AIRCRAFT WAKE TURBULENCE AND ITS DETECTION

AIRCRAFT WAKE TURBULENCE AND ITS DETECTION

Proceedings of a Symposium on Aircraft Wake Turbulence held in Seattle, Washington, September 1-3, 1970. Sponsored jointly by the Flight Sciences Laboratory, Boeing Scientific Research Laboratories and the Air Force Office of Scientific Research

Edited by
John H. Olsen and Arnold Goldburg
Boeing Scientific Laboratories
Seattle, Washington

and

Milton Rogers
Air Force Office of Scientific Research
Arlington, Virginia

 PLENUM PRESS • NEW YORK–LONDON • 1971

Library of Congress Catalog Card Number 70-159027
SBN 306-30541-0

© 1971 Plenum Press, New York
A Division of Plenum Press Publishing Corporation
227 West 17th Street, New York, N.Y. 10011

United Kingdom edition published by Plenum Press, London
A Division of Plenum Publishing Company, Ltd.
Davis House (4th Floor), 8 Scrubs Lane, Harlesden, NW10, 6SE, England

Printed in the United States of America

FOREWORD

        The combination of increasing airport congestion and the ad-
vent of large transports has caused increased interest in aircraft
wake turbulence. A quantitative understanding of the interaction
between an aircraft and the vortex wake of a preceding aircraft is
necessary for planning future high density air traffic patterns and
control systems. The nature of the interaction depends on both the
characteristics of the following aircraft and the characteristics
of the wake. Some of the questions to be answered are: What determines the full characteristics of the vortex wake? What properties
of the following aircraft are important? What is the role of pilot
response? How are the wake characteristics related to the generating aircraft parameters? How does the wake disintegrate and where?

        Many of these questions were addressed at this first Aircraft
Wake Turbulence Symposium sponsored by the Air Force Office of Scientific Research and The Boeing Company. Workers engaged in aerodynamic research, airport operations, and instrument development
came from several countries to present their results and exchange
information. The new results from the meeting provide a current
picture of the state of the knowledge on vortex wakes and their
interactions with other aircraft.

        Phenomena previously regarded as mere curiosities have emerged
as important tools for understanding or controlling vortex wakes.
The new types of instability occurring within the wake may one day
be used for promoting early disintegration of the hazardous twin
vortex structure.

        The influence of the atmospheric variables of stability and
turbulence level on the wake behavior has been examined, but is
not yet fully understood. High turbulence levels are thought to
accelerate wake disintegration by both increasing dissipation and
exciting the natural wake instabilities. High atmospheric stability
causes variations in the descent path of the wake, in some cases the
effect is similar to the spreading in ground effect.

Determining the interaction between the organized vortex wake of one aircraft and the flight of another is the most important practical problem of the symposium. The interaction depends, of course, on the parameters of the generating aircraft as related to the wake structure and on the parameters of the following aircraft. Analytical models have shown that the two most important parameters are the circulation of the vortex and the span of the following aircraft. A suitable control system can significantly reduce motions of the following aircraft.

The three day symposium to discuss these problems was held in September 1970 in Seattle, Washington. Papers presented were either invited or carefully selected from unsolicited submitted works. Three types of papers were accepted: review papers, descriptions of finished research, and descriptions of ongoing research.

The symposium proceedings are arranged according to subject. The main categories are: I. Fundamental Problems; II. Experimental Methods; III. Wake Formation and Character; IV. Stability and Decay of Trailing Vortices; V. Interactions with Vortex Wakes; and finally VI. the Proceedings of the Panel Discussion.

The editors wish to express their appreciation to Colonel W. L. Shields, Jr. (then at AFOSR), Headquarters, United States Air Force, for his constant encouragement and support which brought the Symposium to its successful culmination as a venture jointly sponsored by the Air Force Office of Scientific Research and the Boeing Scientific Research Laboratories. The editors also want to acknowledge the help and counsel of their colleague, Dr. S. C. Crow, in organizing the Program.

The editors wish to thank Mr. Robert Ubell and the staff at Plenum Press for their aid and cooperation in publishing these proceedings as a companion volume to Clear Air Turbulence and Its Detection.

Finally, the editors wish to thank the Symposium Secretary, Mrs. Edna R. Gaston who took care of all correspondence and typing of papers, to Mr. George H. Tweney, the Symposium Arrangements Chairman, and to Mr. Disman W. Peecher, Symposium Arrangements Coordinator.

John H. Olsen and Arnold Goldburg
Seattle, Washington

and

Milton Rogers
Arlington, Virginia

January 1971

CONTENTS

WAKE FORMATION AND CHARACTER

STABILITY AND DECAY

INTERACTIONS WITH VORTEX WAKES

CONTRIBUTORS

F. H. Abernathy, Harvard University, Cambridge, Massachusetts

William H. Andrews, NASA Flight Research Center, Edwards, California

Paolo Baronti, Advanced Technology Laboratories, 400 Jericho Turnpike, Jericho, Long Island, New York

P. L. Bisgood, Royal Aircraft Establishment, Bedford, England

D. Bliss, Massachusetts Institute of Technology, 77 Massachusetts Avenue, Cambridge, Massachusetts

Bernard Caiger, National Aeronautical Establishment, National Research Council of Canada, Montreal Road Laboratories, Ottawa 7, Ontario, Canada

Norman Chigier, NASA - Ames Research Center, 631 Glenbrook Drive, Palo Alto, California

P. M. Condit, The Boeing Company, Everett, Washington

Victor R. Corsiglia, NASA - Ames Research Center, LSA Branch Building, 221-2, Moffett Field, California

Steven C. Crow, The Boeing Company, Seattle, Washington

F. W. Dee, Royal Aircraft Establishment, Bedford, England

Coleman duP. Donaldson, Aeronautical Research Associates of Princeton, Incorporated, 50 Washington Road, Princeton, New Jersey

Calvin C. Easterbrook, Cornell Aeronautical Laboratory, 4455 Genesee Street, Buffalo, New York

Sheldon Elzweig, Advanced Technology Laboratories, 400 Jericho Turnpike, Jericho, Long Island, New York

Leo J. Garodz, National Aviation Facilities Experimental Center, Federal Aviation Administration, Atlantic City, New Jersey

Arnold Goldburg, Boeing Scientific Research Laboratories, Post Office Box 3981, Seattle, Washington

D. G. Gould, National Aeronautical Establishment, National Research Council of Canada, Montreal Road Laboratories, Ottawa 7, Ontario, Canada

J. E. Hackett, Lockheed-Georgia Company, Marietta, Georgia

John C. Houbolt, Aeronautical Research Associates of Princeton, Incorporated, 50 Washington Road, Princeton, New Jersey

Robert M. Huffaker, NASA - Marshall Space Flight Center, S & E-AERO-
AT, Marshall Space Flight Center, Alabama

Robert A. Jacobsen, NASA - Ames Research Center, Moffett Field,
California

A. V. Jelalian, Raytheon Company, 528 Boston Post Road, Sudbury,
Massachusetts

Robert P. Johannes, Air Force Flight Dynamics Laboratory,
Wright-Patterson Air Force Base, Ohio

W. P. Jones, Aerospace Engineering Department, Texas A & M
University, College Station, Texas

Peter F. Jordan, RIAS/Martin Marietta Corporation, 1450 South
Rolling Road, Baltimore, Maryland

William W. Joss, Cornell Aeronautical Laboratory, 4455 Genesee
Street, Buffalo, New York

W. H. Keene, Raytheon Company, 528 Boston Post Road, Sudbury,
Massachusetts

Robert L. Kiang, Stanford Research Institute, 333 Ravenswood
Avenue, Menlo Park, California

R. B. Kiland, Thermo-Systems, Incorporated, 2500 Cleveland Avenue,
North, Saint Paul, Minnesota

M. T. Landahl, Massachusetts Institute of Technology, Cambridge,
Massachusetts

Walter S. Luffsey, Federal Aviation Administration, 800
Indpendence Avenue, S.W., Washington, D. C.

Paul B. MacCready, Jr., Meteorology Research, Incorporated, Post
Office Box 637, Altadena, California

Ralph L. Maltby, Royal Aircraft Establishment, Bedford, England

Barnes W. McCormick, 233 Hammond Building, The Pennsylvania State
University, University Park, Pennsylvania

Joshua Menkes, Department of Aeronautical Engineering Sciences,
University of Colorado, Boulder, Colorado

Derek W. Moore, Imperial College, London SW7, England

Jack N. Nielsen, Nielsen Engineering and Research, Incorporated,
850 Maude Avenue, Mountain View, California

John G. Olin, Thermo-Systems, Incorporated, 2500 Cleveland Avenue,
North, Saint Paul, Minnesota

John H. Olsen, Boeing Scientific Research Laboratories, Post Office
Box 3981, Seattle, Washington

P. R. Owen, Imperial College, London SW7, England

R. Padakannaya, 233 Hammond Building, The Pennsylvania State
University, University Park, Pennsylvania

Patrick C. Parks, NASA - Langley Research Center, Hampton,
Virginia (present address: University of Warwick, Warwick,
England)

Balusu M. Rao, Texas A & M University, College Station, Texas

P. G. Saffman, California Institute of Technology, Pasadena,
California

Richard G. Schwind, Nielsen Engineering and Research, Incorporated,
850 Maude Avenue, Mountain View, California

W. L. Shields, Headquarters, United States Air Force, 1400 Wilson Boulevard, Arlington, Virginia

C. M. Sonnenschein, Raytheon Company, 528 Boston Post Road, Sudbury, Massachusetts

Roger D. Sullivan, Aeronautical Research Associates of Princeton, Incorporated, 50 Washington Road, Princeton, New Jersey

J. G. Theisen, Lockheed-Georgia Company, Marietta, Georgia

Lu Ting, Department of Mathematics, New York University, University Heights, Bronx, New York

Ivar H. Tombach, Meteorology Research, Incorporated, Post Office Box 637, Altadena, California

P. W. Tracy, The Boeing Company, Everett, Washington

Sheila E. Widnall, Massachusetts Institute of Technology, 77 Massachusetts Avenue, Cambridge, Massachusetts

A. Zalay, Massachusetts Institute of Technology, 77 Massachusetts Avenue, Cambridge, Massachusetts

AIRCRAFT WAKES: A NEW LOOK AT A CLASSICAL PROBLEM

W. L. Shields, Colonel, USAF

Headquarters, U.S. Air Force

1.   INTRODUCTION

Aeronautical research has suffered through some hard times during the first decade of the space age; that fact seems to be widely recognized. This recognition is evidenced in many ways: policy statements by the major technical societies, remarks by spokesmen from industry and government, to some extent by re-direction of program efforts and manpower resources, shifting emphasis from space research to aeronautical research. No man-power allocation could illustrate this point better than the re-assignment of Mr. Armstrong to the post of Deputy Associate Ad-ministrator of NASA for Aeronautical Research less than one year from the time he took that "one small step for a man".

In a very direct sense this symposium on aircraft wake tur-bulence also symbolizes this new emphasis. There has indeed been a hiatus in aeronautical research, and it caused a lapse in our national research programs. Overcoming this lapse will not be easy. During the recent lean years of aeronautical research many scientists and engineers have turned to new disciplines. It will not be easy for them to turn again to aeronautical research, nor will it be easy to educate a new generation of aeronautical sci-entists. The limited number of university participants in this symposium, in contrast to government and industry, is evidence of these difficulties.

We can illustrate the need for new aeronautical research with some examples. In 1951 the standard references for both theoret-ical and experimental (we may include operational) calculations of the effects of aircraft - induced vortices were the papers

1

by Spreiter and Sacks (1) and by Bleviss (2). In 1951 the state
of the art in aircraft design was represented by the B-47, a
200,000 pound airplane with a speed of about 500 knots. Twenty
years later the state of the art is represented by the Concorde,
385,000 pounds and Mach 2.2; the C-5A, 728,000 pounds and 500
knots; and the B2707-300, 635,000 pounds and Mach 2.7. Although
there has been some excellent work on a few topics in wake vortex
research since 1951, the same two papers still fairly well represent the extent of our fundamental knowledge.

There are questions other than those related to aircraft design that demand a new look at the aircraft wake problem. For
example, the Air Force is faced with a whole new set of problems
such as high-density mixed traffic operations, aerial refueling,
and air drops from high performance aircraft.

The Federal Aviation Administration must deal with corresponding operational problems in civil air operations. The greater
size and higher performance of civil transports, together with the
increase of traffic density of both the transport and general aviation fleets, have greatly increased the attention given to potential wake hazards. (3-4)

Other groups are concerned with the interaction of aerodynamic
vortices with the environment. For example, the effects of local
meteorological conditions will undoubtedly have major significance
in predicting the motion of aerodynamic vortices. Local weather
modification by means of aerodynamic forces is already a reality
in some limited applications. (5) Pollution problems are now
gaining increased attention too. Aircraft engine smoke is perhaps
the major current consideration but inevitably someone will characterize the aircraft wake as a problem of vorticose pollution.

These considerations, some scientific, some operational, provided the incentive about two years ago for the Air Force Office
of Scientific Research to initiate a program of research in aircraft wakes, and to plan for this symposium. This paper is intended
to suggest a frame of reference for the symposium, first by eluci-
dating some historical aspects of the aircraft wake problem, then
by recounting the genesis of the AFOSR wake turbulence research
program and finally by commenting briefly (and fragmentarily) on
some of the challenging research opportunities contained within the
wake problem. Many of these research opportunities are reflected
in papers that appear on the symposium program.

2. REFLECTIONS ON SOME SEEMINGLY SIMPLE PROBLEMS IN FLUID MECHANICS:

In reviewing the short history of aerodynamic vortices, one
is struck by a pattern of unfolding complexity. The early vis-

ualization of a two-dimensional rollup of a vortex sheet has given way to a complex, three-dimensional picture. In this respect the aerodynamic vortex is similar to some older problems in classical fluid mechanics.

The Karman vortex street is a good example of complexity in a seemingly simple problem. Vortex shedding in the wakes of cylinders was observed long ago, and Karman himself tells of an old Italian painting that shows the phenomenon. (6) Karman's (7) theory for a stable row of vortices provided a relatively simple explanation of observations, however continuing analysis has greatly refined the theory (8) and experiments have shown the structure of the vortex street to be more complicated than the simple flow visualizations suggested. (9-10) Thus, each new research effort reveals new degrees of complexity.

Another example of a seemingly simple problem in classical fluid mechanics is the edgetone, the fluid dynamic sound phenomenon that occurs when a thin, two-dimensional jet impinges upon a sharp edge. It was discovered by Sondhaus in 1854, (11) and many explanations of the phenomenon have been offered since. Nevertheless, each study of the problem has revealed further complexities. It was pointed out by Shields and Karamcheti (12) that almost every edgetone study since 1854 has stated the need for additional study of new aspects of the problem. To date, no satisfactory theory of the apparently simple edgetone has been developed.

The aerodynamic vortex fits this pattern of unfolding complexity. Lanchester was apparently the first to explicate the role of vortex formation in the generation of lift by an airfoil.(13) His drawings and accompanying physical explanation seem simple and straight forward. The first analytical treatment by Prandtl (14) was correspondingly simple. Each subsequent experimental and theoretical study has revealed added complexities, providing an apparently endless list of problems for further research. The two-dimensional problem addressed initially by Lanchester and Prandtl has evolved into a complex three-dimensional problem. The in-flight measurements of McCormick, et al. (15) have shown the actual aircraft vortex to have a complex structure, with several major regions of distinct behavior. Batchelors (16) analysis demonstrates the fundamental importance of three-dimensional considerations. Work by Donaldson and Snedeker (17) has shown how aerodynamic vortices may have a multi-cellular structure, thus introducing a higher order of complexity. Vortex stability, discussed recently by Crow, (18) has also emerged as a major consideration.

Reflecting on this pattern of unfolding complexity in seemingly simple problems, one is not surprised that this symposium addresses

such a wide range of topics related to aerodynamic vortices. A
problem initially limited to the theory of lift has expanded to
encompass hydrodynamic stability, geophysical flows, unsteady
aerodynamics and aircraft operations. Indeed, synergism among
research areas in mechanics has become a familiar pattern. Consider
the fusion of the apparently unrelated subjects of air loads and
structures into the science of aeroelasticity, and the merging of
chemistry, electromagnetism and aerodynamics into magnetogasdy-
namics. It seems not unlikely that vortex research and aerodynamics will follow the pattern and combine to provide new techniques of vortex control; witness the development of the Viggen
airplane that will be discussed in this symposium by Landahl and
Widnall.

3.   GENESIS OF THE AFOSR WAKE TURBULENCE RESEARCH PROGRAM:

The history of aerodynamic vortices associated with generation of lift by airfoils, as we have seen,began early in the history of heavier-than-air flight. Summaries of theoretical and experimental research and practical applications may be found in a
number of papers. (1-15-19-20) These efforts, by and large, kept
pace with other developments in the science of aeronautics, with
the ever-increasing complexity we have noted. However, in this
historical context, aerodynamic vortex research during the past
decade was somewhat of an anomaly. Although numerous questions
arose due to the increased size and speed of modern aircraft, only
a limited amount of research was carried out. Reflecting upon this
situation several years ago, scientists at AFOSR singled out aircraft wake turbulence as a promising area of concentration in re-
juvenating their aerodynamics program. This symposium is partial
evidence of the direction that program is taking.

4.   PROMISING WAKE TURBULENCE RESEARCH TOPICS:

At the outset AFOSR scientists found that research problems
associated with aircraft wake turbulence seemed to fall naturally
into several major categories:

- Formation of trailing vortices

- Decay and breakup of vortices

- Interaction with the environment

- Operational considerations

- Experimental techniques

Each of these categories encompasses theoretical, experimental and practical problems that are very nicely integrated.

Formation of trailing vortices seems to be a promising research area because it contains so many unanswered questions, such as the role of the boundary layers and boundary layer disturbances in vortex rollup, the effect of wing sweep and taper and wing loading on vortex development (with implications for vortex control) and scaling of vortex parameters as functions of time, space and wing geometry. Determination of the appropriate viscosity for turbulent vortices is another important problem. Also, more attention might be given to the birth process of a vortex, with a view toward cancellation of the vortex rollup before or as it occurs. Control of vortices depends upon understanding the basic fluid motions involved in vortex formation and upon developing equipment and techniques for modifying those motions. Finally, vortex control of aircraft will demand a better knowledge of both vortex motions and the interaction of vortices with lifting surfaces. It is not beyond the realm of speculation that vortex considerations might one day be as much a part of aircraft design philosophy as L/D.

Decay and breakup of vortices is a logical area of research because it has both scientific and operational implications. The scientific problems include the nature and origin of instabilities, the significance of external perturbations, and the interaction of paired vortices. The operational implications also seem clear: knowledge of the mechanism of breakup could lead to control of breakup, and reliable avoidance of wake encounters can only be accomplished after more is known about the theory of vortex motion and decay and after means are developed to detect and track a vortex. Also, study of the decay process should be extended to include the final state of a vortex. Furthermore, we may not always be interested in the decay and cancellation of vortices. In geophysical applications, for example, the goal might be a more intense and persistent aerodynamic vortex.

Interaction with the environment in many ways seems to offer the widest range of challenging problems. These include atmospheric effects on vortex decay and breakup, wake energy coupling into natural atmospheric processes, a possible artifical source of clear air turbulence, and even the possibility of weather modification.

Operational considerations appear in each category of wake turbulence research. Assuming that operational needs will include prediction, detection, avoidance, control and induced dissipation of vortices, then these needs will be satisfied only on the basis of improved understanding of the birth, life and death of vortices.

Finally, experimental techniques in wake turbulence research must be improved. One research goal is to increase the validity

of flight and wind tunnel measurements, and to establish the relationship between them in the form of scaling laws. There are substantial difficulties at present in scaling even between aircraft of
similar weight and size. Scaling of vortex parameters from models
to aircraft and from small aircraft to large aircraft presents
some very difficult problems indeed. Another research goal is to
improve the experimental technique of flow visualization. This
has great promise for increasing our understanding of aerodynamic
vortices, and in fact a complete session of this symposium will be
devoted to flow visualization.

This initial categorization of topics in wake turbulence research served fairly well to structure the AFOSR research program.
Although it is now several years old it is still proving useful,
as may be deduced from the symposium program.

5. CONCLUSION

There are several ways in which we might view this symposium
in attempting to judge its potential value. First, it represents
a renaissance of sorts of aeronautical research, after a period
in which space research received relatively greater attention.
Second, it points out opportunities for applying research results
to pressing real-world problems. Finally, we may view this symposium as presenting a menu of challenging basic research topics
in fluid mechanics.

Each participant will probably take a slightly different view
of this meeting. Some will probably be most impressed and en-
couraged by a renewal of emphasis on aeronautical topics. Certainly
there are many current problems in aeronautics that need additional
emphasis. These are not only aerodynamic problems, but also problems in structures, propulsion and control.

On the other hand, some participants will be more impressed
by opportunities to apply research knowledge to practical problems.
There will indeed be many applications of this research, in aircraft
design, in aircraft operations, both enroute and terminal, in flight
safety engineering and possibly in environmental control.

Finally, this symposium will provide a focus for many scientific efforts related to wake turbulence that have not previously
been pulled together, and will develop guidelines for future research. All of the participants should find value in this aspect
of the symposium.

REFERENCES

1. Spreiter, J., and Sacks, A., The rolling up of the railing vortex sheet and its effect on the downwash behind wings. J. Aero. Sci., 18, 1 (1951) p. 21.

2. Bleviss, Z., Theoretical analysis of light plane landing and takeoff accidents due to encountering the wakes of large airplanes, Douglas Aircraft Company, Report SM-18647 (1954).

3. Aviation Week and Space Technology, 25 May 70, p. 31.

4. Flying, November 68, p. 44.

5. OAR Research Review, U.S. Air Force, Office of Aerospace Research, 8, 3 (May-June 69) p. 10.

6. Karman, T. Von, Aerodynamics, Cornell University Press, 1954.

7. Karman, T. Von, Uber den Mechanismus des Widerstandes, den ein bewegter Körper in einer Flussigkeit erfährt. Göttinger Nachrichten, Math. Phys. Klasse, 509 (1911) and 547 (1912).

8. Birkhoff, G., Formation of vortex streets, J. Appl. Phys., 24, 1 (1953) p. 98.

9. Kovasznay, L., Hot wire investigation of the wake behind cylinders at low Reynolds numbers. Proc. Royal Soc., A 198, 174 (1949).

10. Gerrard, J., The three-dimensional structure of the wake of a circular cylinder. J. Fluid Mech., 25, (1966), p. 143.

11. Sondhaus, C., Uber die beim Ausstromen der Luft enstehende Tone, Pogg. Ann., 91, 126 and 214 (1854).

12. Shields, W., and Karamcheti, K., An experimental investigation of the edgetone flow field, Stanford University, Department of Aeronautics and Astronautics, SUDAAR No. 304 (Feb 67).

13. Lanchester, F., Aerodynamics, London, Constable (1911).

14. Prandtl, L., Tragflugeltheorie, Nachr. Ges. Wiss. Gottingen, 107 and 451. (1918)

15. McCormick, B., Tangler, J., and Sherrieb, H., Structure of trailing vortices, J. Aircraft, 5, 3, (1968), p. 260.

16. Batchelor, G., Axial flow in trailing line vortices, J. Fluid Mech., 20 pt. 4 (1964), p. 645.

17. Donaldson, C., and Snedeker, R., Experimental investigation of
    the structure of vortices in simple cylindrical vortex chambers,
    Aeronautical Research Associates of Princeton, Report 47 (1962).

18. Crow, S., Stability theory for a pair of trailing vortices,
    Boeing Scientific Research Laboratories, D 1-82-0918 (Sep 69),
    also presented at AIAA Aerospace Sciences Meeting, (Jan 70).

19. McGowan, W., Trailing vortex hazard, SAE Business Aircraft
    Meeting, April 68, paper no. 680220.

20. Wetmore, J., and Reeder, J., Aircraft vortex wakes in relation
    to terminal operations, NASA TN D-1777 (1963).

THE VELOCITY OF VISCOUS VORTEX RINGS

P. G. Saffman

California Institute of Technology

ABSTRACT

The motion of vortex rings of small cross section is considered. A formula is given for the velocity of a ring in an ideal fluid with an arbitrary distribution of vorticity in the core and an arbitrary circumferential velocity. A definition of velocity is given for the unsteady diffusing ring in a viscous fluid, and the speed is found by the method used for rings in ideal fluids.

The Kelvin formula for the velocity of a circular vortex ring of small cross section in a perfect fluid is

$$U = \frac{\kappa}{4\pi R} \left[\log \frac{8R}{a} - \frac{1}{4} + o\left(\frac{a}{R}\right) \right]. \tag{1}$$

Here R is the radius of the ring, a is the radius of the cross section (circular to leading order) and κ is the circulation about the ring. This formula is for the case when the vorticity in the core is uniform to leading order.

The method used by Lamb (1932) to derive the Kelvin result has been extended to deal with the case when the core vorticity is not uniform and there is in addition an arbitrary axisymmetric circumferential velocity. Details of the proof are given by Saffman (1970). The result is

$$U = \frac{\kappa}{4\pi R} \left[\log \frac{8R}{a} - \frac{1}{2} \right] + \frac{\pi}{\kappa R} \int_o^a v^2 s \, ds - \frac{2\pi}{\kappa R} \int_o^a w^2 s \, ds. \tag{2}$$

The relative error can be shown to be $O(a^2/R^2 \log R/a)$. Here, $v = v(s)$ is the tangential velocity about the center of the core, s is the radial distance from the core center, and $w = w(s)$ is the circumferential velocity around the vortex ring. The slowing down due to the circumferential velocity agrees with that found first, using a perturbation analysis, by Widnall, Bliss and Zalay (these proceedings).

The proof of this formula requires that a steady ring exists, so that the fluid must be perfect and, presumably, w must not be too large. (The critical value of w is not known.) If viscous effects are present, the motion is not steady and the velocity must be defined. A convenient definition is

$$\underset{\sim}{U} = \frac{1}{2} \frac{d}{dt} \int \frac{(\underset{\sim}{r} x \underset{\sim}{\omega} \cdot \underset{\sim}{I})}{I^2} \underset{\sim}{r} \, dV \tag{3}$$

where

$$\underset{\sim}{I} = \frac{1}{2} \int \underset{\sim}{r} \times \underset{\sim}{\omega} \, dV \tag{4}$$

is the impulse of the vorticity distribution. It can be shown (Saffman, 1970) that with this definition of velocity, the speed of an unsteady ring is still given by (2). The values of v and w are leading order given by solution of the diffusion equation for a line vortex.

For the case

$$v = \frac{\kappa}{2\pi s} \left(1 - e^{-s^2/4\nu t} \right), \quad w = 0 , \tag{5}$$

which corresponds to the ring being a circle at time $t = 0$, the velocity of propagation is

$$U = \frac{\kappa}{4\pi R} \left[\log \frac{8R}{\sqrt{4\nu t}} - 0.558 \right] . \tag{6}$$

The difference between this result and that given by Tung and Ting (1967) is stated by Professor Ting to be due to a numerical error.

REFERENCES

Lamb, H., (1932), Hydrodynamics, Cambridge University Press.
Saffman, P. G., (1970), Studies in App. Math. 49, 371.
Tung, C. and Ting, L., (1967), Phys. Fluids, 10, 901.

STUDIES IN THE MOTION AND DECAY OF VORTICES

L. Ting

New York University

ABSTRACT

Solutions of Navier-Stokes equations are constructed as an
asymptotic expansion in terms of a small parameter related to the
Reynolds number of the vortex. A general scheme is presented for
the matching of the inner viscous core of the vortex to the outer
inviscid solution. The singularities in the inviscid theory are
removed and the condition of regularity in the flow field defines
the velocity of the vortex line. This general scheme is applied
to study vortices in two dimensional, axially symmetric and in
three dimensional flow fields. Results for the outer solution
which do or do not require the solution in the viscous core are
obtained separately. Several examples are presented to show
that the solution of this analysis can be identified with that
of classical inviscid theory with the same initial vorticity distribution for the initial instant. They disagree afterwards because
the inviscid theory ignores the diffusion of vorticity in the core.
For a general three dimensional problem, the present scheme of
analysis can be carried out provided the shape of the vortex line
and the three dimensional flow field fulfill a constraint condition.
This condition is automatically fulfilled for the two dimensional
and axially symmetric problems. When the initial velocity of the
vortex is prescribed to be different from that given by the analysis, the necessary modifications to the expansion scheme and the
solution for the subsequent motion are presented.

1. INTRODUCTION

The classical inviscid model of a vortex line moving in an

11

inviscid stream has been employed frequently for the explanation
of many fluid dynamics phenomena[1].   Since a real fluid is viscous,
the inviscid solutions should be identified as special limiting solutions of the viscous theory and the informations which are either
missing or incorrectly provided by the inviscid theory will then be
accounted for by the viscous theory.   A brief review of the essential results and defficiencies of the inviscid theory is
therefore in order.

   The simple case of a two-dimensional incompressible inviscid
flow with a vortex of strength Γ located at the point (X, Y) can be
represented[2] by the sum of a complex potential regular at the
vortex and that of a vortex alone, i.e.

$$W(z) = W^*(z) + \Gamma(2\pi i)^{-1} \ln(z-Z) \tag{1}$$

where $z = x+iy$, $Z = X+iY$ and $W^*(z)$ is regular at $z = Z$.   Based on
the inviscid theory the velocity at the center, $z = Z$ of the vortex
is infinite as $\Gamma/(2\pi|z-Z|)$ and the velocity of the vortex adjusts
instantaneously to the local average velocity or the finite part,

$$\dot{X} - i\dot{Y} = (dW^*/dz)_{z=Z} \tag{2}$$

   For a curved vortex line in a three dimensional flow field,
the velocity $\vec{V}$ in the inviscid theory is a linear superposition of
the velocity $\vec{V}_1$ induced by the vortex line alone and the velocity
$\vec{V}^*$ which is regular along the vortex line, i.e.,

$$\vec{V}(x, y, z, t) = \vec{V}_1(x, y, z, t) + \vec{V}^*(x, y, z, t) \tag{3}$$

Due to the motion of the vortex line, the time, t, enters as a
parameter. $\vec{V}_1$ is given by the Biot-Savat formula[3],

$$\vec{V}_1(P, t) = \frac{\Gamma}{4\pi} \oint \frac{[\vec{P} - \vec{X}(s',t)] \times d\vec{s}'}{|\vec{P} - \vec{X}(s',t)|^3} \tag{4}$$

where s' is the arc length along the vortex line as shown in Fig. 1.
For its behavior near the vortex line, it is convenient to repre-
sent the position vector $\vec{P}$ in curvilinear coordinates s, r and θ, i.e.,

$$\vec{P}(x, y, z) = \vec{X}(s, t) + r\,\hat{r}(s, \theta, t) \tag{5}$$

where r is the shortest distance from the point P to a point $X(s, t)$
on the vortex line.   As $r \to 0$, the integrand becomes singular as $s' \to s$.
With a careful expansion of the integrand near $s'=s$, the following
result is obtained for $r \to 0$,

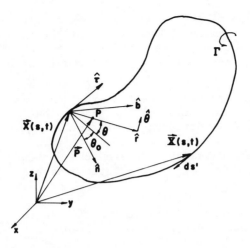

Fig. 1   Curved vortex line in three dimensional stream

$$\vec{V}(P,t) = \vec{V}_1(P,t) + \vec{V}^*(P,t)$$

$$= \frac{\Gamma}{2\pi r}\,\hat{\theta} + \frac{\Gamma}{4\pi R(s,t)}\,\ell n\!\left(\frac{S}{r}\right)\hat{b} + \vec{V}_f(P,t) + \vec{V}^*(P,t) \qquad (6)$$

where $\hat{\theta}$ is the unit circumferential vector, i.e., $\hat{\tau} \times \hat{r}$; $\hat{\tau}$ and $\hat{b}$ are the unit tangential and binormal vector of the vortex line at the point $\vec{X}(s,t)$ and $R(s,t)$ the local radius of curvature. $\vec{V}_f$ and $\vec{V}^*$ are finite when $r \to 0$. S is the arc length of the vortex line or any other reference length. The difference in the definitions of S will of course be absorbed in the term $\vec{V}_f$. For a circular vortex ring[2] S is usually identified as 8R.

For the general case of a curved vortex line in a three di-mensional flow field, the inviscid theory yields a flow field with infinite velocity along the vortex line as given by eq. (6) and an infinite velocity for the vortex line unless a distribution of vorticity over a finite vortical core is assumed.

It is clear that the inviscid theory is inadequate to describe the motion of a circular vortex ring or of a curved vortex line. Furthermore, the infinite velocity at the center of the vortex core which exists also for the two dimensional problem is not physically realistic. The large velocity will be accompanied by a large ve-locity gradient. In the neighborhood of the vortex line, i.e., in the vortical core, the viscous force is no longer small as compared to the inertia term and should therefore be included to describe the actual flow in the same manner as in the boundary layer analy-sis[4,5].

The solution for the motion of a vortex line with a viscous core will be constructed as a solution of the time dependent Navier-Stokes equation with a small parameter ϵ which is defined as the inverse of the square root of the reference Reynolds number $1/\sqrt{Re}$. In this paper, the ratio of the circulation Γ to the kinematic viscosity ν has been chosen as the Reynolds number, i.e., $\epsilon = (\nu/|\Gamma|)^{\frac{1}{2}}$. The selection of the independent variables and the expansion scheme are motivated by the following objectives:

a) At finite distance away from the vortex line, the leading term of the solution (the outer solution) agrees with the inviscid solution.

b) Near the vortex line the circumferential velocity is large and the viscosity terms are comparable to the inertial terms, therefore, the radial variable r is stretched to $\bar{r}\epsilon$ for the inner solution.

c) The condition that the velocity is finite and single valued inside the viscous core together with the matching conditions for the inner and outer solutions should be sufficient to define the velocity of the vortex line.

This general scheme has been successfully applied to the two dimensional problems[6] and to the axially symmetric problems[7].

In both analyses, the leading term of the inner solution, was shown to be symmetric with respect to the axis tangential to the vortex line, i.e., independent of θ. The governing equation yields the diffusion of the vorticity distribution $\bar{\zeta}^{(o)}(\bar{r},t)$ provided an initial data for $\bar{\zeta}^{(o)}$ is given. The governing equation for the leading term of the antisymmetric part $\bar{\psi}^{(1)}(r,t,\theta)$ involves the vorticity distribution $\bar{\zeta}^{(o)}$ as coefficients. In both analyses[6,7], it was assumed that the vortex line was created at the instant t=0 so that the vorticity distribution as given by the similar solution

$$\epsilon^{-2}\,\bar{\zeta}^{(o)}(\bar{r},t) = \Gamma(4\pi\nu\tau)^{-1}\exp(-\eta^2) \qquad (7)$$

where $\eta^2 = \bar{r}^2/(4\Gamma\tau)$, $\tau = t$ for two dimensional problems and $\tau = \int_0^t R_0(t')dt'/R_0(t)$ for axially symmetric problems. With the similar solution, eq. (7), the governing equation for the anti-symmetric part $\bar{\psi}^{(1)}(r,t,\theta)$ is solved and the matching conditions remove the singularity of the outer inviscid solution and defines the velocity of the vortex line.

In the present paper the analyses of the two dimensional and the axially symmetric problem are briefly reviewed in sections 2

and 3.   The effect of initial conditions will then be analyzed.
In particular, the answers to the following questions will be obtained:

i)   What information regarding the motion of the vortex line
can be obtained from the matching condition independent of the
vorticity distribution?

ii)   For a given initial vorticity distribution, how should an
optimum time interval t_o be defined in order that the vorticity
distribution can be approximated by a similar solution originated
at an instant t_o before the initial station.

iii)   For a given initial vorticity distribution, how will the
instantaneous velocity of the vortex line and its subsequent motion
given by the present viscous theory compare with those given by the
inviscid theory[2]?

iv)   If the velocity of the vortex line is initially different
from the value defined by the analysis, what will be the subsequent
motion?

Questions i) and ii) are answered in section 2 and questions iii)
and iv) are answered in section 3.   In section 4, the analysis is
extended to a curved vortex line in a three dimensional flow field.

2.   TWO DIMENSIONAL PROBLEMS

The Navier-Stokes equations for a two dimensional incompressible flow in terms of the stream function ψ and vorticity ζ are

$$\zeta_t + \psi_y \zeta_x - \psi_x \zeta_y = \nu \Delta \zeta \qquad (8)$$

$$\Delta \psi = -\zeta \qquad (9)$$

where Δ is the two dimensional Laplacian operator.   The flow field
away from the vortex point, $X(t)$, $Y(t)$, does not have any steep
velocity gradients.   The stream function ψ may therefore be expanded as a power series of ϵ as follows

$$\psi(t,x,y,\epsilon) = \psi^{(o)}(t,x,y) + \epsilon\,\psi^{(1)}(t,x,y) + \ldots \qquad (10)$$

If the flow field is initially irrotational, eqs. (8 & 9) yield

$$\Delta\psi^{(n)} = 0 \quad \text{and} \quad \zeta^{(n)} \equiv 0 \qquad (11)$$

This means that the flow in the outer region is always a potential
flow to any order of approximation.

Near the vortex point, it is easier to use polar coordinate r, θ with $x = X(t) + r \cos\theta$, $y = Y(t) + r \sin\theta$. The inner solutions will be functions of t, $\bar{r}$, θ and ϵ with $\bar{r} = r/\epsilon$ and will also be expanded in power series of ϵ,

$$\bar{\psi}(t,\bar{r},\theta,\epsilon) = \bar{\psi}^{(o)}(t,\bar{r}) + \epsilon\,\bar{\psi}^{(1)}(t,\bar{r},\theta) + \dots \qquad (12a)$$

$$\bar{\zeta}(t,\bar{r},\theta,\epsilon) = \epsilon^{-2}[\bar{\zeta}^{(o)}(t,\bar{r}) + \epsilon\,\bar{\zeta}^{(1)}(t,\bar{r},\theta) + \dots] \qquad (12b)$$

The independence of the leading term on θ is guided by the matching condition[5] with the outer solution given by eq. (1).   Series (12a & b) will be substituded into eqs. (8 & 9).   A general scheme for deriving the equations for the inner solution will now be stated. The equations designated as E^n, obtained by equating the coefficients of ϵ^n of eq. (8), can be split into two parts E_1^n and E_2^n. The first part depends on θ and the second does not.   By equating each part to zero two equations are obtained.   For the m-th order solution $\bar{\psi}^{(m)}$, the part, $\tilde{\psi}^{(m)}$, which depends on θ is defined by the equation $E_1^{m-2} = 0$ while the part independent of θ, $\bar{\psi}_o^{(m)}$, are governed by the equation $E_2^m = 0$.   This general scheme will be observed for the axially symmetric cases and the general three dimensional cases in the next two sections.   Based on this scheme, the governing equation $E_1^{-1} = 0$ for the leading antisymmetric part $\tilde{\psi}^{(1)}(\bar{r},t,\theta)$ is

$$E_1^{-1} = \bar{\zeta}_{\bar{r}}^{(o)}\,\tilde{\psi}_\theta^{(1)} - \bar{\psi}_{\bar{r}}^{(o)}\,\tilde{\zeta}_\theta^{(1)} = 0 \qquad (13)$$

where $\tilde{\zeta}^{(1)} = \bar{\Delta}\,\tilde{\psi}^{(1)} = (\bar{r}\,\tilde{\psi}_{\bar{r}}^{(1)})_{\bar{r}}/\bar{r} + \tilde{\psi}_{\theta\theta}^{(1)}/\bar{r}^{-2}$.   The governing equation $E_2^o = 0$, for the leading symmetric term $\bar{\psi}^{(o)}(t,\bar{r})$ is

$$E_2^o = \bar{\zeta}_t^{(o)} - \Gamma(\bar{r}\,\bar{\zeta}_{\bar{r}}^{(o)})_{\bar{r}}/\bar{r} = 0 \qquad (14)$$

The boundary condition for the equations are the regularity conditions at $\bar{r} = 0$ and the matching conditions with the outer solution as $\bar{r} \to \infty$.   They are:

$$\bar{\psi}_{\bar{r}}^{(o)}(t,\bar{r}) = 0, \qquad \bar{\zeta}^{(o)}(t,\bar{r}) = \text{finite} \qquad (15a)$$

$$\tilde{\psi}_{\bar{r}}^{(1)}(t,\bar{r},\theta) = 0, \qquad \tilde{\psi}_\theta^{(1)}(t,\bar{r},\theta)/\bar{r} = 0 \quad \text{as } \bar{r} \to 0 \qquad (15b)$$

and $\qquad \bar{v}^{(0)} = - \bar{\psi}_{\bar{r}}^{(0)} \rightarrow \Gamma/(2\pi\bar{r})$

$$\tilde{\psi}^{(1)}(t,\bar{r},\theta) \rightarrow [\psi_y^*(X,Y) - \dot{X}(t)]\bar{r} \, \sin\theta$$

$$+ [\psi_x^*(X,Y) + \dot{Y}(t)]\bar{r} \, \cos\theta \qquad \text{as } \bar{r} \to \infty \qquad (16)$$

Detailed steps in the derivation of eqs. (13-16) can be found in ref. 6 or obtained from section 4 as a special case.

Equation (14) is the well known heat conduction equation and the solution in terms of the initial data $\bar{\zeta}^{(0)}(\bar{r},o)$ is

$$\bar{\zeta}^{(0)}(\bar{r},t) = (2\Gamma t)^{-1} \int_0^\infty \bar{\zeta}^{(0)}(\rho,o) \, e^{-(\rho^2 + \bar{r}^2)/(4\Gamma t)} I_0\left(\frac{\rho\bar{r}}{2\Gamma t}\right)\rho d\rho \qquad (17)$$

where I_0 is the modified Bessel function. From each initial vorticity distribution $\bar{\zeta}^{(0)}(\bar{r},o)$, eq. (13) can be integrated to yield $\tilde{\psi}^{(1)}$ which in turn yields the velocity of the vortex from eq. (16). In the subsection 2.1, the conclusions which can be obtained independent of the initial data $\bar{\zeta}^{(0)}(\bar{r},o)$ will be presented. In the subsection 2.2, the connection of an optimum similar vorticity solution to a prescribed initial data will be described.

2.1 Conclusions Independent of the Initial Vorticity Distribution

As a solution of the heat conduction equation, eq. (14), the total space integral of $\zeta^{(0)}$ is invariant, i.e.,

$$\int_0^\infty \bar{\zeta}^{(0)}(\bar{r},t)(2\pi\bar{r})d\bar{r} = \Gamma \qquad (18)$$

Since the vorticity in the outer solution remains zero, eq. (11), the matching condition requires that

$$\bar{\zeta}^{(0)}(\bar{r},t) = o(\bar{r}^{-\alpha}) \qquad (19)$$

for any positive α. In other words $\bar{\zeta}^{(0)}$ decays exponentially with respect to $\bar{r}$. This is guaranteed by eq. (17) so long as it is true for the initial data. The circumferential velocity $\bar{v}^{(0)}$ is related to vorticity through the relationship $(\bar{v}^{(0)}\bar{r})_{\bar{r}}/\bar{r} = \bar{\zeta}^{(0)}$, therefore,

$$\epsilon^{-1} \ \bar{v}^{(o)} \ (\bar{r}, t) \ = \ \frac{1}{\epsilon \bar{r}} \int_{o}^{\bar{r}} \bar{\zeta}^{(o)} \ (\rho, t) \rho d\rho \qquad\qquad (20)$$

$$\rightarrow \Gamma / (2\pi r) \qquad\qquad \text{as } \bar{r} \rightarrow \infty$$

The matching condition is fulfilled and the singularity of the outer solution at $r = 0$ is removed independent of the vorticity distribution of eq. (17).

The antisymmetric part $\tilde{\psi}^{(1)} (\bar{r}, t, \theta)$ can be resolved into Fourier components with coefficients $\tilde{\psi}_{j1}^{(1)} (\bar{r}, t)$ for $\cos j\theta$ and $\tilde{\psi}_{j2}^{(1)} (\bar{r}, t)$ for $\sin j\theta$. Equation (13) reduces to

$$\frac{\partial^2 \tilde{\psi}_{jk}^{(1)}}{\partial \bar{r}^2} + \frac{1}{\bar{r}} \frac{\partial \tilde{\psi}_{jk}^{(1)}}{\partial \bar{r}} - \left[\frac{j^2}{\bar{r}^2} - \frac{\bar{\zeta}_{\bar{r}}^{(o)}}{\bar{\psi}_{\bar{r}}^{(o)}} \right] \tilde{\psi}_{jk}^{(1)} = 0 \qquad\qquad (21)$$

for $k=1,2$ and $j=1,2,3\ldots$ For eq. (20) and the regularity condition of eq. (15) at $\bar{r} = 0$, the term $\bar{\zeta}_{\bar{r}}^{(o)} / \bar{\psi}_{\bar{r}}^{(o)}$ is of the order of $\bar{r}^{-1}$ and is dominated by the term $j^2 / (\bar{r})^2$ as $\bar{r} \rightarrow 0$. The solution can be written in general as a linear combination of the fundamental solutions which behave as $\bar{r}^j$ and $\bar{r}^{-j}$ respectively, i.e.,

$$\tilde{\psi}_{jk}^{(1)} \sim a_{jk} (t) \ \bar{r}^j + b_{jk} (t) \ \bar{r}^{-j} \qquad \text{as } \bar{r} \rightarrow 0 \qquad\qquad (22)$$

For $j=1$, the regularity condition at $\bar{r}=0$, eq. (15) yields $a_{1k} = b_{1k} = 0$ and, therefore, $\tilde{\psi}_{1k}^{(1)} \equiv 0$. The matching condition of eq. (16) yields

$$\dot{X}(t) = \psi_y^* (X, Y) \quad ; \quad \dot{Y}(t) = - \psi_x^* (X, Y) \qquad\qquad (23)$$

This is the result assumed in the classical inviscid theory.

For $j=2, 3\ldots$, eq. (15) yields only $b_{jk} = 0$ while the matching condition of eq. (16) is homogeneous and yields $a_{jk} = 0$. Therefore, $\tilde{\psi}^{(1)}$ vanishes and there is no contribution to the outer solution, i.e., $\psi^{(1)} \equiv 0$.

2.2 Optimum Similar Vorticity Distribution

For a given initial vorticity distribution $\bar{\zeta}^{(o)} (\bar{r}, o)$, the general integral representation of eq. (17) is not easy to evaluate and furthermore it will be too cumbersome to be in the coefficients of the governing equation for the antisymmetric part. Since $\zeta^{(0)}$ decays exponentially with respect to $\bar{r}$, it can be represented in a decending series in new variables $\bar{t}$ and $\bar{\eta}$ where

$$\bar{t} = t + t_o \quad\text{and}\quad \bar{\eta}^2 = \bar{r}^2 / (4\Gamma\bar{t}) \tag{24}$$

The series solution can be written as[9,10]

$$\zeta^{(0)}(\bar{t},\bar{r}) = \sum_{n=0,1...} C_n (\bar{t})^{-(n+1)} L_n (\bar{\eta}^2) e^{-\bar{\eta}^2} \tag{25}$$

where $L_n (\bar{\eta}^2)$ is the Laguerre polynomial[11]. The coefficients C_n are related directly to the initial data at $t=0$ or $\bar{t}=t_o>0$ by the orthogonal conditions, the results are[10]

$$C_n = 2 (t_o)^{n+1} \int_0^\infty \zeta^{(0)}[0,\bar{\eta} (4\Gamma t_o)^{\frac{1}{2}}] L_n (\bar{\eta}^2)\bar{\eta} d\bar{\eta} \tag{26}$$

In particular, $C_o = 1/(4\pi)$ and the first term in eq. (25) for $\epsilon^{-2} \zeta^{(0)}$ is $\Gamma (4\pi\bar{t})^{-1}/\exp(\bar{\eta}^2)$. It represents the similar solution for an isolated line vortex created at the instant $t = -t_o$. To define the unknown constant t_o, and additional condition of $C_1 =0$ is introduced so that the similar solution will be optimum in the sense that it will differ from the exact solution by $0(\bar{t}^{-3})$ instead of $0(\bar{t}^{-2})$. With $C_1 =0$, t_o is related directly to the initial data, thereby matching its second moment with the optimum solution,

$$t_o = \pi (2\Gamma^2)^{-1} \int_0^\infty \zeta^{(0)}(o,\bar{r})\bar{r}d\bar{r}\cdot \bar{r}^2 \tag{27}$$

Fig. 2 shows the comparison between the exact solution and the optimum similar solution corresponding to an initial profile in the shape of a step function. Good agreement is observed even for finite $\bar{t}$, e.g., for $\xi=\bar{t}/(8t_o)=0.2$.

With $\tilde{\psi}^{(1)}\equiv 0$ and $\psi^{(1)}\equiv 0$, the next order inner solution $\epsilon^2\tilde{\psi}_{jk}^{(2)} (\bar{r},t)$ also satisfies eq. (21) with the same boundary conditions of eq. (15b) while the matching condition as $\bar{r} \to \infty$ is

$$\tilde{\psi}^{(2)} (\bar{r},t,\theta) \to \frac{1}{2} \bar{r}^2 \left\{ \psi_{xx}^* \cos2\theta + \psi_{xy}^* \sin2\theta \right\}_{\text{at } X,Y} \tag{28}$$

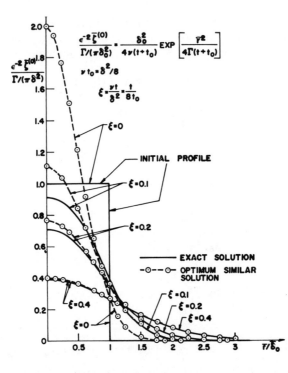

Fig. 2 One-term optimum similar solution and exact solution.

Therefore, only $\widetilde{\psi}_{2k}^{(2)}$ does not vanish. Its solution depends on the vorticity distribution. With a similar solution for $\bar{\zeta}^{(0)}$, the solution for $\widetilde{\psi}_{2k}^{(2)}$ is also similar and $\widetilde{\psi}^{(2)}$ behaves for large $\bar{r}$ as

$$\epsilon^2 \widetilde{\psi}^{(2)} (\bar{t}, \bar{r}, \theta) \rightarrow \left[\overset{*}{\psi}_{xx} \cos 2\theta + \overset{*}{\psi}_{xy} \sin 2\theta \right] \left[\lim_{r \to 0} \left(\frac{r^2}{2} + \frac{35 \epsilon^4 \Gamma^2 \bar{t}^2}{r^2} \right) \right] \qquad (29)$$

Comparison with eq. (28) shows that the term r^{-2} will induce an extra outer solution of quadrupoles of the order of ϵ^4 and are proportional to the local nonuniformity of the flow field ψ^*. These quadrupoles are absent in the classical inviscid theory. Details for the derivations of eqs. (29) are given in ref. 6. The constant 35 for the r^{-2} term in eq. (29) differs from that in ref. 6 due to an error in ref. 6, i.e., using a factor of 1/8 instead of 1/4 in eq. (7) for $\bar{\zeta}^{(0)}$.

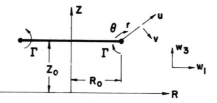

Fig. 3 Circular vortex ring in axially symmetric stream

3. AXIALLY SYMMETRIC PROBLEM

For a circular vortex ring, it is natural to introduce tor-roidal coordinates r, θ which are related to the cylindrical co-ordinates R, Z by the equations $Z = Z_o(t) + r \sin\theta$ and $R = R_o(t) - r \cos\theta$ where Z_o is the location of the ring with radius R_o (see Fig. 3). The same expansion schemes explained in section 2 for the stream function ψ in the outer region and $\bar{\psi}$ in the inner region will be introduced,

$$\psi(t, Z, R, \epsilon) = \psi^{(o)}(t, Z, R) + \epsilon \, \psi^{(1)}(t, Z, R) + \dots \tag{30}$$

and

$$\bar{\psi}(t, \bar{r}, \theta, \epsilon) = \bar{\psi}^{(o)}(t, \bar{r}) + \epsilon \, \bar{\psi}^{(1)}(t, \bar{r}, \theta) + \dots \tag{31}$$

for the outer region the flow will again remain irrotational. The vorticity in the inner core which is tangential to the ring is expanded likewise,

$$\bar{\zeta} = - [(\bar{\psi}_R/R)_R + \bar{\psi}_{ZZ}]/R$$

$$= \epsilon^{-2} \, \bar{\zeta}^{(o)}(t, \bar{r}) + \epsilon^{-1} \, \bar{\zeta}^{(1)}(t, \bar{r}, \theta) + \dots \tag{32}$$

By following the same procedure outlined in section 2, the governing equations for the leading asymmetric part $\widetilde{\bar{\psi}}^{(1)}(t, \bar{r}, \theta)$ in $\bar{\psi}^{(1)}$ and the symmetric part $\bar{\psi}^{(o)}(t, \bar{r})$ are

$$\bar{\zeta}^{(o)}_{\bar{r}} \bar{\psi}^{(o)}_{\theta} - \bar{\psi}^{(o)}_{\bar{r}} \bar{\zeta}^{(1)}_{\theta} = [\bar{r} \, \bar{\zeta}^{(o)}/R_o] \bar{\psi}^{(o)}_{\bar{r}} \sin\theta \tag{33}$$

$$\bar{\zeta}^{(o)}_{t} - \Gamma(\bar{r} \, \bar{\zeta}^{(o)}_{\bar{r}})_{\bar{r}}/\bar{r} = (\dot{R}_o/R_o)[\bar{r}^2 \bar{\zeta}^{(o)}]_{\bar{r}}/(2\bar{r}) \tag{34}$$

By comparing with eqs. (13 & 14) for the two dimensional case, the terms on the right side of eqs. (33) and (34) represent the effect of the curvature of the ring. The boundary condition at $\bar{r}=0$ is the same as eqs. (15a & b). The condition of matching with the outer solution[2], eq. (6), as $\bar{r} \to \infty$ and $r \to 0$ yields,

$$\bar{v}^{(o)} = -\bar{\psi}_{\bar{r}}^{(o)}/R_o \to \Gamma/(2\pi\bar{r}) \tag{35a}$$

$$\tilde{\psi}^{(1)}(t,\bar{r},\theta) \to \Gamma/(4\pi) \; \bar{r} \; \cos\theta\left\{\ell n[8R/r] - 1\right\}$$

$$+ R_o\bar{r}[-(\overset{*}{w}_1 - \dot{R}_o) \sin\theta + (\overset{*}{w}_3 - \dot{Z}) \cos\theta] \tag{35b}$$

where $\overset{*}{w}_1$ and $\overset{*}{w}_3$ are the velocity components along the R and Z axes evaluated at (R_o, Z_o) without the contribution of the vortex ring. To be consistant with the symbols in sections 2 and 3, ψ is the stream function of the velocity components u, v relative to the center of the vortical core. Equations (33) to (35b) are equivalent to those in ref. 7 where u and v were the absolute velocity components.

Similar to section 2, $\tilde{\psi}^{(1)}(t,r,\theta)$ is resolved into Fourier components in θ with coefficient $\tilde{\psi}_{jk}^{(1)}$, eq. (33) reduces to eq. (21) for $k=1,2$ and $j=1,2,3...$ except when $j=1$, $k=1$, i.e.,

$$\frac{\partial^2\tilde{\psi}_{11}^{(1)}}{\partial\bar{r}^2} + \frac{1}{\bar{r}}\frac{\partial\tilde{\psi}_{11}^{(1)}}{\partial\bar{r}} - \left[\frac{1}{\bar{r}^2} - \frac{R_o\bar{\zeta}_{\bar{r}}^{(o)}}{\bar{\psi}_{\bar{r}}^{(o)}}\right]\tilde{\psi}_{11}^{(1)} = -2\bar{r}\bar{\zeta}^{(o)} - \bar{v}^{(o)} \tag{36}$$

3.1   Conclusions Independent of the Initial Vorticity Distribution

Prior to the determination of $\bar{\zeta}^{(o)}$ and $\bar{\psi}^{(o)}$, the conclusions drawn in section 2.1 for $\tilde{\psi}_{jk}^{(1)}$ except $j=1$, $k=1$ are also valid here,

$$\tilde{\psi}_{12}^{(1)} = 0 \tag{37a}$$

$$\tilde{\psi}_{jk}^{(1)} = 0 \qquad k=1,2, \; j=2,3,4.... \tag{37b}$$

The general behaviors for $\bar{\zeta}^{(o)}$ and $\bar{v}^{(o)}$ given by eqs. (18), (19) and (20) remains valid. In particular the behavior of $\bar{v}^{(o)}$ as $\bar{r} \to \infty$ is

$$\tilde{v}^{(o)} = \Gamma/(2\pi\bar{r}) \tag{38}$$

Consequently, the asymptotic behavior of $\tilde{\psi}_{11}^{(1)}$ as $\bar{r} \to \infty$ can be established,

$$\tilde{\psi}_{11}^{(1)}(t,\bar{r}) \to -\Gamma(4\pi)^{-1}\left\{\bar{r}\,\ell n\,\bar{r} + C_1(t)\bar{r} + C_2(t)\bar{r}^{-1} + o(\bar{r}^{-1})\right\} \tag{39}$$

Coefficients $C_1(t)$ and $C_2(t)$ will be related to the boundary condition at $\bar{r} = 0$ after the determination of the vorticity distribution $\bar{\zeta}^{(o)}$. From eq. (37a) the matching condition of eq. (35b) yields

$$\dot{R}_o(t) = w_1^*(t, R_o, Z_o) \tag{40}$$

Eqs. (38) and (39) and the matching conditions of eqs. (35a,b) show that the singularities in the inviscid theory have been removed in general while the axial velocity is given by the formula

$$\dot{Z}_o(t) = w_3^*(t, R_o, Z_o) + \Gamma(4\pi R_o)^{-1}\left\{\ell n(8R/\epsilon) - 1 + C_1(t)\right\} \tag{41}$$

$\dot{Z}_o(t)$ depends on C_1 while the term $C_2(t)\bar{r}^{-1}$ in eq. (39) will induce an extra term in the outer solution of the order ϵ^{-2}.

3.2 Axial Velocity of the Ring and the Vorticity Distribution

Equation (36) is an ordinary differential equation in $\bar{r}$ for $\tilde{\psi}_{11}^{(1)}(t,\bar{r})$ with t appearing as a parameter, therefore, for each prescribed vorticity distribution at each instant $\tilde{\psi}_{11}^{(1)}$ can be determined by eq. (36) and the boundary conditions at $\bar{r} = 0$.

For example, if the initial vorticity distribution is that of a rotating disk (model B in Fig. 4) of radius δ_o, the $\bar{\zeta}^{(o)}$ and $\bar{v}^{(o)}$ distribution at the instant $t=t_o$ is

$$\bar{\zeta}^{(o)} = 2\Omega \qquad \bar{v}^{(o)} = \Omega\bar{r} \qquad \text{for } \bar{r} < \bar{\delta}_o \tag{42a}$$

and

$$\bar{\zeta}^{(o)} = 0 \qquad \bar{v}^{(o)} = \Gamma/(2\pi\bar{r}) \text{ for } \bar{r} > \bar{\delta}_o \tag{42b}$$

where $\Omega = \Gamma/(2\pi\bar{\delta}_o^2)$ and $\bar{\delta}_o = \delta_o/\epsilon$. The solution of eq. (36) with boundary condition at $\bar{r} = 0$ is

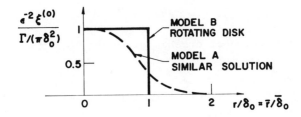

Fig. 4   Two models of initial vorticity distribution

$$\tilde{\psi}_{11}^{1}\,(t_o, \bar{r}) = -\,(5/8)\bar{r}^3\,\Omega \qquad \text{for } 0 \le \bar{r} < \bar{\delta} \qquad\qquad (43a)$$

$$\text{and } \tilde{\psi}_{11}^{1}\,(t_o, \bar{r}) = -\,\Gamma(4\pi)^{-1}\left\{\bar{r}\,\ell n\bar{r} + C_1\bar{r} + C_2\bar{r}^{-1}\right\} \qquad \text{for } \bar{r} > \bar{\delta} \qquad (43b)$$

The continuity of $\tilde{\psi}_{11}^{1}$ and the jump condition of $\tilde{\psi}_{11,\bar{r}}^{(1)}$ at $\bar{r} = \bar{\delta}$ yield,

$$C_1(t_o) = \ell n\bar{\delta} - 3/4 \qquad \text{and} \qquad C_2(t_o) = \bar{\delta}^2/2$$

Eq. (41) yields the axial velocity at $t = t_o$

$$\dot{Z}(t_o) = w_3^*(t, R_o, Z_o) + \Gamma(4\pi R_o)^{-1}\left\{\ell n 8R/\delta_o - 1/4\right\} \qquad (44)$$

With $w_3^* = 0$, eq. (44) is in agreement with the classical inviscid theory (Ref. 2 p. 241) for an isolated vortex ring with vorticity distribution of eq. (42). Due to the diffusion of vorticity governed by eq. (34), eqs. (42) and (44) will not be valid for $t > t_o$. Although eq. (34) involves the unknown function $R_o(t)$, a similar solution can be constructed by the a change of variable t to τ, with t denoting the life time of the line vortex

$$\tau = \int_{o}^{t} R_o(t')\,dt'/R_o(t) \qquad (45)$$

The solution of eq. (34) is given by eq. (7). With this similar solution, $\tilde{\psi}_{11}^{(1)}(t,\bar{r})$ can also be represented by a similar solution

$$\bar{\psi}_{11}^{(1)}(t,\bar{r}) = - (\Gamma^{3/2} \tau^{1/2}/\pi)f(\eta) \tag{46a}$$

and the solution of eq. (33) can be obtained[7] for all t instead of only t_0 as in eqs. (43-44). The asymptotic behavior of $f(\eta)$ for large η is

$$f(\eta) \rightarrow \frac{1}{2}\left[\eta \ell n\eta + 0.442\eta + 1.294\eta^{-1} + o(\eta^{-1})\right] \tag{46b}$$

By comparison with eq. (39) $C_1(t)$ is identified as $0.442 - \frac{1}{2}\ell n(4\Gamma\tau)$ and eq. (41) becomes

$$\dot{Z}_0(t) = w_3^*(t, R_0, Z_0) + \Gamma(4\pi R_0)^{-1}\left\{\ell n[8R/(4\nu\tau)^{\frac{1}{2}}] - 0.558\right\} \tag{47}$$

The coefficients of η and η^{-1} in eq. (46b) and the constant 0.558 in eq. (47) is different from those in ref. 7 due to an error in ref. 7 using a factor of 1/8 instead of 1/4 in eq. (7) for $\bar{\zeta}(o)$.

The η^{-1} term in eq. (46b) for $\epsilon\bar{\psi}_{11}^{(1)}\cos\theta$ calls for an extra term in the outer solution of a doublet distribution along the circular vortex ring with strength $- 1.29\Gamma^2\tau\epsilon^2/\pi$.

For each instant t_0, or τ_0, with given vorticity distribution of eq. (7), (model A of Fig. 4), i.e.,

$$\zeta^{(o)} = \frac{\Gamma}{4\pi\nu\tau_0}\exp[-(r/\delta_0)^2] \quad \text{where} \quad \delta_0 = (4\nu\tau_0)^{\frac{1}{2}} \tag{48}$$

the velocity for the ring can then be defined by inviscid theory. By repeating the analysis carried out for model B of Fig. 4 on p. 241 of Lamb[2] step by step, the instantaneous velocity of the isolated ring is

$$\dot{Z}_0(t_0) = \Gamma[4\pi R_0]^{-1}\left\{\ell n[8R_0/\delta_0] - \frac{1}{2}\left[1 + \ell n2 - \int_0^\infty \ell nx \; e^{-x}dx\right]\right\}$$

$$= \Gamma[4\pi R_0]^{-1}\left\{\ell n[8R_0/\delta_0] - 0.558\right\}$$

The agreement between the present viscous theory and the classical inviscid theory on the instantaneous velocity of the ring with the same vorticity distribution $\bar{\zeta}^{(o)}$ is expected because eq. (33) for $\bar{\psi}^{(1)}$ contains viscosity only implicitly through $\bar{\zeta}^{(o)}$.

The classical inviscid theory is inaccurate for the subsequent motion due to the diffusion of the vortical core. The numerical examples in figs. 5 and 6 demonstrate this point. The motion of a circular vortex ring in a uniform stream U and a doublet at origin is studied. Without the ring it is a flow over a sphere of radius a. The Reynolds number $U a/\nu$ is 6.25×10^4. The strength of the vortex ring Γ is chosen to be $\frac{1}{2}U a$. The initial conditions at $t = 0$ are

$$Z_o(o) = -10a$$

$$\zeta_o(o) = \Gamma (\pi \delta_o^2)^{-1} \left\{ \exp - [r/\delta_o]^2 \right\}$$

Calculations have been carried out for $\delta_o/a = 10^{-2}$, 10^{-3}, and $R_o(o)/a = 0.5, 1$.

For the viscous theory, the velocity components $\dot{R}_o$ and $\dot{Z}_o$ are given by eqs. (40) and (47). The relationship between ζ and t given by eq. (45) requires the knowledge of the ring since its creation, therefore the differential relationship is used, i.e.

$$\tau_t = 1 - \tau \dot{R}_o/R_o \tag{49}$$

with $\tau(o) = \delta_o^2/(4\nu)$.

For the inviscid theory, $\dot{R}$ is also given by eq. (40) and $\dot{Z}$ is given by eq. (47) with $4\nu\tau$ replaced by $\delta^2(t)$. The relationship between δ and t is given by the Helmholtz vortex theorem. From the definition of $\delta(t)$, the radius of the core, it is clear that

$$\zeta(\delta, t)/\zeta(o, t) = e^{-1} \tag{50a}$$

$$\int_o^\delta \zeta(r, t)\, 2\pi r dr = \Gamma[1 - e^{-1}] = \text{const.} \tag{50b}$$

The conservation of mass then gives $2\pi R_o(t)(\pi \delta^2(t)) = \text{const.}$ or

$$\delta(t) = \delta_o[R_o(o)/R_o(t)]^{\frac{1}{2}} \tag{51}$$

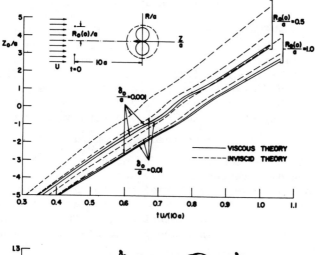

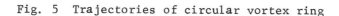

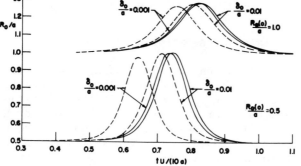

Fig. 5 Trajectories of circular vortex ring

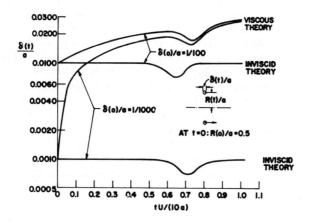

Fig. 6 Variation of the effective size of the vortex core

The condition fulfills eq. (44) without the diffusion term.

Figure 5 shows the differences between the trajectories of the viscous theory and the inviscid theory. The differences are more pronounced for smaller core size. Figure 6 shows the variation of the core size $\delta(t) = (4\nu\tau)^{\frac{1}{2}}$ defined by eq. (50a or b). The size in the inviscid theory does not change except during the passage over the doublet. In the viscous theory, the core radius increases due to diffusion at a higher rate for the core with smaller δ_o.

The trajectories in the inviscid theory for $\delta_o/a = 10^{-2}$ and 10^{-3}

are distinct due to the constant differences in core sizes while the corresponding differences in the viscous theory are much smaller. Similar calculations have been carried out for a circular vortex ring moving over a rigid sphere. For a non-similar initial vorticity distribution an optimum similar solution with the life time defined in section 2.2 can again be introduced. Details of these analyses will be reported elsewhere.

3.3 Extended Asymptotic Solutions

In the governing equation for the leading asymmetric part, eq. (33), time t appears as a parameter. This fact enables the definition of the velocity of the vortex. On the other hand, it represents a special asymptotic solution of the unsteady Navier-Stokes equation since it requires the initial data to be consistant with the solution. An extended asymptotic solution can be con-structed by the introduction of two time scales, the regular time t and the short time variable $\tilde{t}$ with $d\tilde{t}/dt = \epsilon^{-2}$. For two dimensional problems, it has been shown[6] that the average of the extended asymptotic solutions with two time scale over an interval of the regular time scale can be identified as the preceding solutions with only the regular time variable t. The same conclusion can be drawn for the axially symmetric problem. Instead of repeating the derivations for the general statement, an extended solution will be constructed for a special problem, namely, the initial velocity of the ring is prescribed to be different from that of preceding analysis. This phenomenon arises when a fixed vortex ring with life time t_1 is suddenly released or a free isolated vortex ring with life time t_1 encounters a sudden gust with velocity U in the axial direction as shown in fig. 7. The axial velocity of the free stream changes suddenly from 0 for $t < t_1$ to U for $t > t_1$. The extended solution for the inner core with two time variables, t and $\tilde{t} = (t-t_1)/\epsilon^{\delta}$, in the asymmetric part can be expanded in power series of ϵ, i.e.

$$\bar{\psi}(\tilde{t}, t, \bar{r}, \theta, \epsilon) = \bar{\psi}^{(o)}(\bar{r}, t) + \epsilon \bar{\psi}^{(1)}(\tilde{t}, t, \bar{r}, \theta) + \ldots \qquad (52)$$

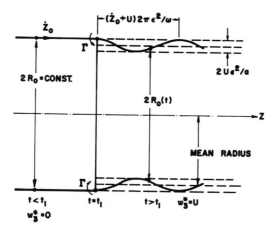

Fig. 7 Motion of circular vortex ring encountering a gust

and similarly for the vorticity function. The governing equation for the leading asymmetric part $\tilde{\psi}^{(1)}(\tilde{t},t,\bar{r},\theta)$ is

$$\tilde{\zeta}_{\tilde{t}} + \bar{\psi}_{\theta}^{(o)}\tilde{\zeta}_{\bar{r}}^{(1)} - \bar{\zeta}_{\bar{r}}^{(o)}\tilde{\psi}_{\theta}^{(o)} = \left[\overline{r\zeta}^{(o)}/R_o\right]\bar{\psi}_{\bar{r}}^{(o)} \sin\theta \qquad (53)$$

$\tilde{\zeta}_{\tilde{t}}$ represents the additional term to eq. (33). Equation (53) is now a partial differential equation in $\tilde{t}$ and $\bar{r}$ and can admit prescribed initial condition. It is easier to remove the inhomogeneous term by considering the deviation from the solution in eq. (46a) for the isolated vortex ring without encountering the gust, i.e.,

$$\tilde{\psi}^{(1)}(\tilde{t},t,\bar{r},\theta) = \psi^{(1)}(\tilde{t},t,\bar{r},\theta) + (-\Gamma^{3/2}\tau^{1/2}/\pi)f(\eta) \cos\theta \qquad (54)$$

with $\tau = t$ for $t \le t_1$. The deviation $\psi^{(1)}$ obeys the homogeneous differential equation of eq. (53) and the additional vorticity $\zeta^{(1)}$ is equal to $-\bar{\Delta\psi}^{(1)}/R_o$. The initial condition and the boundary conditions are,

$$\tilde{t} = 0, \quad \dot{\bar{z}}_o = 0, \quad \dot{\bar{R}}_o = 0 \qquad (55a)$$

$$\bar{r} = 0, \quad \underline{\psi}^{(1)} = 0, \quad \underline{\psi}_{\bar{r}}^{(1)} = 0 \qquad (55b)$$

and $\bar{r} \to \infty$, $\underline{\psi}^{(1)} \to -R_o\bar{r}\left\{[U - \dot{\bar{z}}_o(\tilde{t},t)]\cos\theta - \dot{\bar{R}}_o(\tilde{t},t) \sin\theta\right\}$ \qquad (55c)

The only nonhomogeneous condition of eq. (55c) suggests the dependence of $\underline{\psi}^{(1)}$ on θ, i.e.

$$\underline{\psi}^{(1)} = \underline{\psi}^{(1)}_{11}(\bar{t}, t, \bar{r})\cos\theta + \underline{\psi}^{(1)}_{12}(\bar{t}, t, \bar{r})\sin\theta \tag{56}$$

The governing equations for the Fourier components are:

$$\left[\underline{\varsigma}^{(1)}_{1,k+1}\right]_{\bar{t}} + (-1)^k\left[\bar{\varsigma}^{(0)}_{\bar{r}}\underline{\psi}^{(1)}_{1k} - \bar{\psi}^{(0)}_{\bar{r}}\underline{\varsigma}^{(1)}_{1k}\right]/\bar{r} = 0 \tag{57}$$

for $k=1,2$ with $\varsigma^{(1)}_{13} = \varsigma^{(1)}_{11}$ and $R_o\varsigma^{(1)}_{1k} = -\overset{*}{\psi}{}^{(1)}_{1k}$ and the operator $(*) = [\partial^2/\partial\bar{r}^2 + (1/\bar{r})\partial/\partial\bar{r} - (1/\bar{r}^2)]$. Equation (57) will be solved by the method of separation of variables, i.e.,

$$\underline{\psi}^{(1)}_{1k}(\bar{t}, t, \bar{r}) = (-1)^k V_R(\bar{t}, t) A_k(\bar{r}, t) \qquad \text{for } k=1,2 \tag{58}$$

and eq. (57) becomes

$$(-1)^k (V_k)_{\bar{t}}/V_{k+1} = \left[\bar{\varsigma}^{(0)}_{\bar{r}} A_k + \bar{\psi}^{(0)*}_{\bar{r}} A_k/R_o\right]/(\bar{r}A_{k+1}) = C_k(t) \tag{59}$$

where $A_3 = A_1$ and C_k are separation constants. By representing A_k as a power series, $\bar{r}^\lambda \Sigma\, a_{kn}\bar{r}^{-2n}$, the differential equation for A_k and the boundary condition at $\bar{r}=0$ yield the conditions for nontrivial solution,

$$C_1 C_2 = (64\pi^2 t_1^2)^{-1} \qquad \text{and} \qquad \lambda = 3$$

Due to the separation of variables, the only inhomogeneous boundary condition, eq. (55c), can be split as $\bar{r} \to \infty$,

$$A_1 \to \bar{r}, \qquad V_1 \to U - \dot{\underline{Z}}_o(\bar{t}, t) \tag{60a}$$

and

$$A_2 \to \bar{r}, \qquad V_2 \to \dot{\underline{R}}_o(\bar{t}, t) \tag{60b}$$

Due to symmetry it follows that $k_1 = k_2 = 1/(\delta\pi t_1)$ and $A_1 = A_2$. With $\bar{\varsigma}^{(o)}$ given by the similar solution, A_k can also be represented in a similar solution,

$$A_1 = A_2 = 2(\pi\tau)^{\frac{1}{2}} g(\eta)$$

The differential equation for $g(\eta)$ from eq. (59) is

$$\left\{1 - [1 - \exp(-\eta^2)]/\eta^2\right\}\, [g'' + (1/\eta)g' - g/\eta^2] - 4g\exp(-\eta^2) = 0$$

with the boundary conditions of $g \to \eta$ as $\eta \to \infty$ and $g(o) = 0$. The other half of eq. (59) gives

$$8\pi T_1 (V_1)_{\tilde{t}} = -V_2 \quad \text{and} \quad 8\pi T_1 (V_2)_{\tilde{t}} = V_1$$

The solutions of these equations with the initial conditions of $\underline{Z}_o(o, t_1) = 0$, $\underline{R}_o(o, t_1) = 0$ and with $V_1(o, t_1) = U$, $V_2(o, t_1) = 0$ from eqs. (60a,b), are

$$V_1 = U \cos\omega\tilde{t} \quad \text{and} \quad V_2 = U \sin\omega\tilde{t}$$

and $\dot{\underline{Z}}_o(\tilde{t}, t_1) = U(1 - \cos\omega\tilde{t})$ and $\dot{\underline{R}}_o(\tilde{t}, t_1) = -U \sin\omega\tilde{t}$ $\qquad$ (61)

with $\omega = C_k = 1/(8\pi t_1)$. Equation (61) implies that the velocity components of the ring after receiving the gust at t_1 fluctuate about the undisturbed value with a very small period of $2\pi\epsilon^2/\omega$ or $16\pi t_1 \epsilon^2$. The amplitude of oscillation of the radius of the ring is $U\epsilon^2/\omega$ and the drift of the mean radius is inward by the same amount as indicated by the sketch in fig. 7.

4. THREE DIMENSIONAL PROBLEM

For a curved vortex line in a three dimensional flow field, the flow field at a finite distance away from the core of the vortex line can again be given by the inviscid theory with velocity vector $\vec{V}$ split into two parts as given in eq. (3). The solution is quasi-steady, i.e., the time variable t appears as a parameter due to the motion of the curved vortex line. To study the inner solution, i.e., the solution of the vortical core, and to determine the velocity of the vortex line by the matching with outer solution, it is easier to work directly with the velocity components u,v,w relative to the center of the vortex instead of the single scalar stream function and vorticity which do not exist in the general problem.

The position vector of the center of the viscous vortex is represented as $\underline{\vec{X}}(s,t)$ where s is the arc length measured from a reference point along the vortex line at the initial instant (say t=0). At any instant t, the unit tangential, normal, and binormal vectors $\hat{t}$, $\hat{n}$ and $\hat{b}$ are related to the positive vector function $\underline{\vec{X}}(s,t)$ by the Serret-Frenet formula,

$$\underline{\tilde{X}}_s = \hat{t} \quad , \quad \hat{t}_{\tilde{s}} = k\hat{n} \quad , \quad \hat{b}_{\tilde{s}} = -T\hat{n} \qquad (62a)$$

$$\hat{n}_{\tilde{s}} = -k\hat{t} + T\hat{b} \quad \text{and} \quad \tilde{s}_s = |\vec{X}_s| \qquad (62b)$$

where $\tilde{s}$, k, T are the arc length, curvature and torsion of the curve

at the instant t. For a point P in the viscous core, its position
vector $\vec{P}$ can be written as (see Fig. 1)

$$\vec{P} = \vec{X}(s,t) + r\hat{r} \qquad (63)$$

where $\hat{r}$ lies in the $\hat{b}$, $\hat{n}$ plane of $\vec{X}(s,t)$. The angle between $\hat{r}$ and
$\hat{n}$ is defined as $\theta + \theta_o$ where θ_o is again a function of s and t.
When $\theta_o(s,t)$ is related to $\vec{X}(s,t)$ by $\theta_o(s,t) = -\int_0^s T(s',t)ds'$, the

coordinates s,r,θ become orthogonal and the differential change in
the position vector $\vec{P}$ is expressed by

$$d\vec{P} = \hat{r}h_1\,ds + \hat{r}ds + \hat{\theta}\,rd\theta \qquad (64)$$

with $h_1 = \tilde{s}_s[1-rk\cos(\theta+\theta_o)]$. The velocity of the point P can be
written as the sum of the local velocity of the vortex and the
relative velocity,

$$\vec{V}(s,\bar{r},\theta,t,\epsilon) = \vec{X}_t(s,t,\epsilon) + \bar{u}\hat{r} + \bar{v}\hat{\theta} + \bar{w}\hat{r} \qquad (65)$$

The relative velocity components $\bar{u},\bar{v}$ and $\bar{w}$ are also functions of s,
$\bar{r},\theta,t,\epsilon$ and $\bar{r}$ is the stretched variable r/ϵ. Due to the behavior of
the outer inviscid solution near $r\to0$ given by eq. (6), in which the
leading term is $\Gamma/(2\pi r)$ or $\Gamma/(2\pi\bar{r}\epsilon)$, the matching conditions suggest
the following expansions for $\bar{v},\bar{u}$ and $\bar{w}$

$$\bar{v}(s,\bar{r},\theta,t,\epsilon) = \epsilon^{-1}\bar{v}^{(o)}(s,\bar{r},\theta,t) + \bar{v}^{(1)}(s,\bar{r},\theta,t) + \epsilon\bar{v}^{(2)} +\ldots \quad (66a)$$

$$\bar{u}(s,\bar{r},\theta,t,\epsilon) = \qquad\qquad \bar{u}^{(1)}(s,\bar{r},\theta,t) + \epsilon\bar{u}^{(2)} +\ldots \quad (66b)$$

$$\bar{w}(s,\bar{r}\,\theta,t,\epsilon) = \qquad\qquad \bar{w}^{(1)}(s,\bar{r},\theta,t) + \epsilon\bar{w}^{(2)} +\ldots \quad (66c)$$

In order to later on balance the leading terms of the momentum
equations, the following expansion of pressure $\bar{p}(s,\bar{r},\theta,t,\epsilon)$ is
assumed,

$$\bar{p} = \epsilon^{-2}\bar{p}^{(o)} + \epsilon^{-1}\bar{p}^{(1)} + \bar{p}^{(2)} +\ldots \qquad (67)$$

Since the vortex line is assumed to move with finite velocity the
following expansion is introduced,

$$\vec{X}(s,t,\epsilon) = \vec{X}^{(o)}(s,t) + \epsilon\vec{X}^{(1)}(s,t) +\ldots \qquad (68)$$

The geometric parameters, k^{-1}, T and h_1 can be expanded likewise
and related to $\vec{X}^{(i)}(s,t)$. In particular, the expansions for $\tilde{s}$ and
h_1 are

$$\tilde{s}_s = |\vec{X}_s| = |\vec{X}_s^{(o)}| + \epsilon\vec{X}_s^{(o)}\cdot\vec{X}_s^{(1)}/|\vec{X}_s^{(o)}| +\ldots \qquad (68a)$$

$$h_1 = |\vec{X}_s^{(o)}| + \epsilon\left[\tilde{s}_s^{(1)} - |\vec{X}_s^{(o)}|\bar{r}k^{(o)}\cos(\theta+\theta_o^{(o)})\right] +\ldots \qquad (68b)$$

For the basic equations, $\nabla \cdot \vec{V} = 0$ and the momentum equations $\vec{V}_t + (\vec{V} \cdot \nabla)\vec{V} = -\nabla p + \nu \nabla \cdot \nabla \vec{V}$, the independent variables are transformed to the moving curvilinear coordinates $s, \bar{r}, \theta$ and t and the dependent variables become u, v, w in eq. (65) and p. Without losing any generality, the constant density has been equated to unity. The unknown vector function $\vec{X}(s, t)$ and the associated geometric parameters will be defined later by the matching conditions and the condition of the regularity of the flow field. After the substitution of the power series in ϵ for the variables, eqs. (66-68), into the basic equations, the coefficients of like powers of ϵ in each equation will be equated to zero. The governing equations for the leading and higher order solutions will be obtained systematically

4.1 Governing Equations

The coefficient of ϵ^{-2} in the continuity equation yields

$$\bar{v}_\theta^{(o)} = 0$$

The coefficients of ϵ^{-3} terms in the $\hat{r}$ and $\hat{\theta}$-components of the momentum equation yield respectively,

$$[\bar{v}^{(o)}]^2 / \bar{r} = \bar{p}_{\bar{r}}^{(o)} \quad \text{and} \quad \bar{p}_\theta^{(o)} = 0 \tag{69}$$

The leading terms in the $\hat{r}$-component of the momentum equation are of the order of ϵ^{-2}. The coefficients yield,

$$[\bar{v}^{(o)}/\bar{r}]\bar{w}_\theta^{(1)} = \bar{p}_s^{(o)}/|\vec{X}_s^{(o)}| \tag{70}$$

With $\bar{v}^{(o)}$, $\bar{p}^{(o)}$ and $\vec{X}^{(o)}$ independent of θ, eq. (71) can be integrated with respect to θ,

$$[\bar{v}^{(o)}/\bar{r}][\bar{w}^{(1)}(s, \bar{r}, \theta, t) - \bar{w}^{(1)}(s, \bar{r}, o, t)] = \theta \bar{p}_s^{(o)}/|\vec{X}_s^{(o)}|$$

Since $\bar{w}^{(1)}$ has to be single valued, it follows that $\bar{p}_s^{(o)} = 0$, which in turn yields

$$\bar{w}_\theta^{(1)} = 0 \quad \text{and} \quad \bar{v}_s^{(o)} = 0$$

It should be noted here that once the series expansions of eqs. (66-68) have been assumed, the independence of $\bar{p}^{(o)}$ and $\bar{v}^{(o)}$ on s and θ follows automatically. The governing equations for $\bar{v}^{(o)}(\bar{r}, t)$, however, will be obtained later in the higher order equation in the same manner as in the special cases described in sections 2 and 3.

For the sake of simplifying the higher order equations, the continuity equation will be eliminated. The complete continuity equation can be written as

$$(\bar{r}\bar{u}h_1)_{\bar{r}} + (h_1\bar{v})_\theta + \epsilon\bar{r}[\bar{w}_s + (\vec{X}_{st}\cdot\hat{\tau})] = 0$$

The three variables $\bar{u}, \bar{v}, \bar{w}$ can now be related to two variables $\bar{\psi}$ and $\bar{w}$ as follows,

$$\bar{v} = -\bar{\psi}_{\bar{r}}/h_1 \quad\text{and}\tag{71a}$$

$$\bar{u} = \left\{\bar{\psi}_\theta - \epsilon\int_0^{\bar{r}}\bar{r}d\bar{r}[\vec{X}_{st}\cdot\hat{\tau} + \bar{w}_s]\right\}/(\bar{r}h_1)\tag{71b}$$

The corresponding expansion for $\bar{\psi}(s,\bar{r},\theta,t,\epsilon)$ will be

$$\bar{\psi} = \epsilon^{-1}\bar{\psi}^{(0)}(\bar{r},t) + \bar{\psi}^{(1)}(s,\bar{r},t,\theta) + \epsilon\bar{\psi}^{(2)} +\ldots\tag{72}$$

The relationships between $\bar{u}^{(i)}$, $\bar{v}^{(i)}$, $\bar{w}^{(i)}$ and $\bar{\psi}^{(i)}$ are then obtained from eqs. (71a&b). The first few equations, which replace the continuity equation, are

$$\bar{v}^{(0)}(\bar{r},\theta) = -\bar{\psi}_{\bar{r}}^{(0)}/|\vec{X}_s^{(0)}|\tag{73a}$$

$$\bar{u}^{(1)}(s,\bar{r},\theta,t) = \bar{\psi}_\theta^{(1)}/(\bar{r}|\vec{X}_s^{(0)}|)\tag{73b}$$

$$\bar{v}^{(1)}(s,\bar{r},\theta,t) = -\bar{\psi}_{\bar{r}}^{(1)}/|\vec{X}_s^{(0)}| + h_1^{(1)}\bar{\psi}_{\bar{r}}^{(0)}/|\vec{X}_s^{(0)}|^2\tag{73c}$$

$$\bar{u}^{(2)}(s,\bar{r},\theta,t) = \bar{\psi}_\theta^{(2)}/(\bar{r}|\vec{X}_s^{(0)}|) - \bar{\psi}_\theta^{(1)}h_1^{(1)}/(|\vec{X}_s^{(0)}|^2\bar{r})$$
$$- (\bar{r}|\vec{X}_s^{(0)}|)^{-1}\int_0^{\bar{r}}\bar{r}d\bar{r}[\vec{X}_{st}^{(0)}\cdot\hat{\tau} + \bar{w}_s^{(0)}]\tag{73d}$$

The coefficients of ϵ^{-2} terms in the $\hat{r}$ and $\hat{\theta}$ components of the momentum equation yield,

$$(\bar{v}^{(0)}/\bar{r})\bar{u}_\theta^{(1)} - 2\bar{v}^{(0)}\bar{v}^{(1)}/\bar{r} = -\bar{p}_{\bar{r}}^{(1)}\tag{74a}$$

$$\bar{v}^{(0)}\bar{v}_\theta^{(1)} + \bar{u}^{(1)}[\bar{r}\bar{v}_{\bar{r}}^{(0)} + \bar{v}^{(0)}] = -\bar{p}_\theta^{(1)}\tag{74b}$$

Elimination of $\bar{p}^{(1)}$ from these two equations leads to the governing equation for the leading asymmetric term $\tilde{\psi}^{(1)}$ in $\bar{\psi}^{(1)}$, i.e.,

$$\bar{v}^{(0)}[(\bar{v}_\theta^{(1)}\bar{r})_{\bar{r}} - \bar{u}_{\theta\theta}^{(1)}]/\bar{r} + [\bar{v}_\theta^{(1)}+(\bar{u}^{(1)}\bar{r})_{\bar{r}}]\bar{\zeta}^{(0)}+\bar{u}^{(1)}\bar{r}\bar{\zeta}_{\bar{r}}^{(0)} = 0\tag{75a}$$

or $\bar{\Delta}\bar{\psi}_\theta^{(1)} - [\bar{\zeta}_{\bar{r}}^{(0)}/\bar{v}^{(0)}]\bar{\psi}_\theta^{(1)} = k^{(0)}\{2\bar{\zeta}^{(0)}\bar{r}+\bar{v}^{(0)}\}|\vec{X}_s^{(0)}|\ \sin(\theta+\theta_o)$ (75b)

where $\bar{\zeta}^{(0)}$ has been identified as $(\bar{r}\bar{v}^{(0)})_{\bar{r}}/\bar{r}$ and $\bar{\Delta}=\partial^2/\partial\bar{r}^2+\bar{r}^{-1}\partial/\partial r+\bar{r}^{-2}\partial^2/\partial\theta^2$. The boundary conditions for $\bar{\psi}$ at $\bar{r}=0$ are

$$\bar{u} = \bar{v} = 0 \quad \text{or} \quad \bar{\psi}^{(i)} = \bar{\psi}_{\bar{r}}^{(i)} = 0 \quad \text{for } i=0,1,2... \tag{76}$$

The conditions of matching the inner solution as $\bar{r} \to \infty$ with the outer solution as $r \to 0$ in eq. (6) are

$$\bar{v}^{(o)} \to \Gamma/(2\pi\bar{r}) \tag{77a}$$

$$\bar{\psi}^{(1)} \to |\vec{X}_s^{(o)}| \Gamma k^{(o)} (4\pi)^{-1} \bar{r} \cos(\theta+\theta_o) \, \ell n[s/(\epsilon\bar{r})]$$

$$+ |\vec{X}_s^{(o)}| \bar{r} \left\{ (\vec{V}_o - \vec{X}_t^{(o)}) \cdot \hat{n} \cos(\theta+\theta_o) - (\vec{V}_o - \vec{X}_t^{(o)}) \cdot \hat{b} \sin(\theta+\theta_o) \right\} \tag{77b}$$

and $\quad \bar{w}^{(1)} \to (\vec{V}_o - \vec{X}_t^{(o)}) \cdot \hat{\tau} \tag{77c}$

where $\vec{V}_o$ is the limit of $\vec{V}_f + \vec{V}^*$ as $r \to 0$.

Similar to sections 2 and 3, $\bar{\psi}^{(1)}(t,r,\theta)$ is resolved into Fourier components in $\theta+\theta_o$ with coefficients $\tilde{\psi}_{jk}^{(1)}$ with k=1,2 and j=0,1,2,.... The asymmetric part $\tilde{\psi}^{(1)}(t,r,\theta)$, i.e. $j\neq 0$, is governed by eq. (75b) which yields for each Fourier component,

$$\left\{ \frac{\partial^2}{\partial\bar{r}^2} + \frac{1}{\bar{r}}\frac{\partial}{\partial\bar{r}} - \left[\frac{1}{\bar{r}^2} + \frac{\bar{\zeta}_{\bar{r}}^{(o)}}{\bar{v}^{(o)}} \right] \right\} \tilde{\psi}_{jk}^{(1)} = - k^{(o)} \left\{ 2\bar{\zeta}_{\bar{r}}^{(o)} + \bar{v}^{(o)} \right\} |\vec{X}_s^{(o)}| \delta_1 {}_j \delta_{1k} \tag{78}$$

Prior to the determination of $\bar{\zeta}^{(o)}$ and $\bar{v}^{(o)}$, the following conclusions can be drawn from eq. (78) and the boundary conditions in the same manner as in sections 2 and 3,

$$\tilde{\psi}_{12}^{(1)} = 0 \quad \text{and} \quad \tilde{\psi}_{jk}^{(1)} = 0 \quad \text{for } k=1,2, \; j=2,3,... \tag{79a}$$

and $\tilde{\psi}_{11}^{(1)} \to -\Gamma(4\pi)^{-1} k^{(o)} |\vec{X}_s^{(o)}| \left\{ \bar{r}\ell n\bar{r} + C_1 \bar{r} + C_2 \bar{r}^{-1} + o(\bar{r}^{-1}) \right\} \tag{79b}$

Coefficients C_1 and C_2 are unknown functions of t and s to be determined after the solution of eq. (78) with given $\bar{\zeta}^{(o)}$ and $\bar{v}^{(o)}$. It is clear that the present viscous analysis removes the singularities of the inviscid theory. The matching condition of eq. (77b) gives

$$\vec{X}_t^{(o)}(s,t) \cdot \hat{n} = \vec{V}_o(s,t) \cdot \hat{n} \quad \text{and} \tag{80a}$$

$$\vec{X}_t^{(o)}(s,t) \cdot \hat{b} = \vec{V}_o(s,t) \cdot \hat{b} + \Gamma k^{(o)}(4\pi)^{-1} \left\{ \ell nS/\epsilon + C_1(t,s) \right\} \tag{80b}$$

The component $\vec{X}_t^{(o)} \cdot \hat{\tau}$ will not influence the motion and the configuration of the vortex line. $\vec{V}_o(s,t)$ is the value of $\vec{V}_f + \vec{V}^*$ at r=0, i.e., at $\vec{X}^{(o)}$ where $\vec{V}_f$ and $\vec{V}^*$ are defined by eqs. (3,4,5). Similar to the case of a circular vortex ring in section 3 the velocity component normal to the plane of oscillation of the vortex line, $\vec{X}_t^{(o)} \cdot \hat{b}$, which depends on $C_1(t,s)$, can be defined only after the determination of the leading symmetric part of the inner solution $\bar{v}^{(o)}(t,\bar{r})$.

4.2 Decay of Vorticity and Similar Solution

The governing equation for the leading symmetric part can be obtained by the general procedure outlined in section 2. The coefficients of ϵ^{-1} terms in the θ-component of the momentum equation are equated to zero. The equation is integrated with respect to θ from 0 to 2π. By making use of the single valuedness of the physical quantities and of the eqs. (73a-d) and (75a), the result is

$$\frac{\partial \bar{v}^{(o)}}{\partial t} - \Gamma\left[\frac{1}{\bar{r}}\frac{\partial}{\partial \bar{r}} \bar{r} \frac{\partial}{\partial \bar{r}} - \frac{1}{\bar{r}^2}\right]\bar{v}^{(o)} = \frac{(\bar{r}\bar{v}^{(o)})_{\bar{r}}}{\bar{r}^2|\vec{X}_s^{(o)}|} \int_o^{\bar{r}} (\vec{X}_{st}^{(o)}\cdot\hat{\tau}+\bar{w}_s^{(1)})\bar{r}d\bar{r} \qquad (81)$$

Since $\bar{v}^{(o)}$ is a function of t and $\bar{r}$ only, the integral over $|\vec{X}_s^{(o)}|$ has to be a function of $\bar{r}$ and t, say $\bar{r}^2 g(t,\bar{r})$, i.e.,

$$\int_o^{\bar{r}} (\vec{X}_{st}^{(o)}\cdot\hat{\tau} + \bar{w}_s^{(1)})\bar{r}d\bar{r} = \bar{r}^2 g(t,\bar{r})|\vec{X}_s^{(o)}| \qquad (82)$$

By noting that $|\vec{X}_s| = \tilde{s}_s$ from eq. (68a) and that $\vec{X}_{st}\cdot\hat{\tau} = (\tilde{s}_s\hat{\tau})_t\cdot\hat{\tau} = \tilde{s}_{ts}$ eq. (82) can be integrated with respect to s along the total length S of the closed vortex line. The result is

$$g(t) = \frac{1}{2} \dot{S}(t)/S(t) \qquad (83)$$

Substitution of eqs. (82) and (83) into eq. (81) yields

$$\frac{\partial \bar{v}^{(o)}}{\partial t} - \Gamma\frac{\partial}{\partial \bar{r}} \left\{\frac{1}{\bar{r}}\frac{\partial}{\partial \bar{r}}\left[\bar{r}\bar{v}^{(o)}\right]\right\} = \frac{\dot{S}(t)}{2S(t)}\left[\bar{r}\bar{v}^{(o)}\right]_{\bar{r}}$$

With $[\bar{r}\bar{v}^{(o)}]_{\bar{r}}$ identified as $\bar{\zeta}^{(o)}$, the leading vorticity component tangential to the vortex line, the equation becomes

$$\frac{\partial \bar{\zeta}^{(o)}}{\partial t} - \frac{\Gamma}{\bar{r}}\frac{\partial}{\partial \bar{r}} \bar{r} \frac{\partial \bar{\zeta}^{(o)}}{\partial \bar{r}} = \frac{\dot{S}(t)}{S(t)}\left[\bar{\zeta}^{(o)} + \frac{1}{2} \bar{r}\bar{\zeta}_{\bar{r}}^{(o)}\right] \qquad (84)$$

It should be pointed out here that with $k^{(o)}(s,t)$ reduced to $1/R_o(t)$, $|\vec{X}_s^{(o)}|$ to $R_o(t)/R_o(o)$ and $\dot{S}(t)/S(t)$ to $\dot{R}_o(t)/R_o(t)$; the the governing equations for the three dimensional problem, eqs. (75b), (78) and (84) are identified with eqs. (33), (36) and (34) respectively for the axially symmetric problem. With $k^{(o)}$ and $\dot{S}(t)$ equal to zero and $|\vec{X}_s^{(o)}|$ equal to unity, these equations are identified with eqs. (13), (21) and (24) for the two dimensional problem.

In particular with $R_0(t)/R_0(t)$ replaced by $S(t)/S(t)$ eq. (34) is the same as eq. (84), therefore, the similar solution for eq. (84) is again given by eq. (7), i.e.,

$$\zeta^{(0)} = (4\pi\tau_3)^{-1} \exp(-\eta^2) \qquad (85a)$$

where $\eta^2 = \bar{r}^2/(4\Gamma\tau_3)$ and $\tau_3 = \int_0^t S(t')dt'/S(t)$ (85b)

The corresponding similar solution, $f(\eta)$, for eq. (78) is the same as eq. (45), i.e.

$$\tilde{\psi}_{11}^{(1)}(t,\bar{r}) = -(\Gamma^{3/2}\tau_3^{1/2}/\pi) \, k^{(0)} |\vec{X}_s^{(0)}| f(\eta) \qquad (86)$$

The asymptotic behavior of $f(\eta)$ is the same as eq. (46), only the definition of η or τ_3 is changed by eq. (85b). By comparison with eq. (79b) $C_1(t,s)$ is identified as $0.442 - \frac{1}{2}\ln(4\Gamma\tau_3)$ and eq. (80b) becomes

$$\vec{X}_t^{(0)}(s,t)\cdot\hat{b} = \vec{V}_0(s,t)\cdot\vec{b} + \Gamma k^{(0)}(4\pi)^{-1}\left\{\ln[S/(4\nu\tau_3)^{\frac{1}{2}}] + 0.442\right\} \qquad (87)$$

The constant 0.442 is an invariant number for the similar solution. The constant -0.558 in eq. (47) is meaningful only for circular vortex rings where w_3^* is identified as $\vec{V}^*\cdot\vec{b}$ while $\vec{V}_f\cdot\vec{b}$ combined with the $\ln S$ term as $\ln(8R_0)-1$.

The comparisons between the viscous and the inviscid theories made in section 3 apply also in the present case. Since eq. (78) does not contain any viscous term explicitly, the velocity of the vortex line given in the present analysis will agree with inviscid theory for each instant provided the same vorticity distribution is assumed for the inviscid theory. Present analysis yields the correct time history of the vorticity diffusion which the inviscid theory fails to do.

4.3   Condition of Compatibility

It should be noted that eq. (82) yields $\bar{w}_{s\bar{r}}^{(1)}=0$. With the boundary condition of $\bar{w}^{(1)}=0$ at $\bar{r}=0$, it follows that

$$\bar{w}_s^{(1)} = 0 \quad \text{for all } \bar{r} \quad \text{and} \quad \tilde{s}(s,t) = sS(t)/S(o). \qquad (88)$$

Equation (88) says that the vortex line will move and deform in such a way that its arc length will change in direct proportion to the total length of the vortex line at each instant. This is a very strong geometrical constraint. Indeed, the matching condition for $\bar{w}^{(1)}$ as $\bar{r} \to \infty$ and eq. (88) yield

$$[(\vec{V}_o(s,t) - \vec{X}_t^{(o)}) \cdot \hat{\tau}]_s = 0$$

which in turn can be written as with the aid of eq. (80a)

$$[\bar{V}_o(s,t)]_{ss} \cdot \hat{\tau} + [\bar{V}_o(s,t)]_s \cdot k\hat{n} = 0 \qquad (89)$$

Since $\bar{V}_o(s,t)$ is the value of $\vec{V}_f + \vec{V}^*$ in eq. (6) at $\vec{X}(s,t)$, eq. (89) imposes a compatability condition on the shape of the vortex line and the prescribed three dimensional flow field. Condition (85) is always fulfilled for two dimensional or axially symmetric problems which are independent of s.

For a three dimensional problem, the initial data, i.e., the initial position of the vortex line $\vec{X}(s,o)$ and the outer flow field, may not fulfill eq. (89). In this case, the special asymptotic expansions of this section should be extended by introducing multiple time scale as done in section 3.3. Whether this extended solution will approach the special asymptotic solution fulfilling eq. (89) as time increases is still an open question and is currently under invesitgation.

Conclusions

By the assumption of a stretching and inner-outer expansion scheme, special asymptotic solutions of Navier-Stokes equations have been obtained. The leading term in the outer solution is given by the classical inviscid theory. The inner solution removes the singularities of the inviscid theory at the vortex line regardless of the vorticity distribution in the viscous core.

The velocity of the vertex line, except in the two dimensional case, depends on the vorticity distribution. For the same vorticity distribution at each instant the present theory and the classical inviscid theory yield the same velocity for the vortex line; however, the latter fails to account for subsequent diffusion of vorticity in the small vortical core. In the present analysis the governing equation for the vorticity distribution includes both the diffusion and curvature effects. A similar solution can be constructed and the curvature effect is included in a transformation of the time variable. For a given initial vorticity distribution, it is shown that the solution will approach a similar solution for large time. An similar solution optimum in this asymptotic sense is defined in section 2.2. Its two parameters, the total strength Γ and the life time at the initial station t_o are defined by the matching of the total integral of the initial vorticity distribution and its second moment.

For a curved vortex line in a three dimensional flow field, the present scheme leads to a compatibility condition on the shape of the vortex line and the local three dimensional flow velocity. This condition is always fulfilled in the two dimensional or axially asymmetric cases. The condition defining the velocity of the vortex line can of course be looked upon as a condition of compatibility caused by the expansion scheme. When the initial conditions do not agree with the condition of compatibility the expansion scheme should be generalized or extended. An example of the extended asymptotic solution is given in section 3.3.

ACKNOWLEDGEMENT

This research is supported by AFOSR Grant 67-1062C. The author wishes to acknowledge the collaborations of Professors Max Gunzburger and Susan J.N. Shaw of the Department of Mathematics and Dr. Huai-Chu Wang of the Aerospace Laboratory, New York University.

REFERENCES

1. Prandtl, L. Essentials of Fluid Dynamics, Hafner Publ, N.Y., 1952.
2. Lamb, H., Hydrodynamics, 6th ed., Dover, N.Y., 1932.
3. Prandtl, L. and Tietjens, O.G., Fundamentals of Hydro and Aeromechanics, Dover, N.Y., 1934.
4. van Dyke, M., Perturbation Methods in Fluid Dynamics, Academic Press, New York, 1964.
5. Cole, J.D., Perturbation Methods in Applied Mathematics, Blaisdell Publ., Toronto, 1968.
6. Ting, L. and Tung, C., Motion and Decay of a Vertex in a Non-uniform Stream, Phys. Fluids vol. 8, pp. 1039-1051, June 1965.
7. Tung, C. and Ting, L., Motion and Decay of a Vortex Rings, Phys. Fluids vol. 10, pp. 901-910, May 1967.
8. Carslaw, H.S. and Jaeger, J.C., Conduction of Heat in Solid, Oxford Univ. Press., London, 1959.
9. Ting, L. and Chen, S., Perturbation Solutions and Asymptotic Solutions in Boundary Layer Theory, J. Eng. Math. vol. 1, pp. 327-340, October 1967.
10. Kleinstein, G. and Ting, L., "Optimum One-term Solutions for Heat Conduction Problems" N.Y.U. report AA-68-43, to appear in ZaMM, 1970
11. Magnus, W., Oberhettinger, F. and Soni, R.P., Formulas and Theorems for Special Functions, Springer-Verlag, N.Y., 1966.

TRANSPORT OF A VORTEX WAKE IN A STABLY STRATIFIED ATMOSPHERE

I. H. Tombach

Meteorology Research, Inc.

ABSTRACT

Atmospheric stratification affects the downward motion of an aircraft vortex wake and may influence the persistence and stability of the vortex pair configuration. Observations of actual wakes have shown significant variation in the distance to which they descend and in their lifetimes under different degrees of atmospheric stability. This behavior has been modeled analytically in this paper as a pair of infinite vortices in an inviscid, compressible, stably stratified atmosphere with entrainment characterized by a single parameter which is related to the difference between the density in a particular region of the wake and that external to the wake. It has been found that the motion of such a vortex system is governed by a parameter Q which depends on the initial circulation and vortex spacing, on the atmospheric stability, and on the entrainment parameter.

The nature of the transport follows one of two patterns, depending on whether Q is less than or greater than a critical value Q_{crit}. If $Q < Q_{crit}$, the circulation decreases more rapidly than the momentum and the vortices separate as they descend to an equilibrium level. If $Q > Q_{crit}$, the momentum of the vortices decreases more rapidly than the circulation and, after an initial period of slow divergence, the vortices attempt to converge as they descend. In both cases, the descent takes place in a well-defined characteristic time which depends solely on the atmospheric density gradient.

1. INTRODUCTION

The transport of the trailing vortex wake of an aircraft is frequently modeled by the laminar potential flow solution for a parallel pair of infinite line vortices.[1] This solution indicates that, in the absence of ground effect, the two vortices will descent indefinitely at a constant velocity while maintaining a constant separation. Observations of actual wakes[2,3] have shown, however, that the vortices often stop their descent at some level below the aircraft and then remain near that level until they dissipate. The amount of descent varies from time to time, with atmospheric stability apparently a dominant factor in determining its magnitude. The variation can be quite great. For example, the wake from an aircraft has been observed to descend twenty times as far in the neutrally stable atmosphere over the ocean as in the stable air over land.[4]

A theoretical analysis of the detailed behavior of the wake under these atmospheric conditions is extremely difficult because of the great number of variables involved. Dominant factors involved in a description of the motion include the atmospheric stability, the initial properties of the wake, and the turbulence in the atmosphere and in the wake, which govern the rate at which the wake entrains the surrounding atmosphere and the rate and manner in which the entrained fluid is mixed into the wake. This paper represents an effort to create a simplified approximate model of the dominant effects of atmospheric stability on the transport of a trailing vortex wake, in which only the major aspects of the wake motion are considered and many of the second order effects are ignored. In this model, the wake is represented by a pair of parallel infinite vortices embedded in a nearly elliptical cylinder of accompanying fluid which moves through an inviscid, compressible, stably stratified atmosphere. The effects of entrainment and buoyancy on the circulation are represented by a simplified single parameter model. The approach taken is similar in many respects to that used by Turner.[5]

2. THE EQUATIONS OF MOTION

A schematic portrayal of the vortex pair is shown in Fig. 1. The vorticity is assumed to be concentrated into two contrarotating vortex cores whose diameters are small compared to their separation and each of which has a circulation of magnitude Γ. Fluid within the nearly elliptical cylinder marked by the dashed lines is assumed to be convected together with the vortices, as in the potential flow solution,[6] hence the dashed line is a streamline separating the flow field into two regions: the external atmosphere (whose properties are represented by ρ and T), and the fluid accompanying the vortices (with average properties ρ' and T'). Since

the vortices entrain the ambient air, this separation cannot hold
rigorously, but it seems a reasonable approximation if the entrainment rate and the descent velocity of the vortex system are sufficiently slow so that, at any particular instant, a near equilibrium state exists and the elliptical shape holds. This assumption
of geometric similarity, in which the scale of the wake profile is
b, is not valid though for large well mixed wakes in a strongly
stable atmosphere, for which a collapse in the vertical extent of
the wake due to gravity, may occur.[7]

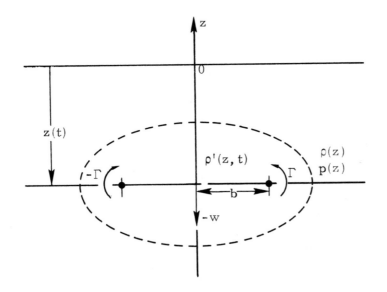

Fig. 1. Schematic cross section of vortex wake

Initially, at $t = 0$, the fluid properties within the vortex
system are the same as those external to it. The properties at the
initial level, $z = 0$, are represented by the subscript $(\ )_o$. It is
assumed that the extreme variations encountered for the density are
small, i.e., the Boussinesq assumption that $\left|\dfrac{\rho(z)-\rho_o}{\rho_o}\right| \ll 1$ is made,
which holds as long as the descent distance is small compared to the
scale height of the atmosphere, viz., as long as $|z| \ll 10$ km.

Using these assumptions, the consequences of which will be discussed later, the equations of motion for the vortex system follow
easily. The downward velocity of the entire vortex system is
given by

$$w = -\frac{\Gamma}{4\pi b} \tag{1}$$

and the overall vertical momentum of the motion (sometimes called
the impulse, and defined as the time integral of all the external
forces required to achieve this state of motion[8]) is

$$M = -2\rho' \Gamma b$$

$$\simeq -2\rho_o \Gamma b \tag{2}$$

The momentum is affected by the buoyancy which the accompanying
fluid acquires, so the momentum conservation equation is

$$\frac{dM}{dt} = g \ V \ (\rho-\rho')$$

$$= \rho_o F \tag{3}$$

where

$$F = g \ V \ \frac{\rho-\rho'}{\rho_o} \tag{4}$$

is a measure of the buoyancy, and the cross-sectional area of the
accompanying oval cylinder of fluid is $V = qb^2$. The value of the
factor q can be computed numerically for the potential flow solution, and is approximately 11.62.

The mass of fluid in the oval cylinder changes only if there
is entrainment of the ambient air. As the cylinder descends, the
volume occupied by the accompanying fluid changes due to compressi-
bility, as well as entrainment, hence the mass conservation equation is

$$\frac{d}{dt} (\rho'V) = \rho\left(\frac{dV}{dt} - \frac{gV}{a^2} w\right) \tag{5}$$

where a is the speed of sound in the ambient air. It is assumed
here that there is no heat transfer to the vortex system, except
for that which is convected along with the entrained mass, so that
changes in the cylinder volume would occur adiabatically in the
absence of entrainment. The term gVw/a^2 in Eq. (5) is the adiabatic rate of change of volume, which is subtracted from the total
rate of volume change to compute the increase in volume due to entrainment.

It is convenient at this point to introduce

$$G = -\frac{g}{\rho_o} \frac{d\rho}{dz} - \frac{g^2}{a_o^2} \tag{6}$$

as a measure of the stability of the atmospheric stratification.
G is the difference between the actual atmospheric density gradient
and the adiabatic one, multiplied by $- g/\rho_0$. Over the range of
altitude a wake travels, it is consistent with previous approximations to assume G to be a constant. Positive values of G correspond to stable stratification.

Some manipulation of Eq. (5) gives then the mass conservation
equation as

$$\frac{dF}{dt} = - \, GVw \tag{7}$$

assuming again that density changes are small over the distance
traveled.

When the vortex system has acquired buoyancy, the circulation
about each vortex is altered by the buoyancy. The circulation
equation for inhomogeneous fluids (the Bjerkness equation) is[9]

$$\frac{d\Gamma}{dt} = - \oint_C \frac{1}{\rho} \, \nabla p \cdot \underline{d\ell} \tag{8}$$

where C is the contour along which the circulation is computed.
For the present application, take C as shown in Fig. 2, so that
legs 1 through 3 are entirely in the ambient atmosphere and leg 4
bisects the oval along the line of symmetry between the vortices.
Then, assuming the density outside the oval to be of an undisturbed

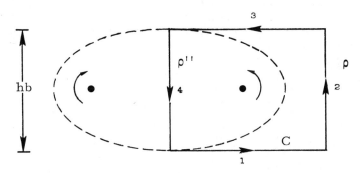

Fig. 2.   Circuit used to compute rate of change of circulation

constant value ρ over the distance corresponding to the height of
the oval, and the density along leg 4 inside the oval to be ρ'',
and noting that $\nabla p = - \rho g$ in the vertical direction, evaluation of
the line integral in Eq. (8) gives, taking legs 1 through 4 in

order

$$\frac{d\Gamma}{dt} = 0 + \frac{\rho g}{\rho} hb + 0 - \frac{\rho g}{\rho''} hb$$

$$= - ghb \frac{\rho - \rho''}{\rho''} \tag{9}$$

where hb is the height of the oval. The density ρ'' along the
dividing line need not be the same as the average over the whole
oval, ρ', and the difference between ρ'' and ρ' will depend on the
entrainment rate and the internal mixing rate. Hence the rate of
change of circulation depends on the "aspect ratio" of the oval,
the buoyancy and entrainment, the internal mixing, and the scale
of the vortex system. This change in circulation is assumed to
affect the wake as though it took place as a decrease in the circulation of the concentrated vortices, although in actuality the
vorticity would be distributed over the volume of the wake.

Since buoyancy will tend to make the upper surface of the oval
cylinder unstable while stabilizing the lower surface, and since
the descending cylinder is a bluff body, it is likely that most of
the entrained mass of ambient fluid will be drawn in at the upper
side of the oval. Because of the rotation of the vortices, it will
then initially circulate downward near the dividing plane between
the two vortices while mixing with the wake fluid. It is useful to
interpret the mixing in terms of a length by noting that the
density ρ'' along this dividing plane will be formally the same
density as that which would be obtained by adiabatic compression of
a parcel of ambient air taken from a level some distance Δz above
the level of the vortex system. Thus Δz can be considered as a
dimension which characterizes the rate of entrainment and the rate
at which the entrained air mixes with the fluid already within the
cylinder.

Relating $\rho''(z)$ to $\rho(z+\Delta z)$ using the adiabatic density gradient, and $\rho(z)$ to $\rho(z+\Delta z)$ using the actual density gradient, and
using the usual density approximation, Eq. (9) can thus be written

$$\frac{d\Gamma}{dt} = - Ghb\Delta z \ . \tag{10}$$

At this point, it is necessary to assume some element of the
entrainment process in order to determine the behavior of Δz. It
is instructive to consider some representative ways in which ρ''
could vary as the wake descends. One extreme sample of possible
behavior is the case with $\Delta z = 0$, which would occur if the internal
mixing was sufficiently weak so that the entrained fluid was still
at ambient density along the dividing plane, and thus circulation
would remain constant as the wake descended. There would still be

some buoyancy though because, although $\Delta z = 0$ implies that ρ'', the density along the dividing plane, will be equal to the local ambient density ρ, it is likely that the mean internal density ρ' will still be less than ρ. If Δz is greater than zero, then there is mixing between the entrained fluid and the buoyant fluid already within the oval and thus ρ'' is less than the ambient density. As will be shown later, an increase in Δz also corresponds to an increased entrainment rate, which would be expected to follow from the increased mixing.

For a realistic case with finite entrainment, it seems valid to assume that, if the heretofore hypothesized geometric similarity of the wake configuration is reasonable, then Δz should probably also scale in the same way as all the other dimensions. Hence, it is assumed that $\Delta z = mb$, where m is an unknown constant. (The vortex half-spacing b is not the only length scale which could be used, but it is probably the dominant one for Δz. If Δz is artificially visualized as the distance above the wake from which the fluid along the dividing plane was supplied, such an assumption seems plausible.) A similar assumption, simplified for an incompressible fluid, was used by Turner[5] and experiments by Woodward[10] with thermals (which are like buoyant entraining vortex rings) seem to provide some experimental substantiation for it.

Using the above assumption and incorporating the constants h and m into one constant $s - hm$ allows Eq. (10) to be written as

$$\frac{d\Gamma}{dt} = - sGb^2 \ . \tag{11}$$

Eqs. (1), (2), (3), (7), and (11) are the five equations which will be solved in the next section for the five unknowns M, F, Γ, b, and w in terms of the initial conditions, the stability of the atmosphere, G, and the parameter s.

3. THE SOLUTION FOR THE MOTION

It is useful to non-dimensionalize all of the variables, so let

$$t = \left(\frac{8\pi}{q}\right)^{1/2} G^{-1/2} \tau \tag{12}$$

$$F = Bf \tag{13}$$

$$M = \rho_o \left(\frac{8\pi}{q}\right)^{1/2} BG^{-1/2} m \tag{14}$$

$$\Gamma = \frac{1}{2}\left(\frac{8\pi}{q}\right)^{1/2}(3s)^{1/3}\,B^{2/3}\,G^{-1/6}\,\gamma \qquad (15)$$

$$b = (3s)^{-1/3}\,B^{1/3}\,G^{-1/3}\,\beta \qquad (16)$$

and $$\qquad z = (3s)^{2/3}\,q^{-1}\,B^{1/3}\,G^{-1/3}\,\zeta \qquad (17)$$

where τ, f, m, γ, β, and ζ are the non-dimensional variables corresponding to t, F, M, Γ, b, and z, respectively. The constant B will be chosen later. Note that for τ to be real and meaningful G must be positive and non-zero, i.e., the density gradient must be stable. (Solution for $G = 0$ will not be considered here.)

In terms of the non-dimensional variables, the equations of motion become

$$\frac{d\zeta}{d\tau} = -\frac{\gamma}{\beta} \qquad (18)$$

$$m = \gamma\beta \qquad (19)$$

$$\frac{dm}{d\tau} = f \qquad (20)$$

$$\frac{df}{d\tau} = -\beta^2\,\frac{d\zeta}{d\tau} \qquad (21)$$

and $$\qquad \frac{d\gamma}{d\tau} = -\frac{2}{3}\,\beta^2 \qquad (22)$$

The solutions, using the initial conditions $\gamma(\tau = 0) = \gamma_0$, $m(\tau = 0) = m_0$, $\beta(\tau = 0) = \beta_0$, $f(\tau = 0) = 0$, and $\zeta(\tau = 0) = 0$ are easily found to be

$$m = m_0 \cos \tau \qquad (23)$$

$$f = -m_0 \sin \tau \qquad (24)$$

$$\gamma = m_0^{2/3}\left\{Q - (\tau + \sin \tau \cos \tau)\right\}^{1/3} \qquad (25)$$

$$\beta = -m_0^{1/3}\,\frac{\cos \tau}{\left\{Q - (\tau + \sin \tau \cos \tau)\right\}^{1/3}} \qquad (26)$$

and $$\qquad \frac{d\zeta}{d\tau} = m_0^{1/3}\,\frac{\left\{Q - (\tau + \sin \tau \cos \tau)\right\}^{2/3}}{\cos \tau} \qquad (27)$$

where $Q = \gamma_o^3/m_o^2$. The non-dimensional descent distance ζ is obtained by a numerical integration of Eq. (27). It should be recalled that m_o is negative for the usual wake.

Although m and f depend only on the time τ, the other variables γ, β, and $d\zeta/d\tau$ depend also on the initial parameter $Q = \gamma_o^3/m_o^2$. In particular, the solutions differ depending on whether this initial parameter is less than, equal to, or greater than $\pi/2$. If $Q < \pi/2$, the circulation decays to zero before the momentum; if $Q = \pi/2$, the circulation and momentum reach zero simultaneously; and if $Q > \pi/2$, the momentum decays more rapidly than the circulation. Since all the solutions are expressed in terms of some power of m_o, reference to Eqs. (12) through (17) gives the interesting result that B can be chosen arbitrarily.

4. DISCUSSION OF THE MOTION

The solutions given by Eqs. (23) through (27) are plotted in Fig. 3 for a representative value of $Q < \pi/2$, in Fig. 4 for $Q = \pi/2$, and in Fig. 5 for $Q > \pi/2$. Also shown on these figures is the numerically computed curve for the non-dimensional descent ζ.

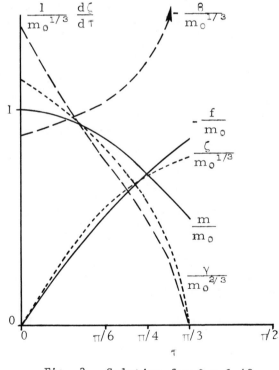

Fig. 3. Solution for $Q = 1.48$

Some major characteristics of the solution can be obtained
directly from the non-dimensionalization. In particular, Eq. (12)
shows that the time scale of the motion is proportional to $G^{-1/2}$,
hence the greater the atmospheric stability, the shorter the actual
time which is required for the motions shown in Figs. 3, 4, and 5
to take place. Similarly, Eq. (17 indicates that the descent will
be less if the stability is greater, which is further reinforced
by the decrease in Q as the stability increases. (See Eq. (28)
below.)

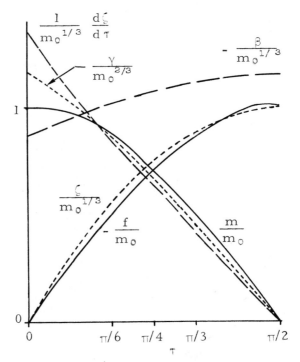

Fig. 4.   Solution for $Q = \pi/2$

Interpretation of some aspects of the motion is easier in
terms of the physical variables. In particular, the initial condition parameter is

$$Q = \frac{\gamma_0^3}{m_0^2} = \frac{2}{3sG^{1/2}} \left(\frac{q}{8\pi}\right)^{1/2} \frac{\Gamma_0}{b_0^2} \qquad (28)$$

so a decrease in stability will increase Q, as will an increase in
initial circulation or a decrease in the initial vortex spacing.

The case with Q < π/2 might be called a "weak vortex" or strong stability case. Here the buoyancy kills the circulation quite rapidly, so that at some time τ < π/2 the circulation becomes zero while m is still non-zero. This requires that the spacing between the vortices become very large (at τ = π/3 for the example in Fig. 3). Overall, the behavior qualitatively resembles that of a vortex pair in ground effect.

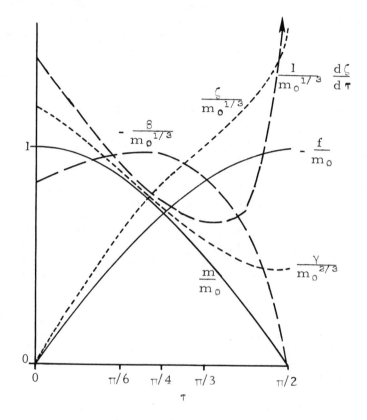

Fig. 5. Solution for Q = 1.66

If the initial circulation is increased until Q = π/2, then the situation shown in Fig. 4 prevails. The circulation and momentum both decay to zero in the same time, at which point the wake is a motionless, buoyant cylinder of scale b = 4/3 b_o. Subsequent solution as the wake buoys upward after τ = π/2 is not straight-forward since it is not reasonable to assume that s is a constant through the last stages of decay and the start of the buoyancy-created circulation. For times τ < π/2, the solution is qualita-tively the same as that in Fig. 3, with the circulation, momentum, and velocity decreasing and the buoyancy and separation increasing.

For the strong vortex or weak stability case, in which $Q > \pi/2$, the initial behavior is again the same, as shown in Fig. 5. But, as the momentum decays more rapidly than the circulation, the vortices begin to close together and the velocity of descent begins to increase. In a real wake, although a certain amount of shrinking of the scale is possible since the pressure increases as the wake descends, any substantial decrease in scale would require a loss of fluid from the wake. It is unlikely that such a loss of fluid could take place in an orderly manner, and thus the shrinking portion of the solution is probably unstable and may result in a collapse of the ordered motion.

The rate of entrainment by the vortex system can be calculated directly from the equations of motion, and is best expressed as the rate of change of wake mass per unit distance of descent:

$$- \frac{d}{dz} (\rho'V) = \rho V \left| \frac{\rho_o g}{a_o^2} \frac{1}{\rho} - \frac{4\pi F}{\Gamma^2} + \frac{\pi sG}{\rho_o} \frac{(-M)^3}{\Gamma^5} \right| \qquad (29)$$

(Recall that for the descending wake M and z are negative.) Eq. (29) indicates that, for any given set of local wake properties, the entrainment rate increases with increasing s and G, but decreases with increasing F. Hence, overall buoyancy (F) tends to reduce entrainment, possibly as a result of the stabilization of the rotating vortices by the light fluid, while buoyancy between the vortices (s) increases entrainment and acts as a destabilizing influence.

The relationship between increasing s and increasing entrainment, when applied to Eq. (28), shows that increasing values of Q correspond to decreasing entrainment if all other parameters remain fixed. This now helps explain the two distinct characteristics of the solutions for Q less than $\pi/2$ and greater than $\pi/2$. In Fig. 3, with $Q < \pi/2$, the effect of the strong combination of stability and entrainment relative to vortex strength is such as to cause the wake to stop its descent and to grow. In Fig. 5, where $Q > \pi/2$, the weak mixing and entrainment result in buoyancy eventually controlling the vortex motion by decreasing the wake momentum. This results in detrainment of fluid from the oval. (Note that detrainment is possible for s > 0 if F/Γ^2 is large enough.)

These aspects of the wake motion are displayed in a different manner in Fig. 6, in which the vortex separation is plotted as a function of the distance of descent. The vertical distance scales inversely with s, so that increasing entrainment results in a greater descent due to the decreased buoyancy. Counterbalancing this, increased entrainment manifests itself in greater vortex separation, since Q decreases with increasing s, and consequently

in decreased descent velocity and decreased descent. Consideration
of increases in stability is easier and the result is always a
decreased descent.

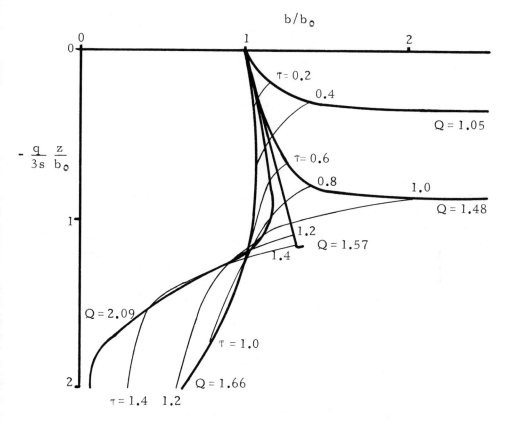

Fig. 6. Variation of vortex spacing with descent distance. The
thin lines connect points of equal τ

For most modern aircraft, ranging from the DC-3 to the Boeing
747, the ratio Γ_o/b_o^2 is about 1 s^{-1}. For a moderately stable atmosphere, $G^{1/2}$ is about 0.02 s^{-1}. The constant q is approximately
12 and s is probably of the order of 10. Insertion of these numbers
into Eq. (28) gives Q $\sim$ 2, hence real wakes appear to fall near the
range of values of Q considered in Fig. 6. For the same atmospheric
conditions, the time $\tau = \pi/2$ over which the motion takes place corresponds to about 100 seconds.

5. CONCLUSIONS

To ascertain the validity of the solutions which have been obtained, it is useful to look again at some of the assumptions which

were used in the analysis. It can be shown that the Boussinesq
assumption that the overall density differences are small results
in approximately correct equations of motion only if the following
inequality is satisfied:

$$\frac{w^2}{a_o{}^2} \ll \left| \frac{\rho - \rho'}{\rho'} \right| \ll 1 \; . \tag{30}$$

Initially, when the wake is moving downward with no buoyancy, this
condition is violated and thus the very start of the motion is
modeled incorrectly. This is probably a relatively unimportant
error.

It was also assumed that the wake cross-sectional shape is
geometrically similar at all times and that the parameter Δz was
proportional to the scale of this profile. The similarity assumption is probably valid as long as the wake is not too far from
potential flow conditions, in that it is not very buoyant and the
entrainment is not too great. The proper approach to use toward
entrainment is not clear and experiments such as those performed
by Woodward[10] are the only recourse. The assumption used in the
present work gives reasonable results but, being coupled to the
similarity assumption, it breaks down whenever the similarity does.
Consequently, this model is unable to show the expected vertical
oscillation of the wake about an equilibrium level.

The obtained solution is thus most accurate for describing
the wake descent to the equilibrium level. Since one of the
methods by which the aircraft wake decays is due to a breakup of
the parallel configuration because of vortex interactions, the
determination of the change in vortex spacing gives a clue to the
stability of the wake. In particular, one might expect the case
in which the vortices try to converge ($Q > \pi/2$ to be particularly
unstable and the lifetime of the organized wake to be quite short.

It is apparent that, in the presence of a stable atmospheric
density gradient, the transport of the vortex wake is quite different from that predicted by the potential flow solution. The
above model suggests some ways in which atmospheric stability and
the entrainment mechanism act to influence the overall motion.

Unfortunately, it is not possible to evaluate the validity of
the obtained results by comparison with existing experiment. Except for some measurements which are not sufficiently complete to
provide the necessary detail,[2,3] the author knows of no published
data in which the stability of the atmosphere was measured along
with the descent and growth of the wake. Such experimental data
are necessary to determine both the validity of the model and to

ascertain the value of s. In the absence of detailed data, all that can be said is that the behavior predicted by the model is intuitively reasonable and the numbers presented are of the right magnitude.

ACKNOWLEDGMENT

The author is indebted to the Air Force Office of Scientific Research for support of this research.

REFERENCES

1. Prandtl, L., and O. G. Tietjens, Fundamentals of Hydro- and Aeromechanics. Dover Publications, New York, 1957, 00 209-212.

2. Smith, T. B., and P. B. MacCready, Jr., "Aircraft Wakes and Diffusion Enhancement." Meteorology Research, Inc., Alta-dena, Calif., Final Report to Dugway Proving Ground, Dugway, Utah, Cont. DA-42-007-CML-545, 1963.

3. Andrews, W. H., "Flight Evaluation of the Wing Vortex Wake Generated by Large Jet Transports." Paper presented at Symp. on Aircraft Wake Turbulence, Seattle, September 1-3, 1970.

4. MacCready, P. B., Jr., "An Assessment of Dominant Mechanisms in Wake Decay." Paper presented at Symp. on Aircraft Wake Turbulence, Seattle, September 1-3, 1970.

5. Turner, J. S., "A Comparison between Buoyant Vortex Rings and Vortex Pairs." J. Fluid Mech., 7 (1960), 419-432.

6. Lamb, Sir Horace, Hydrodynamics. Dover Publications, New York, 1945, pp 221-222.

7. Schooley, A., "Airplane Contrails Related to Vertical Wake Collapse." Bull. Amer. Meteor. Soc., 50 (1969), 719.

8. Lamb, op. cit., pp 161-162.

9. Prandtl, op. cit., pp 194-195.

10. Woodward, B., "The Motion in and around Isolated Thermals." Quart. J. Roy. Meteor. Soc., 85 (1959), 144-151.

ADDENDUM

After the above paper was written, the work of Scorer and Davenport[*] on the same problem was brought to the author's attention. They solve exactly the same equations of motion, but assume that Γ is a constant (equivalent to Δz or $s = 0$ in the present solution) so their solution is the limiting one for $Q \gg \pi/2$, in which the vortices converge as they descend. They maintain that this configuration is stable up to a certain time (equal to $\tau = 0.66$ in the present paper) because the vorticity generated along the buoyant upper surface is detrained and so no instability is produced in the vortices. (This also justifies the assumed constant circulation.) After this time, according to their model, some of the external fluid and vorticity is mixed into the vortex oval and eventually results in its destabilization and destruction.

---

[*]Scorer, R. S., and L. J. Davenport, "Contrails and Aircraft Downwash." J. Fluid Mech., 43 (1970), 451-464.

SPLIT-FILM ANEMOMETER SENSORS FOR THREE-DIMENSIONAL VELOCITY-VECTOR MEASUREMENT

J. G. Olin and R. B. Kiland

Thermo-Systems Inc.

ABSTRACT

A probe for the fast-response measurement of the total air velocity vector is described. In the past, velocity-vector measurement has been achieved with a probe consisting of three, orthogonal, hot-film sensors. However, this measurement requires an a priori knowledge of the octant of the velocity vector. The new probe has a similar configuration but uses three split-film sensors to provide automatic octant indication. Each hot-film sensor consists of a 6-mil diameter, 80-mil long cylindrical quartz rod coated with a 1000 Å platinum film. The film of this new sensor is axially segmented, or split, with two splits $180°$ apart. Each split film element is electrically heated to the same constant temperature by a separate electronic constant-temperature anemometer control system. The non-uniformity of heat flux around the sensor is detected by comparing the individual heat flux to each split-film element. For Reynolds numbers less than approximately 5×10^4, the heat flux to the upstream split film is greater than the downstream. This is used to detect the sense of the velocity component normal to the sensor. Such information from all three orthogonal sensors yields the octant of the air velocity vector. The sum of the heat fluxes to the two split-film elements of a sensor is invariant with azimuthal angle. This permits the use of standard techniques to measure the magnitude and direction of the velocity vector within the detected octant.

Experimental data and empirical correlations are presented for the variation of heat flux with velocity magnitude and two directional angles - the yaw and azimuthal angles to the sensor.

The heat-flux ratio of the downstream to the upstream split-film element has a simple quadratic dependence on azimuthal angle. The sum of the heat fluxes to the two elements of a sensor is correlated in the typical fashion with an effective cooling velocity. Other probe configurations using split-film sensors are described - the rugged three-dimensional probe, the boundary-layer probe, and the horizontal wind-vector probe.

1. PRINCIPLE OF OPERATION

Constant-temperature thermal anemometers — hot-film and hot-wire anemometers — have been widely used to measure fluid velocity and turbulence parameters. They combine the important advantages of extremely fast response, high sensitivity, wide dynamic range, and good spatial resolution. Measurements of the instantaneous air velocity vector in three-dimensional flow field has been made successfully in several applications with a probe consisting of three, orthogonal, 6-mil diameter, hot-film sensors.* However, this probe has the same response in each of the eight octants formed by the axes of the three sensors. Hence, it yields a non-unique solution for the instantaneous velocity vector in applications where the octant of the velocity vector is unknown before the experiment. The new probe described herein eliminates this ambiguity by using split-film sensors to precisely determine the octant of the velocity vector. It also advances the state-of-the-art in the supporting structure for sensors; a unique cantilevered, or single-ended, sensor support significantly reduces support interference and increases strength.

The principle of operation of split-film sensors is based on the well-known nonuniformity of the heat flux distribution around a constant-temperature cylinder in cross flow[1,2], as shown on Figure 1. Such a constant-temperature cylinder is constructed by coating a cylindrical quartz substrate with a thin metallic film, as shown on Figure 2. The film is electrically heated to a constant temperature by an electronic constant-temperature anemometer control system. The new concept is to split this metallic film into halves, $180°$ apart, as shown in Figure 1, and to operate each of the two split-film elements at the same temperature with separate anemometer control systems. The nonuniformity in heat flux is detected by comparing the individual heat flux to each split-film element. For a Reynolds number less than about 5×10^4, the heat flux to the upstream split-film element Q_1 (the one containing the upstream stagnation line) is always greater than the downstream element Q_2. This is experimentally verified, as shown by the calibration of the heat-flux ratio Q_2/Q_1 versus the azimuthal angle θ on Figures 12 - 14. This information is used to detect

_____
*Thermo-Systems Model 1294 Probe

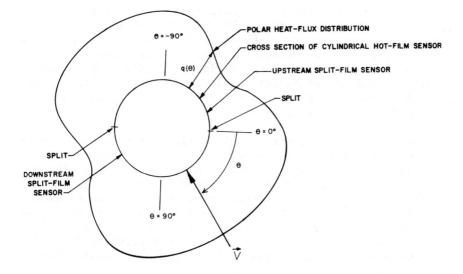

Figure 1. Non-Uniform Heat-Flux Distribution Around A
Constant-Temperature Cylinder In Cross Flow

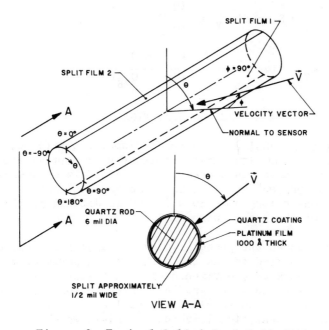

Figure 2. Typical Cylindrical Split-Film
Anemometer Sensor

the sense of the velocity-vector component normal to the split-film sensor. The difference $Q_1 - Q_2$ or, alternatively, the ratio Q_2/Q_1 for each sensor can be used to give quantitative information on the direction of the velocity vector. We describe herein a "standard" data-reduction procedure in which this information is not needed. The direction and magnitude of the velocity vector within the detected octant is determined with a method applicable to standard non-split sensors.

The array of three, orthogonal, split-film sensors for three-dimensional velocity-vector detection is shown on Figure 3. For simplicity of explanation, we assume in the remainder of the text that $Q_1 = Q_2$ at the two splits of a sensor. In practice this is not true because of slight differences in the area and resistance of the two split-film elements. These differences are evaluated during sensor calibration and are used to determine a near-unity correction factor R, such that $Q_1 = R\,Q_2$ at the two splits. With the above assumption, the signs (± 1) of the orthogonal velocity components are:

$$S_X = \text{Sign of } V_X = \frac{(Q_{1C} - Q_{2C})}{|Q_{1C} - Q_{2C}|},$$

$$S_Y = \text{Sign of } V_Y = \frac{(Q_{1A} - Q_{2A})}{|Q_{1A} - Q_{2A}|}, \text{ and} \tag{1}$$

$$S_Z = \text{Sign of } V_Z = \frac{(Q_{1B} - Q_{2B})}{|Q_{1B} - Q_{2B}|}.$$

2. CONSTRUCTION

Each sensor consists of a 6-mil diameter quartz rod coated with a thin platinum film with a thickness of approximately 1000 Å. The entire sensor is coated with a thin film of quartz for environmental protection. The total sensor length is 200 mils. The active sensor length is 80 mils and is defined by plating the extremities of the sensor with gold.

Figure 4 shows the probe. To avoid support interference, the sensors are supported at one end with long supports of minimum diameter. Results of experiments show that support interference is very small with this design. All three sensors fit into a 0.3 - inch diameter sphere yielding excellent spatial resolution. The probe has a pneumatically-actuated shield which can be operated remotely. This important feature prevents possible sensor breakage and contamination when measurements are not being made. Another new feature is that the calibration of the sensors can be

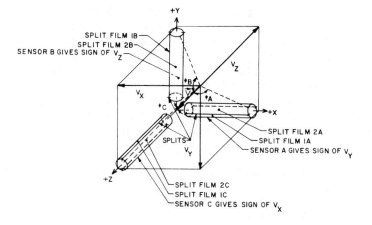

PROBE SHANK OF 1080 PROBE IS IN THE $(-V_X, -V_Y, -V_Z)$ OCTANT

Figure 3. Three Orthogonal Split-Film Sensors
For Total Vector Measurement

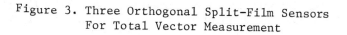

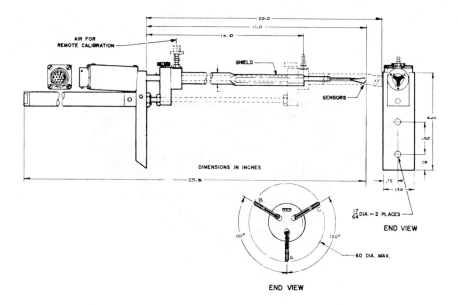

Figure 4. Shielded 1080 Probe

checked remotely. A solenoid valve on the probe can be adjusted
to release air through a sonic orifice at the base of the shield.
This precisely meters the air. The air is then ducted through
the shield and past the sensors to give a calibration check point.
A nozzle flattens the flow profile before the air flows past the
sensors. A small thermocouple is mounted on the probe near the
sensor array to measure ambient temperature, which is used to correct
for possible changes in ambient air temperature and density. The
thermocouple is designed for maximum frequency response. At least
25 Hz is achievable. The cold thermocouple junction is electronic.

The total vector anemometer system* is shown on Figure 5. It
consists of power supply, constant-temperature anemometer control
module, cable, and probe. The single control module operates all
six channels, or split-film elements, of the probe. The common-
ality of many electronic components facilitated compact packaging
of the control module. The system has seven analog outputs – six
velocity outputs and one temperature output.

It is important that the temperatures T_{o1} and T_{o2} of the two
split film elements are very close to identical. If this is not
true, heat will be conducted from the hotter sensor to the colder
sensor. This can cause erroneous octant indication, and, if strong
differences in sensor temperature exist, one sensor can power the
other by conductive interaction to such an extent that one bridge
turns off. A "Balance" potentiometer on the control module allows
fine adjustment of the temperature of the two split-film elements
of a sensor to achieve the necessary temperature equality of the
two elements.

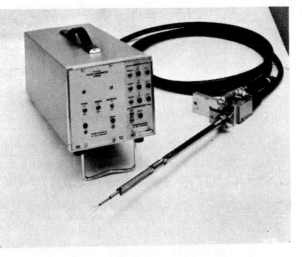

Figure 5. Total Vector Anemometer
System

---

*Thermo-Systems' Model 1080 System

3.   EXPERIMENTAL DATA AND CORRELATIONS

The experimental objective was to calibrate the probe in air over a wide velocity range for the complete solid angle (4π steradians). Actual data was obtained in the velocity range of 0.1 fps to 400 fps over the complete solid angle ($-90^{\circ} \leq$ yaw angle $\phi \leq 90^{\circ}$ and $0 \leq$ azimuthal angle $\theta \leq 360^{\circ}$). Data shown in this paper is in the range of 3 fps to 300 fps. Correlations were found for most of the variables. With the exception of mutual interference between the three sensors, which was found to be small, the performance of a single sensor determines the performance of the total probe. A single sensor with dimensions and a support configuration identical to that when mounted on the total probe (Figure 4) was used in the calibration experiments.

Experiments were conducted by holding the sensor fixed and moving a jet of compressed air around the sensor over the complete solid angle. Thermo-Systems' Model 1125 Calibrator was modified for these tests. The jet diameter was 1.250 inches. The sensor was nominally 1 inch from the mouth of the jet. The orientation of the jet axis was precisely controlled and measured by using two rotary fixtures with orthogonal axes. The flow rate through the jet was accurately measured with a precision Δp cell. Ambient temperature and pressure were recorded during the experiments to convert raw velocity data into a "standard" velocity U_S referred to standard pressure P_S = 29.92 in. Hg and standard temperature T_S = 530°F. In the calibration experiments the ambient pressure P was about 29 in. Hg, the ambient temperature T was about 75°F, and the sensor temperature was about 550°F (an overheat ratio of 1.5).

The frequency response of velocity detection was measured during the calibration experiments. It is nominally 1000 Hz.

The heat flux to a given split-film element, say, element number 1, is

$$Q_1/\Delta T = K_1 \, E_1^2, \text{ and} \tag{2}$$

the total heat flux Q_T to the two split-film elements of a sensor is

$$Q_T/\Delta T = Q_1/\Delta T + Q_2/\Delta T = K_1 \, E_1^2 + K_2 \, E_2^2, \tag{3}$$

where:

$$\Delta T = T_o - T,$$

$$E = \text{Bridge voltage,}$$

$$K = C \cdot \frac{R_{sensor}}{(R_{sensor} + R_{bridge})^2}, \tag{4}$$

In Eq. (4), R_{sensor} is the resistance of the split-film element, R_{bridge} is the resistance of the series resistor in the bridge circuit, and C is a near-unity correction factor chosen to make Q_T independent of θ. The factor C is determined during sensor calibration and accounts for slight differences between the two split-film elements.

The relationship between velocity magnitude U and heat flux is basic to any thermal anemometer. The heat flux to a split-film element varies with the direction of the velocity vector, i.e., with the azimuthal angle θ and the yaw angle ϕ, shown on Figure 2. Figure 6 shows that the total heat flux Q_T is essentially invariant with θ. The small dependency on θ appears to be caused by support interference. In other words, the total sensor acts as if it were not split. This important fact greatly simplifies the standard data-reduction procedure because it eliminates dependence on the azimuthal angle.

The calibration curve for the total heat flux Q_T at a yaw angle ϕ of $0°$ is shown on Figure 7. The correlation for this curve is shown on Figure 8. We used a King's-law relationship (without the constant term) in two segments:

$$Q_T/\Delta T = B \ (U_S)^{1/n}, \tag{5}$$

where:

$$B = 0.177, \ n = 3 \ \text{for 3 sfps} \leq U_S \leq 19 \ \text{sfps and}$$

$$B = 0.0125, \ n = 2.25 \ \text{for 19 sfps} \leq U_S \leq \approx 300 \ \text{fps.}$$

The data showing the variation of the total heat flux Q_T with the yaw angle ϕ is shown in Figure 9. This ϕ dependence is correlated by using the concept of the "effective velocity" U_{Eff}. U_{Eff} is the velocity that will give the same heat flux at $\phi = 0$ as the actual velocity U_S gives at the actual value of ϕ ($-90° \leq \phi \leq 90°$). Its purpose is to embody all of the ϕ-dependence of Q_T.

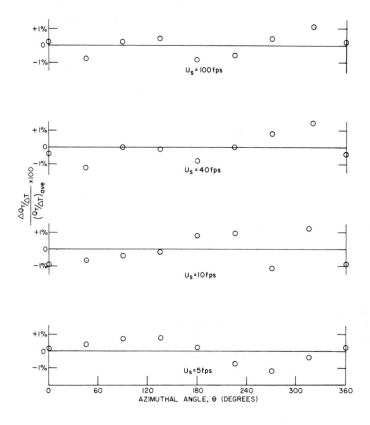

Figure 6. Variation of Total Heat Flux
with Azimuthal Angle θ

Typically[3,4], the following correlation for U_{Eff} is used:

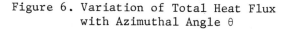

$$U^2_{Eff}/U^2_S = \cos^2 \phi + k^2_1 \sin^2 \phi. \qquad (6)$$

The factor k_1 is usually taken as a constant with a value of about 0.2. We tried to correlate the data of Figure 9 with Eq. (6) but found that k_1 depends on U_S and ϕ, albeit weakly. The major reasons for the variability of k_1 are: (1) the wide dynamic range of U_S (about three orders-of-magnitude) and (2) the single-ended sensor support interference which causes a slightly asymmetric ϕ-distribution which is different for $+ \phi$ and $- \phi$. The correlation

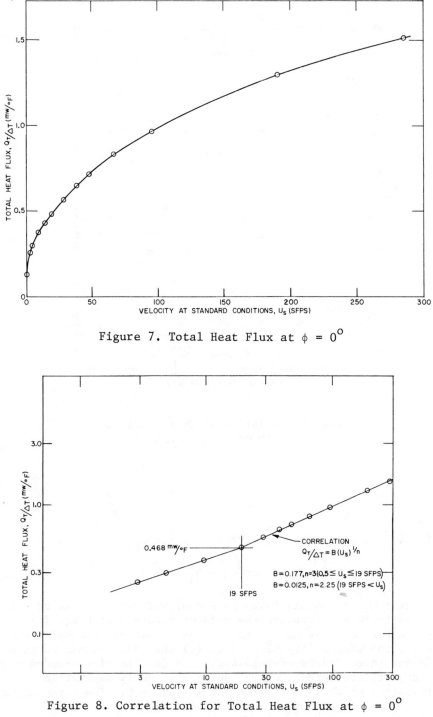

Figure 7. Total Heat Flux at $\phi = 0^{\circ}$

Figure 8. Correlation for Total Heat Flux at $\phi = 0^{\circ}$

we used for U_{Eff} is

$$U_{Eff}^2 / U_S^2 = \cos^2 \phi + \alpha(U_S, \phi) \, k^2 (U_S) \, \sin^2 \phi, \tag{7}$$

where:

$$\alpha(U_S, \phi) = 1 + \beta(U_S) \cos^2 \phi. \tag{8}$$

The values of $\beta(U_S)$ and $k(U_S)$ are given in Table 1. Comparison between this correlation and the data is shown on Figures 10 and 11.

TABLE 1.   Parameters Used in the Correlation for Effective Velocity Given by Eq. (7)

ϕ		U_S (sfps)		
		0 to 20	20 to 30	$\geq$ 30
$k(U_S)$	Neg.	$0.90 \, U_S^{-1/2}$		
	Pos.		0.20	0.20
$\beta(U_S)$	Neg.	0	$-12 + 0.6 \, U_S$	$5 + 0.033 \, U_S$
	Pos.	0	$-22 + 1.1 \, U_S$	$9 + 0.067 \, U_S$

The complete correlation for the total heat flux to each of the three sensors now becomes:

$$Q_T / \Delta T = B \cdot U_{Eff}^{1/n} \, (U_S, \phi). \tag{9}$$

The effective velocity is based on standard conditions (P_S and T_S). The solution to the data-reduction procedure is the velocity U_S at standard conditions. Since hot-film sensors measure mass velocity ρV, where ρ is the gas density the actual velocity, U is found by making the following simple density correction:

$$U = \frac{\rho_S}{\rho} U_S = \left(\frac{P_S}{P}\right) \left(\frac{T}{T_S}\right) U_S. \tag{10}$$

The instantaneous value of ambient temperature T is measured by the thermocouple on the probe. P is the local barometric pressure.

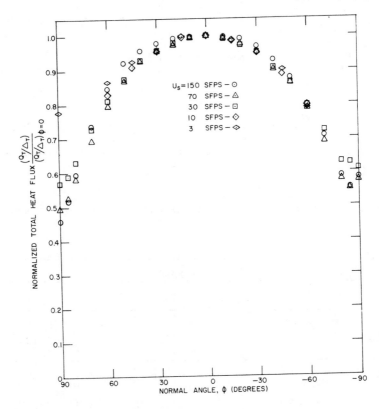

Figure 9. Variation of Total Heat Flux
with Yaw Angle ϕ

Two trigonometric relationships are useful in the data-reduction procedure:

$$\sin^2\phi_A + \sin^2\phi_B + \sin^2\phi_C = 1 \text{ and}$$

$$\cos^2\phi_A + \cos^2\phi_B + \cos^2\phi_C = 2, \tag{11}$$

where ϕ_A, ϕ_B, and ϕ_C are the yaw angles for the three sensors.

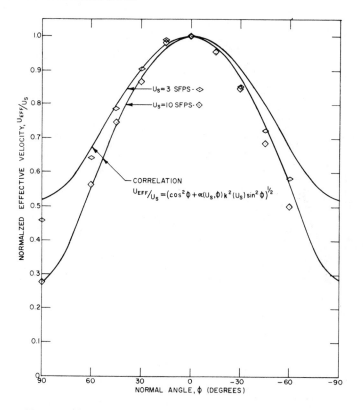

Figure 10. Effective Velocity At Low Velocities

4.   STANDARD DATA–REDUCTION PROCEDURE

The octant of the velocity vector and the signs of the orthogonal velocity components are found from Eq. (1). This also tells
whether the ϕ's are negative or positive in Table 1. Since the octant is known, we assume the convention that the ϕ's are positive.

The exact solution for the velocity vector $\vec{V}$ — U, ϕ_A, ϕ_B,
and ϕ_C — requires iteration. The first iterative approximation
to $\vec{V}$ is found by assuming that $\alpha(U_s,\phi)\, k^2\, (U_s) = \text{constant} = k_1^2$.
A good assumption is that $k_1^2 = 0.04$. The effective velocities
$U_{Eff,A}^{(1)}$, $U_{Eff,B}^{(1)}$, and $U_{Eff,C}^{(1)}$ for the three sensors are determined
from (See Eq. 9):

$$U_{Eff}^{(1)} = (\frac{Q_T}{B^{(1)}\Delta T})^{n^{(1)}}.$$ (12)

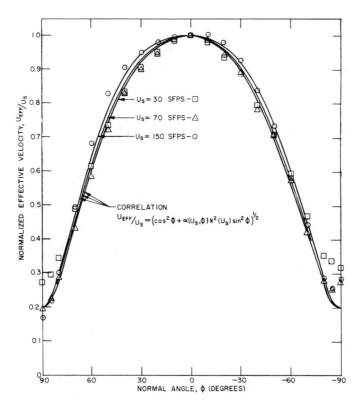

Figure 11. Effective Velocity at High Velocities

One of the two sets of values for B and n in Eq. (5) are assumed, say, $B^{(1)}$ and $n^{(1)}$. Using Eqs. (6), (10), and (11), we get the first approximation to the velocity magnitude:

$$U^{(1)} = (\frac{P_S}{P})(\frac{T}{T_S}) \; [\frac{U_{Eff,A}^{(1)^2} + U_{Eff,B}^{(1)^2} + U_{Eff,C}^{(1)^2}}{2 + k_1^2}]^{1/2} \qquad (13)$$

The yaw angles are found from Eqs. (6) and (11). For example,

$$\phi_A^{(1)} = arc \; sin \; [\frac{1 - U_{Eff,A}^{(1)^2}/U_S^{(1)^2}}{1 - k_1^2}]^{1/2}. \qquad (14)$$

$\phi_B^{(1)}$ and $\phi_C^{(1)}$ are found similarly. The final solution for the first approximation, in terms of orthogonal velocity components, is:

$$U_X^{(1)} = S_X U^{(1)} \sin \phi_A^{(1)},$$

$$U_Y^{(1)} = S_Y U^{(1)} \sin \phi_B^{(1)}, \text{ and} \tag{15}$$

$$U_Z^{(1)} = S_Z U^{(1)} \sin \phi_C^{(1)}.$$

The second approximation $\vec{V}^{(2)}$ is found by choosing the proper values of B and n based on the first iteration solution, computing the final values of U_{Eff}, and then using the following relationship (see Eqs. (7) and (11)):

$$U^{(2)} = (\frac{P_S}{P})(\frac{T}{T_S}). \tag{16}$$

$$[\frac{U_{Eff,A}^2 + U_{Eff,B}^2 + U_{Eff,C}^2}{2 + k^{(1)^2}(\alpha_A^{(1)} \sin^2\phi_A^{(1)} + \alpha_B^{(1)} \sin^2\phi_B^{(1)} + \alpha_C^{(1)} \sin^2\phi_C^{(1)})}]^{1/2} \text{ and}$$

$$\phi_A^{(2)} = \arc\sin [\frac{1 - U_{Eff,A}^2 / U_S^{(2)^2}}{1 - \alpha_A^{(1)} k^{(1)2}}]^{1/2}. \tag{17}$$

$\phi_B^{(2)}$ and $\phi_C^{(2)}$ are found similarly. Further approximations follow the same procedure[5]. The iteration converges very rapidly. Only two or three iterations are necessary to be within one percent of the final solution.

The accuracy of the data and correlations is best assessed by comparing the actual velocity in a test with the values of $\vec{V}$ calculated from the raw data with the above procedure. Several tests were conducted with the probe shown on Figure 4 for a velocity range of 3 sfps to 300 sfps and over the complete solid angle. The results of these tests are:

(1) Accuracy of velocity magnitude U: $\pm$ 3% of reading and
(2) Accuracy of two independent yaw angles ϕ over the complete solid angle (4π steradians): $\pm$ 3°.

We feel that this level of accuracy is very good for an anemometer having complete three-dimensional measuring capability.

5. VARIATION OF HEAT FLUX WITH AZIMUTHAL ANGLE

The ratio Q_2/Q_1 of the heat flux to the downstream split film to the upstream varies with the azimuthal angle θ, as shown on Figures 12 and 13. This variation is caused by the nonuniform heat-flux around the sensor, shown on Figure 1. At the splits, Q_2/Q_1 has maximum sensitivity to θ. This is important for accurate octant detection. The sensitivity decreases as the velocity U decreases and the magnitude of the yaw angle ϕ increases. In addition to its obvious importance in octant detection, Q_2/Q_1 bears quantitative information on the velocity vector $\vec{V}$. Q_2/Q_1 varies with U, ϕ, and θ, and therefore the ratios $(Q_2/Q_1)_A$, $(Q_2/Q_1)_B$, and $(Q_2/Q_1)_C$ for the three sensors can be used to uniquely determine $\vec{V}$. Since Q_2/Q_1 varies most strongly with θ, it probably is most useful for determining the directional angles θ_A, θ_B, and θ_C.

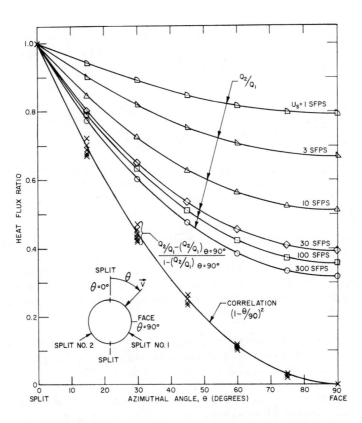

Figure 12.   Variation of Heat Flux Ratio with
Velocity U_S at $\phi = 0°$

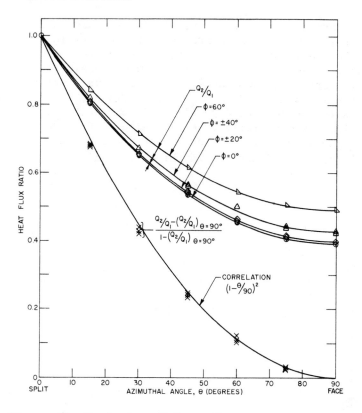

Figure 13. Variation of Heat Flux Ratio with Normal
Angle ϕ at U_S = 30 sfps

Figures 12 and 13 also show the correlation of Q_2/Q_1 with θ:

$$\frac{Q_2/Q_1 - (Q_2/Q_1)_{\theta = 90^\circ}}{1 - (Q_2/Q_1)_{\theta = 90^\circ}} = \left(1 - \frac{\theta}{90}\right)^2 : \qquad (18)$$

This simple quadratic correlation is quite good over a wide range
of U and θ. Improved accuracy can be attained over a more limited
range. In Eq. (18), the U-and ϕ-dependencies are buried in the
parameter $(Q_2/Q_1)_{\theta = 90^\circ}$. The dependence of $(Q_2/Q_1)_{\theta = 90^\circ}$ on U_S
is not simple, and the θ-dependence is very complicated, especially
since it is different for $+ \phi$ and $- \phi$. We made attempts to find
good algebraic correlations, but enjoyed little success. Hybrid
data-reduction procedures, combining parts of the standard data-reduction procedure and the Q_2/Q_1 data, did not prove to be superior

to the standard data-reduction procedure described in the previous
section, primarily because of present difficulties in achieving the
extremely low tolerances required in mounting the sensors. We conclude that the best use of Q_2/Q_1 is in two-dimensional flows where
ϕ is nearly zero.   In such cases, if U_S is known a priori or is in
a narrow range, then $(Q_2/Q_1)_{\theta = 90^\circ}$ is a constant in Eq. (18), and
the Q_2/Q_1 data can be used directly to determine the directional
angle θ. This is especially true in the range $U_S \geq 30$ sfps, where
$(Q_2/Q_1)_{\theta = 90^\circ}$ is nearly constant. This application is used in the
boundary-layer probe described in the next section.

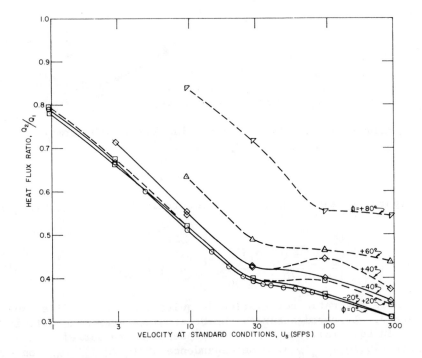

Figure 14.   Heat Flux Ratio at $\theta = 90^\circ$

6.   OTHER SPLIT FILM SENSORS

We have investigated several configurations of split-film
sensors for measuring three-dimensional and two-dimensional flows.
Of these, three configurations have proved most practical and best
suited for use in fluid-mechanics research.

Rugged Three-dimensional Probe

Figure 15 shows a probe[6] consisting of two, orthogonal, 60-mil
diameter quartz rods coated with a platinum film. The large diameter makes the sensor very rugged. They can withstand human
touch, impact by large particles, shock, and vibration. The large
diameter has the additional advantage of reducing contamination by
smaller airborne particles. The probe has application to instantaneous, three-dimensional flow-field mapping in jet-engine inlets,
around rotating helicopter blades, and in similar severe environ-
ments.

Each cylindrical sensor has four axial splits. Theoretically,
three split-film elements would be the minimum required for unique
velocity-vector detection. Four split-film elements are required
to eliminate the effect of an unexpected flow phenomenon, which we
now describe. In our calibration tests (10 sfps to 300 sfps), we
found that the heat flux from the downstream side of the sensor
increased from its value at a yaw angle of $\phi = 0$ as the magnitude

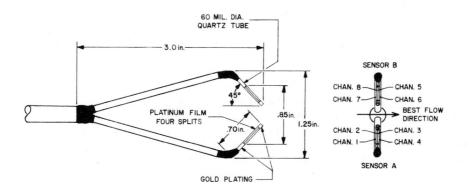

Figure 15. Rugged Three-Dimensional Probe

of ϕ increased. This resulted in an unexpected doubly-humped ϕ-distribution much different than the typical distribution shown on Figure 9. This phenomenon affected the heat flux to the most downstream split-film element at velocities as low as 10 sfps (Re_d = 310). It even affected the total heat flux - the sum of the heat fluxes from the four split-film elements - when the velocity exceeded 60 sfps (Re_d >1900). To avoid these difficulties, data from the two (or possibly one) most upstream split-film elements are used in data reduction. Data from the others is neglected. It was found that four splits, instead of three, were necessary to avoid the double-valued ϕ-distribution of heat flux. Support interference is not responsible for this phenomenon. Rather, the secondary flows associated with the three-dimensional flow field over a yawed cylinder seems to be causitive. As yet, a detailed explanation of this phenomenon is not known by the authors.

Boundary-Layer Probe

Figure 16 shows a probe for two-dimensional boundary-layer measurements. The sensor diameter is 6 mils or 2 mils. It is split axially with two splits 180° apart - the same as the sensors of the three-dimensional probe shown in Figure 2. The splits are in a plane parallel to the wall - the most sensitive orientation for detecting the directional angle θ. The sensor is single-ended for minimum support interference. The support construction also permits measurements very close to the wall.

Since the sensor is relatively insensitive to small velocity components parallel to the sensor axis (see Figure 9), it measures the instantaneous velocity vector $\vec{V}$ in the plane normal to the sensor axis. The sum of the heat fluxes $Q_T = Q_1 + Q_2$ gives the velocity magnitude U, and the ratio Q_2/Q_1 gives the directional angle θ, as described in the preceeding section. Turbulence parameters - intensity, Reynolds stress, probability density, etc. - are calculated from $\vec{V}$. Flows with either high or low turbulence intensities can be measured. Simple expressions for turbulence intensity and Reynolds stress are obtainable from the previously described correlations[7]. Turbulence parameters in boundary layers are often measured with the conventional cross-wire probe. The split-film sensor has the following advantages over the cross-wire: (1) it can get much closer to the wall, (2) it has much better spatial resolution, and (3) it can measure large-scale turbulence unambiguously over the azimuthal angle range of $\pm 90^o$, whereas the cross-wire probe is limited to $\pm 45^o$.

Horizontal Wind-Vector Probe

A cylindrical sensor with three (the minimum number required) axial split-film elements, each 120° apart, can be used to measure the velocity vector anywhere in the plane perpendicular to the sensor axis. If the sensor axis is vertical it becomes a horizontal

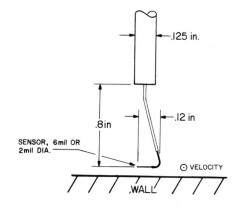

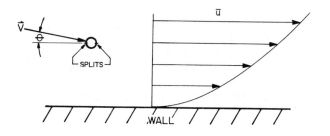

Figure 16. Split-Film Boundary-Layer Probe

velocity-vector probe. In meteorological and micrometeorological applications, it provides high temporal and spatial resolution measurement of horizontal wind speed and direction. Sensors of this type having a diameter of 60 mils, with and without teflon coating, have been successfully conducted in wind tunnels and in flow facilities simulating the expected meteorological environment on the planet Mars.

7. CONCLUSION

A new concept - split-film anemometer sensors - has been developed for three-dimensional air velocity-vector measurement. A probe consisting of three, orthogonal, 6-mil, cylindrical sensors, each with two axial splits 180° apart, offers a combination of the following advantages:

(1) Three-dimensional measurement in the complete solid
    angle (4π steradians) over the nominal velocity
    range of 0.5 to 500 fps.,

(2) High accuracy -

        Velocity magnitude: ± 3% of reading, and

        Two independent yaw angles: ± 3°,

(3) High frequency response: 1000 Hz, and

(4) Good spatial resolution: 0.3 inches.

Support interference is minimized by supporting the sensors on only
one end with supports of small diameter. The probe has a thermocouple for measuring ambient temperature, a retractable shield, and
a means for remote calibration of the sensors. The data-reduction
procedure uses algrebraic correlations for the calibration data and
is relatively straightforward and computer-compatible.

The probe has application to many three-dimensional flow fields,
including:

(1) Aircraft wakes,

(2) Flow around objects in wind tunnels or in the
    earth's boundary layer,

(3) Rotating propellers or blades, and

(4) Micrometeorology.

The Federal Aviation Administration is planning to tower-mount
ninety such probes for detailed flow-field mapping of aircraft
wing-tip vortices.

Several other types of split-film probes have been developed
for specific applications. Large split-film sensors have the much
desired property of strength and directional sensitivity over the
complete 360° azimuthal angle. The use of a small split-film sensor for velocity-vector and turbulence measurements in boundary
layers is particularly attractive. Properly exploited, split-
film sensors should become a powerful flow-measurement tool.

REFERENCES

1. Schmidt, E. and Wenner, K., "Heat Transfer Over The
    Circumference Of A Heated Cylinder In Transverse Flow",
    NACA TM 1050, Oct. 1943.

2. Lorisch, W. from M. ten Bosch, "Die Wärmeüber-fragung",
    3rd ed., Springer-verlag, Berlin, 1936.

3. Champagne, F., Sleicher G., and Wehrman, O., "Turbulence
    Measurements With Inclined Hot Wires", Parts I and II,
    Journal of Fluid Mechanics, Vol. 28, p. 153 (Part I) and
    p. 177 (Part II), 1967.

4. Hinze, J. O., Turbulence, McGraw-Hill Book Company, Inc.,
    New York, 1959, p. 103.

5. "Operating and Service Manual for Model 1080 Total Vector
    Anemometer", Thermo-Systems Inc., St. Paul, Minn., 1970.

6. Olin, J. G., and Kiland, R. B., "Rugged Thermal Anemometer
    for Measurement of Three-Directional Fluid Velocity Vector",
    Thermo-Systems Inc., Research Report No. 3, Jan. 28, 1970.

7. "Measuring Velocity Components with Constant Temperature
    Anemometers", Thermo-Systems Inc., St. Paul, Minn., 1970.

SUB-SCALE MODELING OF AIRCRAFT TRAILING VORTICES

Robert L. Kiang

Stanford Research Institute

ABSTRACT

Most experimental studies of trailing vortices have been conducted with one of the following three methods: (1) wind-tunnel
study, (2) intercepting the vortices shed by one airplane with
another airplane, (3) measurements of the vortices shed by a low-
flying airplane with ground-based instruments. The wind-tunnel
method is suitable for the study of vortex formation, but not the
long-time decay process. The second method would have been the
most realistic approach, but unfortunately, the difficulty in coordinating the two traveling airplanes plus the various unknown
effects of the atmosphere have so far precluded any consistent and
useful data. The third kind of study inevitably includes the ground
effect in addition to all the unpredictable atmospheric effects.
The data obtained, though useful to the air-traffic control near the
runways, are nevertheless hard to analyze; therefore it is difficult
to separate the various mechanisms responsible for the vortex decay.

This paper describes a pilot study of trailing vortices using
a sub-scale model in a more controllable laboratory environment.
The essential feature of this laboratory facility is to have a
vortex-generating wing moving along a pair of elevated rails so that
the vortices shed by the wing remain relatively fixed with respect
to ground-based instrumentation and can be observed and measured
throughout their entire life-span. The vortices are rendered visible by two methods: by smoke traces and by drifting soap bubbles.
A 16-mm movie camera is used to record the visible vortices. The
smoke-traced vortices give better qualitative pictures, but the
soap-bubble pictures are more suitable for quantitative studies.
Hot-wire anemometer probes are also used to study the turbulence

nature of the vortices. Ease of operation, repeatability, and
versatility are some of the obvious advantages of this experimental
setup.

Geometrically similar wings of different sizes running at
various speeds and angles of attack have been tested. The vortices
generated by these model wings all contain high levels of turbulence.
They generally last for a few seconds. A scaling study is also
described based on existing theories that predict the decay rate of
a turbulent line vortex. None of the theories examined is capable
of correlating the results of the sub-scale experiments (chord
Reynolds number of the order of 10^5) and those of full-scale airplanes (chord Reynolds number ranging from 10^6 to 10^8). This indicates that a satisfactory theory of turbulent vortex decay is still
lacking.

1. INTRODUCTION

A lifting wing is known to shed a vortex sheet behind its
trailing edge. Spreiter and Sacks[1] have shown that this vortex
sheet will quickly roll up into two counter-rotating vortices. The
vortex pair, with its organized motion, can persist for a relatively
long time. In fact, the trailing vortices shed by a heavy aircraft
can last for several minutes, posing a significant safety hazard to
any smaller airplanes that should happen to intercept them. From
the air-traffic control standpoint, an accurate prediction of the
vortex motion would be most desirable. The ability to do that, however, hinges on a thorough understanding of the mechanisms responsible for the vortex decay. This understanding is still lacking.

Owing to the turbulent nature of the flow, theoretical investigations in this field have been meager,[2,3] and a satisfactory theory
of turbulent vortex is yet to be developed. On the other hand, experimental studies of the trailing vortices have mostly been conducted with one of the following three methods: (1) wind-tunnel
tests, (2) intercepting the vortices shed by one airplane with
another airplane, and (3) measurements of the vortices shed by a
low-flying airplane with ground-based instruments. Quantitative
data of very high accuracy can be obtained from the wind-tunnel
method. But the limited dimensions of the test section precludes
any far-field studies. This limitation restricts the wind-tunnel
to more or less the study of vortex formation rather than the long-
time decay process. The second method would have been the most realistic approach, but unfortunately, the difficulty in coordinating
the two traveling airplanes plus the various unknown effects of the
atmosphere have so far precluded any consistent and useful data.
The third kind of study inevitably includes the ground effect in
addition to all the unpredictable atmospheric effects. The data
obtained, though useful to the air-traffic control near the runways,

are nevertheless hard to analyze; therefore it is difficult to separate the various mechanisms responsible for the vortex decay.

It becomes quite obvious that some means of obtaining quantitative vortex data in the far field are needed, preferably without the influences of the ground and the atmospheric inhomogeneities. A novel experimental facility has therefore been built that allows one to observe and measure the trailing vortices throughout their entire life spans. The following sections describe the facility, the experimental methods, and the kind of data that have been obtained from these experiments.

2. EXPERIMENTAL FACILITIES

The essential feature of the facility, called the "moving-wing facility," is to have a vortex-generating model wing moving along a pair of elevated rails so that the vortices shed by the wing remain relatively fixed with respect to ground-based instrumentation and therefore can be easily measured. The entire facility is housed indoors, thus eliminating all the unwanted atmospheric disturbances.

Figure 1 shows an over-all view of this facility. The pair of steel rails is 30 feet long and is elevated 3 feet from the ground. Underneath the wing is a carriage, which slides along the rails supported by four sleeve bearings. The wing itself is an NACA 0012, rectangular, wooden wing with an aspect ratio of 6. The model wing shown in the figure has a 5-foot span. Another geometrically similar wing with a span of 8 feet is also available for the experiments. As can be seen in the figure, the wing is mounted at a negative angle of attack. This allows the generated vortex pair to drift upward rather than downward so as to minimize the ground effect.

The wing/carriage assembly is towed by a steel cable that is

WING SPAN (ft)	WING SPEED (fps)	CHORD Re
5	33	1.7×10^5
	47	2.4×10^5
8	33	2.7×10^5

FIGURE 1   OVER-ALL VIEW OF THE MOVING-WING FACILITY

TABLE 1   EXPERIMENTAL SPEEDS OF THE MOVING WING

wound onto a drum driven by a 5-horsepower motor. At the start of
each experimental run, the wing sits at the right-hand end of the
rail. It is then accelerated by the towing cable in a sling-shot
fashion. At about 10 feet down the track, the cable is automatically released and the wing/carriage assembly is allowed to coast
with nearly constant speed through the 8-foot test section. After
that, the assembly is stopped by a frictional brake system. The
coasting speed of the wing through the test section is measured by
two magnetic pickups situated 5 feet apart alongside the rails. The
variations of the wing speed within the test section was measured to
be less than 5 percent. Table 1 lists the speeds of the wings and
their corresponding chord Reynolds numbers used in the tests.

Some of the auxiliary facilities are also shown in Figure 1.
The round disc on the left-hand side of the figure is a "clock."
It rotates at exactly 2 revolutions per second during the experiment
so that the exposure time of the photographic equipments can be
checked. A bubbled machine can be seen sitting on top of a high
stand on the left-hand side of the figure. Soap bubbles are used
as one of the two flow-visualization methods, the other being smoke.
Another indispensable piece of equipment used in the experiment is
an 8-x-8-x-3-foot light box, which houses a column of flood lights
so that a sheet of light about a foot wide is cast onto the middle
portion of the test section. The light box is situated to the left
of the bubble-machine stand, outside the scope of Figure 1.

3. STUDIES USING SMOKE AND HOT-WIRE ANEMOMETRY

A smoke generator was built which consists of a hot plate
enclosed in a sealed container. The container is slightly pressur-
ized with nitrogen so that the smoke generated inside the container
is forced out of a smoke pipe and into the test section. Model
train smoke liquid is selected for generating smoke, primarily
because of its nontoxic property.

The smoke-traced vortex is photographed by a 16-mm movie
camera at 64 frames per second. Figures 2 (a)-(f) show a few
reprints from these movie frames. The vortex seen in these figures
is shed by the 5-foot-span wing set at a (negative) angle of attack,
α, of $8°$. Wing velocity, V, is 33 fps. The sense of rotation of
the vortex is counterclockwise. Clearly visible in these figures
is the dark core at the center of the vortex. Since the smoke rises
in a relatively thin sheet, this dark core indicates the existence
of a strong axial flow.

Constant-temperature hot-wire probes are used in conjunction
with the smoke. They are mounted on the horizontal rods seen in
Figure 2. The flow velocity measured by the hot-wire probe is recorded both on a magnetic tape recorder and a visicorder. A typical

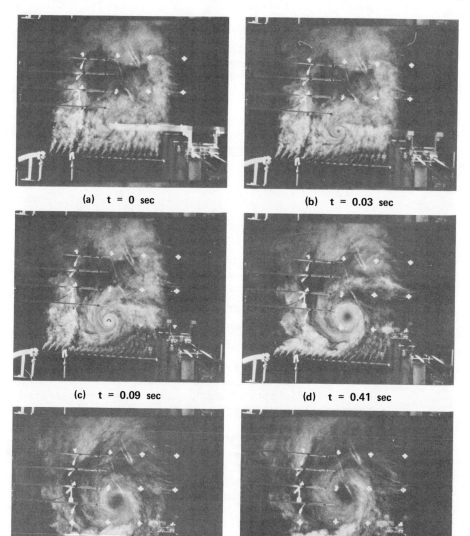

(a) t = 0 sec (b) t = 0.03 sec

(c) t = 0.09 sec (d) t = 0.41 sec

(e) t = 0.72 sec (f) t = 1.03 sec

FIGURE 2   SMOKE-TRACED TRAILING VORTEX
S (span) = 5 ft, α = 8°, V = 33 fps

velocity trace is shown in Figure 3. Several observations are made
from the study of these smoke pictures and velocity traces:

   (1) The most striking feature of the velocity trace is the
        manifestation of a high level of turbulence. In fact, the

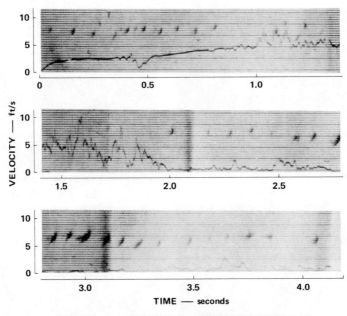

FIGURE 3   HOT-WIRE ANEMOMETER TRACE

turbulence level quite often exceeds 50 percent. This
makes one wonder whether the ordinary theory of turbulence
where quadratic perturbation quantities are neglected is
a reasonable approximation of the physical phenomenon.

(2)   Correlating the smoke pictures and the velocity traces,
it is noticed that the most turbulent part of the trace
usually coincides with the time when the vortex center
is nearest to the probe. This indicates that the vortex
core is highly turbulent.

(3)   The trailing vortices generated by the two model wings
last only a few seconds as compared with several minutes
of the full-sized trailing vortices.

4.   STUDIES USING SOAP BUBBLES

The study of trailing vortices using smoke and hot-wire anemo-
metry has several shortcomings:

(1)   Smoke dissipates too fast. Quite often, the smoke traces
are hardly identifiable when the velocity trace indicates
that the vortex is only half-way through its life span.

(2)   Residual room currents, low as they may be, are sufficient

to cause slight irregularities in the path of the vortex center. This makes the correlation of data from different but otherwise identical runs extremely difficult.

(3) The single-wire anemometer probe measures the combination of the tangential velocity, v_θ, and the radial velocity, v_r. With the axial component also present, only a three-dimensional probe will be able to resolve all three velocity components.

Because of these shortcomings, it was thought that if each of the individual smoke particles were resolvable by the photoplate, then the mean velocity of the swirling flow could be measured directly from the projected movie frames. This would then eliminate the necessity of correlating data from different runs. The use of soap bubbles was thought to be a way of accomplishing this.

A bubble machine was built which consists of a hollow circular cylinder with rows of holes drilled on its wall. An air pipe, which delivers a row of small jets blowing radially outward, is inserted longitudinally into the hollow cylinder. The cylinder is made to rotate slowly around its axis. It first exposes one row of holes to the soap solution contained in a trough, filling each hole with a film of soap solution. Then, as this row of holes comes into alignment with the jets, soap bubbles are blown out. In general, the size of the holes determines the size of the bubbles. The soap solution used is purchased from a theatrical supply agency. The soap bubbles generated from this solution have extremely high resistence to bursting. For instance, stray bubbles are often found in the experiment room up to twenty minutes after each experiment.

The bubble machine generates bubbles about 1/4 inch in diameter. Under the gravitational field, these bubbles acquire a terminal velocity of about 2 inches per second, from which the characteristic response time of these bubbles to a step change in air velocity is calculated to be about 6 milliseconds. This is entirely adequate for the measurement of the mean velocities of the vortex.

As shown in Figure 1, the bubble machine is placed on a high stand. As the bubbles drift down to the test section, the model wing is shot down the rails. Again, the bubble-traced vortex is photographed by a 16-mm movie camera at both 64 and 2 frames per second. The pictures taken at 2 frames per second have an exposure time of 1/4 second; the illuminated bubbles then show up as velocity vectors. Reprints from these movie frames are shown in Figure 4.

The vertical displacement of the vortex center is easily measured from these bubble pictures, and the results are shown in Figure 5 and Table 2. It is seen from Figure 5 that the upward motion of the vortex center slows down as time progresses and seems to approach

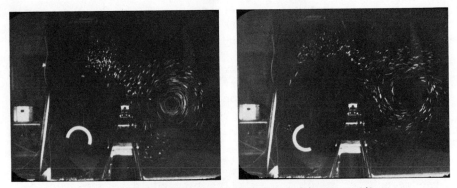

(a) t = 1/2 sec   (b) t = 2-1/2 sec

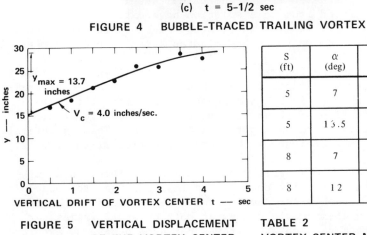

(c) t = 5-1/2 sec

FIGURE 4   BUBBLE-TRACED TRAILING VORTEX

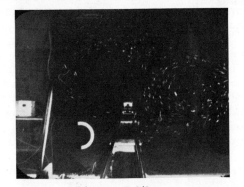

y_{max} = 13.7 inches

V_c = 4.0 inches/sec.

VERTICAL DRIFT OF VORTEX CENTER t — sec

FIGURE 5   VERTICAL DISPLACEMENT
            OF THE VORTEX CENTER

S (ft)	α (deg)	V_c (ips)	y_{max} (inches)
5	7	3.4	9
5	13.5	3.8	14
8	7	4.0	14
8	12	5.8	16

TABLE 2
VORTEX-CENTER MOVEMENTS FOR
VARIOUS WING CONFIGURATIONS

an asymptotic height. It should be noted that this behavior is purely the result of the decay of the vortex pair, since neither the ground effect nor the lapse rate should have a significant influence on this experiment. The slight scattering of data shown

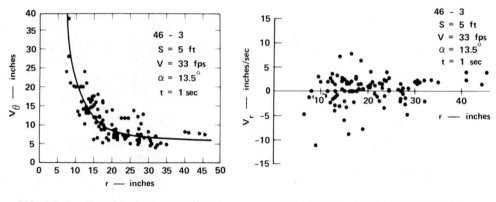

FIGURE 6   TANGENTIAL VELOCITY   FIGURE 7   RADIAL VELOCITY

in Figure 5 is primarily due to the uncertainty in determining the exact location of the vortex center.   Table 2 shows clearly that a larger wing at a higher angle of attack will create a stronger vortex pair.

Each of the velocity vectors shown in Figure 4(a) for instance, is decomposed into its tangential and radial components, v_θ and v_r. These velocity components are then plotted as a function of their radial distance from the vortex center, r.   Typical results are shown in Figures 6 and 7.   The deviation of the data points from their mean is about the same in both v_θ and v_r diagrams.   During the data reduction, the evidence is such that the scattering of the data is more a reflection of the turbulent nature of the flow than some experimental scatter.   Again, it can be seen from Figure 6 that a turbulence level of 50 percent is fairly common, which is consistent with the previous hot-wire anemometer measurements.

Figures 8(a)-(e) show the variation of the mean tangential-velocity profile with time for the two model wings at different angles of attack and running speeds.   This kind of quantitative data will be very useful for comparison with theories.

After an elapsed time of about 3 to 5 seconds in these moving-wing experiments, the vortex center is no longer identifiable, such as in Figure 4(c).   But it is still possible to represent the vortex decay information quantitatively in another fashion, that is, via the speed distribution function.   What must be done is to measure the length, which represents the speed, of each bubble trace and count the number of bubbles that have speeds within a certain speed range.   A representative plot of the speed distribution functions is shown in Figure 9.   The ordinate in this figure is a normalized dis-tribution function.

Vortex movie was also taken from the side view.   In those

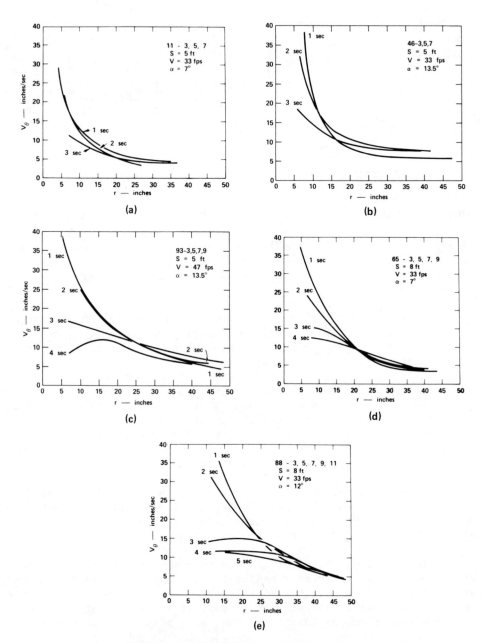

FIGURE 8   TANGENTIAL VELOCITY PROFILES

frames, bubbles that are situated inside the vortex core are seen to have very high axial velocities in the direction of the moving wing.

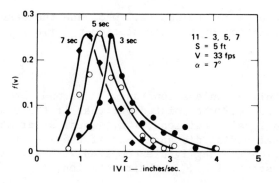

FIGURE 9   SPEED DISTRIBUTION FUNCTIONS

5.   THE SCALING STUDY

Searching for a reasonable scaling law so that the experimental results obtained by the model can be applied to the full-sized vortices, a simple scaling study was made based on existing theories of turbulent vortex.

A straight forward dimensional analysis of a decaying vortex pair in a uniform fluid medium gives the following nondimensional equation:

$$\frac{v_\theta r}{\Gamma_0} = f\left(\frac{r^2}{\varepsilon t} , \frac{r}{s'}\right) , \tag{1}$$

where

v_θ = tangential velocity of the vortex,

r = radial distance from the vortex center,

Γ_0 = initial circulation of a line-vortex,

ε = eddy viscosity,

t = time,

s' = spacing between the two vortex centers,

and f denotes an unknown functional relationship. In Eq. (1), ε is assumed to be very large compared to the laminar kinematic viscosity, so the latter can be neglected.

There exist in the literature two theories of the turbulent

line vortex. One is due to Squire,[2] who made the assumption that ε is a linear function of the circulation Γ_o, i.e.,

$$\varepsilon = a\Gamma_o \quad , \tag{2}$$

where "a" is a dimensionless constant. A second theory is due to Hoffmann and Joubert,[3] who pointed out that if the eddy viscosity depends only on local conditions, i.e., local shear stress, local density and local radial distance, and if the inertia terms in the equation of motion are negligible compared with Reynolds stresses, then ε is a constant, i.e., in their notation,

$$\varepsilon = \nu_T = \text{constant} \quad . \tag{3}$$

For the vortices generated by a lifting wing, e.g., an elliptically loaded wing, it is possible to relate the circulation Γ_o to the various parameters of the generation wing:

$$\Gamma_o = \frac{4}{\pi \rho V} \frac{W}{S} = \frac{2}{\pi} VC_L \frac{S}{AR} \quad , \tag{4}$$

where

ρ = density of air,

V = forward velocity of the wing,

W = weight of the airplane,

S = span of the wing,

C_L = lift coefficient, and

AR = aspect ratio.

Substituting Eq. (4) into Eq. (1) and using either Eq. (2) or (3) gives the following equations:

$$\frac{\pi}{2} AR \frac{r}{S} \frac{v_\theta}{VC_L} = f\left(\frac{r^2}{S^2} \frac{\pi}{2a} AR \frac{S}{VC_L t} , \frac{r}{S} \frac{S}{s'}\right) \quad , \text{ Squire } , \tag{5}$$

$$\frac{\pi}{2} AR \frac{r}{S} \frac{v_\theta}{VC_L} = F\left(\frac{r^2}{S^2} \frac{S^2}{\nu_T t} , \frac{r}{S} \frac{S}{s'}\right) \quad , \text{ Hoffmann-Joubert.} \tag{6}$$

Introducing the following dimensionless variables

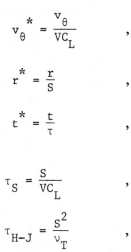

$$v_\theta^* = \frac{v_\theta}{VC_L} \quad ,$$

$$r^* = \frac{r}{S} \quad ,$$

$$t^* = \frac{t}{\tau} \quad ,$$

where

$$\tau_S = \frac{S}{VC_L} \quad ,$$

$$\tau_{H-J} = \frac{S^2}{\nu_T} \quad ,$$

and subscripts S and H-J denote the expressions for the characteristic time τ that are due to Squire's and Hoffmann-Joubert's theories, respectively, Eqs. (5) and (6) then become

$$\frac{\pi}{2} AR\ r^*\ v_\theta^* = f\left(\frac{\pi}{2}\frac{AR}{a}\frac{r^{*2}}{t^*}\ ,\ \frac{S}{s'}r^*\right) \quad ,\ \text{Squire}, \tag{7}$$

$$\frac{\pi}{2} AR\ r^*\ v_\theta^* = F\left(\frac{r^{*2}}{t^*}\ ,\ \frac{S}{s'}r^*\right) \quad ,\ \text{Hoffmann-Joubert}. \tag{8}$$

Being a dimensionless equation, Eq. (7) should describe the behavior of a vortex pair regardless of its size if the underlying theory is correct. The same argument also applies to Eq. (8). Having the experimental data from the moving-wing experiments makes it possible to test the above theories by applying Eq. (7), or Eq. (8), to both the model wing and a number of full-sized airplanes.

The one single parameter that characterizes the decay of a vortex pair is its overall decay time (which is not rigorously defined). The moving-wing experiments show that the vortices generated by the 5-foot-span wing have a decay time on the order of 3 seconds. Let subscripts 1 and 2 denote the model wing and a full-sized airplane, respectively; the two dimensionless times should be equal:

$$\frac{t_1}{\tau_1} = \frac{t_2}{\tau_2} \quad .$$

	S (ft)	V (fps)	t_S (sec)	$t_{H\text{-}J}$ (sec)
Model Wing	5	30	3	3
Cherokee	30	150	3	100
707	150	800	3	2700 (45 min)
747	200	800	5	4800 (80 min)

TABLE 3   RESULTS OF SCALING STUDY

Since τ_1 and τ_2 are combinations of their respective wing parameters, they are known quantities. As mentioned earlier, t_1 is equal to 3 seconds; t_2, which now represents the vortex decay time of the full-sized airplane, can thus be calculated. The calculated decay time for a light airplane (represented by a Piper Cherokee), a heavy passenger plane (represented by a Boeing 707), and a jumbo jet (represented by a Boeing 747) are given in Table 3.

The analysis here is concerned only with orders of magnitude; hence, the differences in aspect ratio, lift coefficient, and ratio of S to s' between the model wing and any one of the three airplanes are ignored. From Table 3, it is seen that according to Squire's and Hoffmann-Joubert's theories, the vortex pair shed by a 747 would last for 5 seconds and 80 minutes, respectively; both numbers are far from realistic. The inability of either theory to correlate the laboratory results and the full-sized airplanes indicates that a satisfactory theory of turbulent vortex decay is still lacking.

6.   DISCUSSION

It is obvious that the moving-wing experiments have quite a number of merits. Among them are ease of operation, low cost, repeatability, and versatility. However, there is also an obvious limitation: the finite length of the track introduces the effects of starting and stopping vortices. Olsen[4] has observed in his water-tank experiment that the effect of the starting vortex is neglegible, but the sudden stop of the wing creates a hydraulic jump-like disturbance that propagates backward along the vortex core. Up to the time when the gross core-instability develops that will quickly destroy the vortex, this disturbance will have traveled a distance of about 30 chord-lengths. If Olsen's result can be scaled directly to the moving-wing setup, then the effect of the stopping vortex will indeed be present at the test section shortly after the moving wing is brought to a stop. Some means of alleviating this effect, such as lengthening the track, will have to be considered.

REFERENCES

1.   Spreiter, J. R., and Sacks, A. H., "The Rolling Up of the Trailing Vortex Sheet and Its Effect on the Downwash behind Wings," J. Aero. Sci., Vol. 18, No. 1, Jan. 1951, pp. 21-32.

2.   Squire, H. B., "On the Growth of a Vortex in Turbulent Flow," Aero. Res. Counc. London, Paper No. 16, 1954, p. 666.

3.   Hoffmann, E. R., and Joubert, P. N., "Turbulent Line Vortices," J. Fl. Mech., Vol. 16, 1963, p. 395.

4.   Olsen, J. H., "Results of Trailing Vortex Studies in a Towing Tank." Appears in this same Symposium Proceedings.

THE UTILITY OF DOPPLER RADAR IN THE STUDY

OF AIRCRAFT WING-TIP VORTICES

Calvin C. Easterbrook and William W. Joss

Cornell Aeronautical Laboratory, Inc.

ABSTRACT

The feasibility of utilizing a special purpose Doppler radar to investigate air motions in aircraft wakes is explored and tested. Two similar experimental methods are undertaken to investigate the potential value of the Doppler radar approach. The first method involves injection of radar reflecting chaff into the wing-tip vortex of an aircraft in flight, and subsequently recording the Doppler spectrum of the return from the chaff packet as obtained by the radar looking normal to the flight path. The second method consists of Doppler measurements in the wake behind aircraft on the approach to an airport during snow conditions, where natural snow crystals and flakes assume the role of radar reflecting tracers of air motion. The power spectra derived from both techniques reflect the distribution of velocities of scatterers (chaff elements or snow crystals) in the direction of the radar, weighted by the spatial distribution of the scatterers contained in the radar sensitive volume.

1. INTRODUCTION

The prediction of position, strength, and other characteristics of wing-tip vortices generated by aircraft in flight has assumed increasing importance with the ever-widening disparity in size and weight among aircraft operating in the same airspace. Furthermore, it is recognized that existing analytical techniques have been so simplified that they are of little value in accurate predictions. What appears to be needed at the present time are improved experimental techniques that will provide reliable wake measurements for

97

prediction purposes, and, at the same time, supply more detailed information to be used in advancing the understanding of the formation and decay of aircraft wakes.

Direct probing methods currently being used for full-scale vortex measurements suffer from several disadvantages. The main problem with direct measurement is the probe and probe-carrier interaction with the vortex. There are very definite indications that disturbances in a vortex produced by the probe, be it an aircraft or tower-mounted sensor, may lead to rapid change and eventual breakup of the vortex. Another serious drawback to direct probing is the difficulty in locating the vortex and in establishing a position of the measurement relative to the vortex position. Finally, ground-based direct probes are made with the vortex either approaching or in ground effect. This severely limits the scope of the measurement program.

It is the purpose of this paper to introduce a remote, indirect measurement technique that avoids the problems listed above. The method involves the use of microwave Doppler radar to measure the velocity distribution of suitable tracers directly injected or otherwise imbedded in an aircraft generated vortex. Interpretation of the radar data depends to a large extent on the assumption that air motions in a vortex remain axially symmetric with the axis aligned in the direction of the flight path for a significant portion of the vortex life span. If this assumption is valid, then measurement of the single component velocity field either normal to or parallel with the flight path should produce some very useful results.

2.   TRACER-VORTEX INTERACTION

It is well-known that a force must be continuously applied to a body in order to make it follow a curved (i.e., inwardly accelerated) path. In the case of a parcel of air traveling around a nearly circular path in a vortex, this force is supplied by the existing pressure gradient. If a test particle, with a density different than air, were injected into such a vortex pattern, its trajectory would not follow that of the air. That is, if the particle were heavier than air, it would tend to be thrown outward. If it were lighter than air, say a helium-filled balloon, it would be forced inward. At first glance, then, it would seem that solid chaff particles, many times heavier than air, could never remain close to the center of a vortex. However, this conclusion is not true in general, as is demonstrated through observation of naturally-occurring vortices. In particular, tornadoes, dust devils, and waterspouts can pick up and entrain large solid objects or particles for long lengths of time. In addition to drawing these particles inward, the vortices are often seen to accelerate the objects axially. These two motions in the vortex, i.e., the radial inflow and motion along the axis, are intimately connected. Unfortunately, they are often ignored in a simplified model of the flow.

In studying the interaction of chaff particles with the flow in a wing-tip vortex, the radial inflow becomes a crucial factor. It is this inflow, and the associated drag it produces on the chaff particles, which allows the particles to become entrained by the wing-tip vortex. This phenomenon is well-illustrated, for instance, by the behavior of dense smoke particles in a vortex.

The centrifugal force acting on a particle can be related to the observed motion of the particle as follows:

$$\frac{F}{W} = \frac{V_\theta^2}{rg} \tag{1}$$

Here, F is the centrifugal force, W, the weight of the particle, V_θ, the instantaneous tangential velocity, r, the instantaneous radius of curvature, and g, the acceleration of gravity.

The centrifugal force, in a steady circular motion, is just balanced by the drag force produced by the relative inward motion of the air, i.e.,

$$F = D = 1/2 \rho V_R^2 C_D A \tag{2}$$

Here, D is the drag, ρ, the air density, V_R, the relative (inward radial component) velocity, C_D, a drag coefficient and A, an appropriate area. This expression can be simplified by relating the relative velocity to the terminal velocity of the particle. That is, when a particle is falling at its terminal velocity, its drag just equals its weight. If it moves at twice the terminal velocity, its drag is four times its weight, i.e., the drag increases as the square of the velocity. Therefore, the drag can be given by

$$D = W \left(\frac{V_R}{V_t} \right)^2 \tag{3}$$

where V_t is the terminal velocity.

Combining equations (2) and (3)

$$F = D = W\left(\frac{V_R}{V_t}\right)^2 \tag{4}$$

Combining equations (4) and (1)

$$\frac{F}{W} = \frac{V_\theta^{\ 2}}{rg} = \left(\frac{V_R}{V_t}\right)^2 \tag{5}$$

This expression relates the tangential and inward radial velocity components to the required particle terminal velocity for a circular motion. However, this criterion must be supplemented by a further criterion dealing with the stability of such a balance of forces. That is, unless the particle is in stable equilibrium, a uniform circular motion will never result.

The existence of a mechanism to provide the required stable orbits observed in practice is suggested by the experimental work of Ying and Chang.[1] They show that in a laboratory-generated vor-tex, the maximum radial inflow velocity occurs at a larger radius than the maximum tangential velocity. This situation is illustrated in Figure 1.

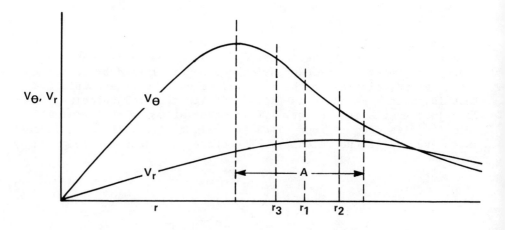

Figure 1   TANGENTIAL AND RADIAL VELOCITIES IN A VORTEX AS A FUNCTION OF RADIUS (YING AND CHANG)

Consider the behavior of a particle placed in region A. If at r_1, the terminal velocity is such that equation (5) is satisfied, the forces on the particle will be instantaneously balanced. Now, suppose the particle moves to a larger radius, r_2. We see that the tangential velocity V_θ decreases; hence, the centrifugal force decreases. (This decrease is further enhanced by the increase in radius.) At the same time, the radial inflow velocity, and the drag on the particle, increase. Therefore, as the centrifugal force decreases and the (inwardly directed) drag increases, the particle experiences a net inward, or restoring, force. The converse is true if the particle had moved to a smaller radius, say r_3. Region A may not necessarily be the only region where such a stable orbit is possible. In fact, using the analytical expressions given by Ying and Chang, one finds stable orbits at any radius where the particle's terminal velocity is such that equation (5) is satisfied. Although the expressions of Ying and Chang may be slightly inaccurate, it is encouraging that stable orbits may be found over large regions in the vortex.

In summary, the chaff particles are held in the vortex by the inward directed radial velocity component. As long as the centrifugal force decreases more rapidly than the drag due to the inward velocity as the radius increases, the balance of forces is stable and the chaff particles should rotate steadily about the vortex.

3.   TRACER MOTION AND THE DOPPLER SPECTRUM

Having shown that it is at least reasonable to expect chaff or similar tracers entrained in a vortex to remain near the core for a significant length of time, it is instructive to investigate the nature of the Doppler signature that would be produced by such a radar target.

The Doppler spectrum represents the power received per unit increment of relative velocity between the radar and the target. For a distributed target consisting of random scatterers, the received spectrum is directly proportional to the number of scatterers in the radar sensitive volume with a given velocity increment in the direction of the radar.

For simplicity, it is assumed that the motion in a vortex is solid rotation and that the radar views this motion at right angles to the axis. The tangential velocity of the air is then given by

$$V_\theta = \omega r \qquad\qquad (6)$$

and the velocity at a given point as measured by the radar is

$$V_r = \omega r \cos \vartheta, \qquad\qquad (7)$$

where θ is measured from a cross beam reference passing through
the axis of the vortex. It is further assumed that the vortex is
uniformly filled with scatterers and that there are n particles
per unit volume. The total number of particles in a single quadrant of unit length of vortex is then given by

$$N = n \int_0^{\pi/2} \int_0^a r\, dr\, d\theta \tag{8}$$

where a is the outer radius of the radar target. The integral (8)
can be rewritten in terms of the velocities V_θ and V_r through
equations (6) and (7). Making this transformation and splitting
the integration into two parts covering an inner radius a_o and the
outer radius a, equation (8) becomes

$$N = \int_0^{\omega a_o} \int_{\omega a_o}^{\omega a} F(V_r, V_\theta)\, dV_\theta\, dV_r + \int_{\omega a_o}^{\omega a} \int_{V_r}^{\omega a} F(V_r, V_\theta)\, dV_\theta\, dV_r \tag{9}$$

$$\text{where } F(V_r, V_\theta) = \frac{n}{\omega^2 \sqrt{1 - V_r^2 / V_\theta^2}} \tag{10}$$

It is seen that (9) can be expressed in terms of the Doppler
velocity only, i.e.,

$$N = \int_0^{\omega a} N(V_r)\, dV_r \tag{11}$$

$$\text{where } N(V_r) = \int_{\omega a_o}^{\omega a} F(V_r, V_\theta)\, dV_\theta \qquad o \le V_r \le \omega a_o \tag{12}$$

$$\text{and } N(V_r) = \int_{V_r}^{\omega a} F(V_r, V_\theta)\, dV_\theta \qquad \omega a_o \le V_r \le \omega a \tag{13}$$

Now the function $N(V_r)$ represents the total number of scatterers with unit velocity increment in the direction of the radar and is thus equivalent to the received power spectrum. Substituting (10) in (12) and (13) we have

$$N(V_r) = \frac{n}{\omega^2} \int_A^{\omega a} \frac{V_\theta dV_\theta}{\sqrt{V_\theta^2 - V_r^2}} \tag{14}$$

$$\text{or } N(V_r) = \frac{na}{\omega} \left[\sqrt{1 - \left(\frac{V_r}{\omega a}\right)^2} - \sqrt{\left(\frac{A}{\omega a}\right)^2 - \left(\frac{V_r}{\omega a}\right)^2} \right] \tag{15}$$

where $A = \omega a_o$ for $o \leq V_r \leq \omega a_o$

or $A = V_r$ for $\omega a_o < V_r \leq \omega a$

Using equation (15) and the appropriate values of A, two important Doppler signatures can be obtained. It must be pointed out that (15) gives only half the spectrum since the integration was carried out over only one quadrant. However, based on the assumption of radial symmetry, the other half of the spectrum is an image of the first.

By setting $A = \omega a$, equation (15) gives the power spectrum for a uniform distribution of scatterers throughout the vortex core. This spectrum is shown in Figure 2-1. Setting $A = \omega a_o$ and computing $N(V_r)$ for V_r in range o to ωa_o and then setting $A = V_r^o$ in range ωa_o to ωa gives the power spectrum for target distribution in a ring and is shown in Figure 2-2. These cases are important ones since any other chaff distribution can be approximated utilizing a combination of the two.

While the above analysis treats a much simplified case, deviations from the solid rotation concept that may exist in a real vortex are not likely to change the gross features of the spectrum. A single peaked response will imply a uniformly filled vortex and a double peak will strongly suggest a ring distribution. In all cases, the maximum half-width of the spectrum will be a measure of the maximum tangential velocity of the tracers in the radar sensitive volume. This may or may not coincide with the maximum velocity in the vortex, depending upon how the tracers are distributed.

4. THE CHAFF INJECTION EXPERIMENT

Some simple experiments were planned with the hope of getting some indication of the validity of the preceding arguments. The experimental effort of necessity had to be modest and was possible only through the use of existing equipment.

An operational Doppler weather radar was available as well as a research aircraft equipped with wing-tip racks designed for cloud seeding work. It was a simple task to make chaff dispensers that would fit the wing-tip racks and be fired by the same electrical circuits used to fire silver iodide flares. Figure 3 shows the Aztec aircraft with wing-tip rack and Figure 4 shows an exploded view of a chaff dispenser tube. A very small powder charge was mated to an electrical detonator and used to eject about 150,000 dipole chaff elements from a tube. In the experiments conducted, three such tubes were loaded in the rack and fired in rapid sequence (machine gun fashion). The radar used is an X-band unit with a peak power output of only eight kilowatts. The antenna beamwidth is approximately one degree to the half power points and the trans- mitted pulse length is 1/2 microsecond giving a maximum range resolution of 75 meters.

The geometry of the chaff experiments is shown in Figure 5. The chaff string ejected from the wing-tip of the cooperating air- craft was viewed at right angles to the flight path, the flight path being perpendicular to the plane of the paper in Figure 5. The Doppler spectrum of the radar return was displayed on a comb- filter type spectrum analyzer with a velocity resolution of 0.1 meters per second. The spectrum was recorded by photographing the analyzer display with a 35 mm camera every two to three seconds after chaff ejection. The flash contacts of the camera were coupled to a recorder so that an accurate time history of the spectrum could be reconstructed from the photographs.

The chaff was dropped when the aircraft flight path was exactly normal to the radar beam. This was accomplished by optically track- ing the aircraft while constantly monitoring the Doppler spectrum of the skin return. When the signal approached the zero velocity point on the spectrum display, the operator in the aircraft was signaled by radio to fire the chaff. The antenna was stopped at this point and the picture sequence started. In less than two seconds, the aircraft moves out of the radar beam, leaving the chaff packet imbedded in the vortex producing a very large signal.

Three experiments were conducted in this way with two of them producing excellent results. In the unsuccessful attempt, the chaff almost missed the radar beam and was quickly lost. However, the initial signature was identical to the other two.

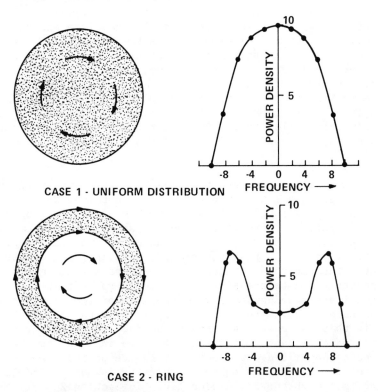

CASE 1 - UNIFORM DISTRIBUTION

CASE 2 - RING

Figure 2   COMPUTED POWER SPECTRA FOR TWO CHAFF DISTRIBUTIONS
IN SOLID ROTATION

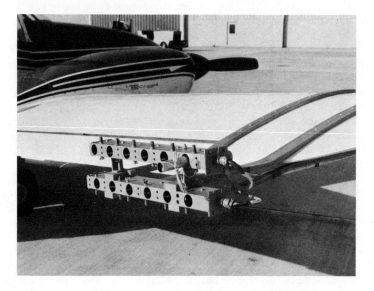

Figure 3   CLOUD SEEDING FLARE RACK WITH CHAFF DISPENSER IN PLACE

Figure 6 shows a series of spectra obtained from one of the experiments. The first frame (16A) shows the aircraft skin-return just at the moment of chaff release. The next frame is about two seconds later and is from the chaff alone, the aircraft signal having moved to the left out of the display. The parameters displayed are voltage in the Y axis and velocity in the X direction. The velocity scale is approximately two meters per second per large division with zero velocity at the center. The constant offset in all the spectra is caused by a wind drift component away from the radar.

The individual spectra and the time history shown in Figure 6 show some rather interesting behavior. For instance, the initial chaff return very definitely shows the double peaked form predicted for a ring dispersion. Frames 20B and 21A show similar forms with reduced peak separation. The maximum tangential velocity depicted by frame 17A is about 15 feet per second, decaying steadily from that point. The tendency for decreasing spectrum width and a shift of the spectral form to the single peak seems to indicate that the chaff particles are being fed toward the axis of the vortex rather than being thrown out. Another possible interpretation of the data, however, is that the chaff is already outside of the velocity maximum at frame 17A and that the chaff is moving away from the core and being deposited outside the influence of the wake. Unfortunately, the solution to this ambiguity is not available at present, although the discussion given in Section 2 would appear to make a good case for the inflow interpretation.

The rates of decay of the maximum recorded velocities for the two usable experiments are plotted in Figure 7. The chaff appears to remain under the influence of the wake for almost a full minute. Actually, the indicated rate of decay may be influenced by drifting of the vortex out of the radar beam. This drift is certainly expected from the frequency offset of the signal and is confirmed by the decreasing area under the spectrum. If all the chaff particles remained in the sensitive volume for the entire period, the area under the spectrum would remain constant. Therefore, it is quite possible that the measured decay rates as deduced from the photographs are too optimistic and that appreciable chaff may well be still retained in the vortex core at the 60-second point.

The chaff used in the experiments described have a terminal velocity of about two feet per second and appear to establish stable orbits at a maximum tangential velocity of 16 feet per second. Smaller diameter chaff are available with terminal velocities as low as 0.5 feet per second. From the stability criterion given previously, this light chaff could be expected to stabilize in orbits with velocities as high as 250 feet per second. Therefore, it is not unreasonable to believe that the chaff-Doppler method could be extended to measurements of vortices generated by large aircraft.

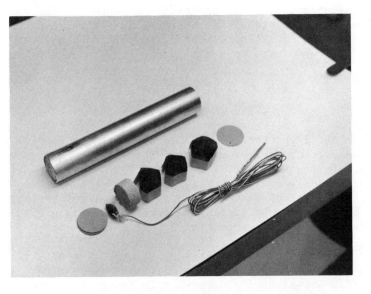

Figure 4   EXPLODED VIEW OF CHAFF DISPENSER

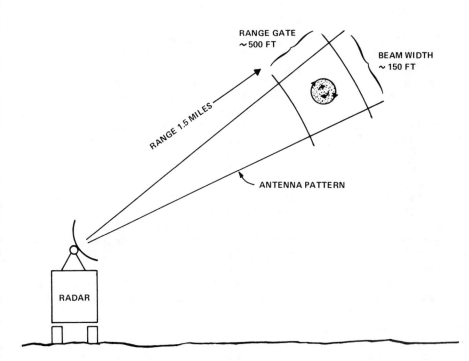

Figure 5   GEOMETRY OF THE CHAFF EJECTION EXPERIMENTS

The main deficiency in the radar, as shown by the data obtained, is the lack of automatic antenna tracking. If tracking had been available, the chaff targets would have remained centered in the radar sensitive volume, thus eliminating this source of apparent velocity decay.

5. WAKE MEASUREMENTS IN SNOW

The chaff drop experiments indicate that the Doppler radar method has a great deal of promise. However, the requirement for a cooperating aircraft, the mechanics of ejecting chaff, and the rather transient nature of the chaff experiments dictate a somewhat slow rate of accumulation of data. Therefore, it was decided to try and make similar measurements using natural snow as the tracer and non-cooperating aircraft.

The radar was subsequently set up in the vicinity of the approach to the main ILS runway at Buffalo International Airport. The site was established about one mile inside the outer marker at a range of two miles from the localizer center line. The antenna and range gate were set to bracket the approach corridor, with the radar looking normal to the flight path. Viewing geometry for this experiment is given in Figure 8.

Estimates of signal intensity made prior to the snow measurements indicated that the signal to noise ratio would be adequate at 1.5 miles range when the visibility was reduced to 3/4 of a mile in snow. However, the actual data were marginal due to the generally low signal levels and the high signal-to-noise ratio. The noise referred to here is produced by the background volume of snow that is not interacting with the wake in the radar sensitive volume. It was found, during the course of the measurements, that the range gate width had to be extended to two microseconds (1000 feet) to ensure the wake remaining in the sensitive volume.

In spite of the low signal-to-noise ratio, wakes were detected in the snow behind aircraft on the approach. Figure 9 shows some examples of spectra taken in two different measurement sequences. In the first frame of sequence 1, the generating aircraft is seen on the right as it proceeds past the radar beam, leaving the wake signature on the left. The photos following are taken at about five-second intervals. The wake spectra are seen to decrease in width as in the chaff experiments but in this case, the signal does not slowly disappear. The spectrum simply reverts back to the undisturbed snow condition existing before the passage of the aircraft. This background spectrum is a measure of the turbulence level in the snow. The velocity scale in the photos of Figure 9 is five times larger than in the chaff photos, making one large-scale division equivalent to ten meters per second.

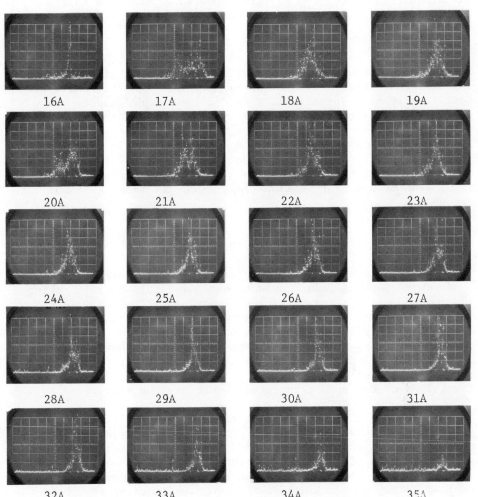

Figure 6 SEQUENCE OF PHOTOGRAPHS OF DOPPLER SPECTRA
SHOWING THE TIME HISTORY OF THE DECAY

Initial velocity measurements taken from the two sequences shown are approximately 26 to 30 feet per second. These values appear to be much too low for the type of aircraft involved (BAC 111). The difficulty here appears to be one of signal level. The volume of the high velocity part of the vortex is too small to involve enough snowflakes to be recorded above the background. However, the measured spectra do indicate that a larger volume but lower velocity part of the vortex is indeed recorded.

While the snow measurements are not nearly as quantitative as the chaff experiments, the results do suggest that the method is feasible with increased sensitivity. The required sensitivity could easily be obtained with a specially-designed radar. The working distance could be reduced by an order of magnitude. The snow-target cross section could be increased by increasing the radar frequency and the background snow signal could be reduced appreciably by reducing the size of the sensitive volume. Minimizing the sensitive volume would necessitate compensating for vortex motion by precision tracking of the range gate and the antenna.

6.   CONCLUSIONS

The results of a simple theoretical analysis suggest that a microwave Doppler radar could become a valuable tool in making measurements of motions in wing-tip vortices. The preliminary experiments involving actual radar measurements of chaff injected into a tip vortex tend to support the analytical work. Similar measurements made of aircraft wakes in snow also show promise. While the experiments were not highly quantitative, the results are certainly encouraging. This is especially true if one considers the rather modest effort required to acquire the measurements.

It is felt that both techniques described could provide very useful quantitative measurements of maximum tangential velocities, decay rates and even axial velocities if the vortices were probed along the axis. Certain improvements to the radar would be required. The chaff measurement method would be advanced a great deal by adding even a single axis monopulse tracking system to the radar. The snow measurement technique could be greatly improved by increasing the sensitivity and the addition of automatic track-ing as suggested in the previous section. Antenna and range gate tracking in the snow case would have to be designed to maximize the width of the spectrum rather than the signal intensity as in normal tracking systems.

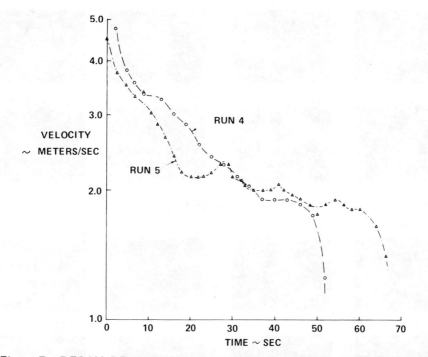

Figure 7   DECAY OF MAXIMUM TANGENTIAL VELOCITIES COMPUTED
FROM THE TIME HISTORIES OF DOPPLER SPECTRA

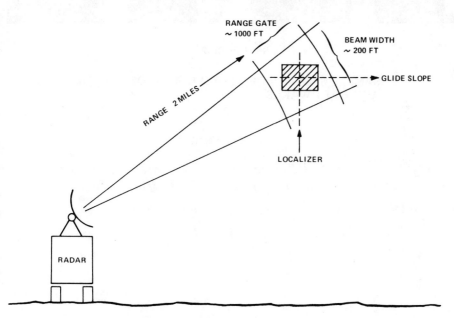

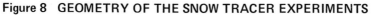

Figure 8   GEOMETRY OF THE SNOW TRACER EXPERIMENTS

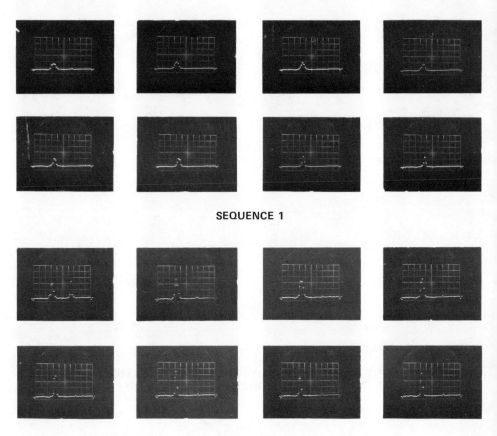

SEQUENCE 1

SEQUENCE 2

Figure 9   SAMPLE SPECTRA GENERATED BY SNOW IMBEDDED IN
AIRCRAFT WAKES

7.   REFERENCES

1. Ying, S. J. and C. C. Chang, 1970: Exploration Model Study of
    Tornado-like Vortex Dynamics. J. Atmos. Sci., 27, 3-14.

APPLICATION OF LASER DOPPLER SYSTEMS TO VORTEX MEASUREMENT AND DETECTION

R. M. Huffaker, A. V. Jelalian, W. H. Keene, and

C. M. Sonnenschein

NASA/Marshall Space Flight Center and Raytheon Company

ABSTRACT

A laser Doppler system for the measurement of atmospheric wind velocity and turbulence has been developed. This system utilizes the Doppler frequency shift undergone by a beam of radiation when scattered by particles suspended in the flows. From the measurement of this difference frequency between the scattered and the reference laser light and knowledge of the geometry of the system, the velocity is directly determined. A three-dimensional system has been developed for wind tunnel and jet type flow studies. Using this system, measurements of velocity have been made in wind tunnels and in jet flows with velocities in excess of Mach 2 and compared with theory and hot wire instrumentation. The agreement was good.

Using the concept proven in wind tunnel aerodynamic applications, the feasibility of extending the technique to the measurement of atmospheric wind velocity and turbulence was demonstrated. A one-dimensional research unit has been developed. Using this one-dimensional laser Doppler system, measurements of atmospheric wind velocity have been made and good comparisons with simultaneous cup-type anemometer measurements has been demonstrated. These comparisons in mean wind velocity will be presented.

Comparisons with cup and hot wire anemometer data with the laser Doppler system for measurement of the time history of the wind velocity and the statistical properties of the wind velocity and the statistical properties of the wind velocity fluctuation will also be presented. Comparisons for different types of wind conditions will be presented.

This Doppler system has also been applied to the problem of detecting the presence of an aircraft trailing vortex. Results of the test program will be presented. The results showed that when the vortex was visually sighted and known to be in the sensitive scattering volume of the laser Doppler system, the velocity distribution of the vortex was measured. Consideration for a three-dimensional research system for monitoring the presence and velocity structure of aircraft trailing vortices will be given. Operational consideration for an airport warning system will also be presented.

Approximately four years ago, the NASA/Marshall Space Flight Center initiated the development of a laser Doppler technique for the measurement of gas velocity. This technique utilizes the Doppler frequency shift undergone by a beam of radiation when scattered by particles suspended in the flow. From the measurement of this difference frequency between the scattered and the reference laser light and knowledge of the geometry of the system, the velocity is directly determined. A three-dimensional system has been developed for wind tunnel and jet type flow studies. Using this system, measurements of velocity have been made in wind tunnels and in jet flows with velocities in excess of Mach 2.0 and compared with other velocity instrumentation. Measurements of jet shear layer turbulence have been made with the Doppler system and compared with theory and hot wire instrumentation.

Using the laser Doppler concept proven in wind tunnel aerodynamic applications, the feasibility was demonstrated for extending the technique to the measurement of atmospheric wind velocity and turbulence by utilizing a one-dimensional $10.6\,\mu m$ laser Doppler velocimeter. Subsequently this one-dimensional research unit was applied to the problem of detecting the presence of an aircraft trailing vortex.

This paper discusses the results of a test program utilizing a CW CO_2 Laser Doppler Velocimeter System that has demonstrated the capability of remotely detecting wing tip vortices by the measurement of Doppler shifted backscatter from atmospheric aerosols naturally suspended in the atmosphere.

Before getting into the experimental results, a review of some of the theory associated with an optical heterodyne system is in order.

Figure (1) illustrates the expression for the S/N ratio of an optical heterodyne system. Here we may observe that this S/N is dependent upon the detector quantum efficiency Q, the received signal power (P_R), the reference local oscillator power P_{LO}, the electronic bandwidth (B) which is determined by the differential atmospheric velocity over the sampled volume and some noise terms. The noise terms are dependent upon the reference local oscillator, optical noise background, and the electronic noise figure of the receiver. By increasing the local oscillator power obtained directly from the laser, one may make the S/N expression independent of the background and arrive at the quantum noise limited S/N expression generally used.

The electronic bandwidth associated with the S/N expression is typically determined by the amount of differential velocity over the path being sampled. Figure (2) illustrates the typical expression of the Doppler shift for a coaxial backscatter system. Here we may observe that the Doppler shift is equal to twice the relative target velocity divided by the transmission wavelength and multiplied by the angle between the relative velocity vectors. Correspondingly the Doppler spectral spread may be shown to be directly related to the differential velocity over the path.

For the laser system utilized during the vortex measurements a focused coaxial optical system was utilized. In this configuration range resolution is obtained by focusing the optical system at a known range and accepting the major portion of the return signal from the depth of field of the optics. Figure (3) illustrates the S/N equation for the focused coaxial system. Here we may observe that the S/N is related to the detector quantum efficiency, here denoted as η, the transmitted power P_T, the atmospheric backscatter coefficient $B(\pi)$, the wavelength λ, $w(R)$ the atmosphere path weighting function, and inversely related to Planck's Constant, the transmitter frequency and the electronic bandwidth.

Referring to Figure (4) we may observe that the atmospheric path weighting function in turn is dependent upon the propagation path R, wavelength λ, optics diameter D, and optics focal length f.

Evaluation of this expression over the depth of field of the optical telescope results in approximately 50% of the returned heterodyne signal occurring from this volume and results in the S/N ratio being independent of range and optics diameter.

Having discussed some of the theory associated with the instrumentation, we will now discuss the trailing vortex detection experiment.

The S/N power at the output of the receiver for a monochrometer source is:

$$S/N = \frac{Q \, P_R \, P_{L.O.}}{B \, h \, f \left(P_{L.O.} + P_N + P_{AMP} \right)}$$

where:

Q = detector quantum efficiency

P_R = received signal power

$P_{L.O.}$ = local oscillator signal power

P_N = equivalent optical noise power

P_{AMP} = equivalent noise figure power of post detection amplifier

B = electronic bandwidth

h = Planck's Constant

f = transmission frequency

for:

$P_{L.O.} \gg P_N + P_{AMP}$ the S/N expression becomes:

$$S/N = \frac{Q \, P_R}{Bhf}$$

Figure 1   S/N Expression

The equation for the Doppler frequency shift (Fd) may be expressed as:

$$Fd = \frac{2V}{\lambda} \cos \theta$$

where:

V = relative velocity

λ = transmission wavelength

θ = pointing angle

and the Doppler spectral bandwidth (Δfd) caused by the uncertainty of wind velocity (ΔV) over the sampled volume is:

$$\Delta fd = \frac{2 \, \Delta V}{\lambda} \cos \theta$$

Figure 2   Bandwidth Expression

S/N equation for a focused coaxial system

The S/N equation for a focused coaxial system over a path R_1 to R_2 is:

$$S/N = \frac{nP_T\beta(\pi)\,\lambda}{2h\upsilon B}\sum_{R_1}^{R_2} w(R)$$

where:

η = quantum efficiency

P_T = transmitted power

$\beta(\pi)$ = atmospheric backscatter coefficient

λ = operating wavelength

h = Planck's Constant

υ = operating frequency

$w(R)$ = atmospheric path weighting function

R = range

Figure 3 S/N Equation for a Focused Coaxial System

S/N Equation For A Focused Coaxial System

The S/N equation for a focused coaxial system over a path R_1 to R_2 is:

$$S/N = \frac{nP_T\,\beta(\pi)\,\lambda}{2h\upsilon B}\sum_{R_1}^{R_2} w(R)$$

where:

$$w(R) = \frac{dR}{\dfrac{4\,\lambda R^2}{\pi D^2}\left(1 + \dfrac{\pi D^2}{4\,\lambda R}\right)^2 \left(1 - \dfrac{R}{f}\right)^2}$$

In the near field of a focused coaxial system approximately 50% of the received heterodyne signal occurs from the depth of field of the optical system and therefore the S/N equation becomes

$$S/N\Big|_{\text{Depth of field}} = \frac{nP_T\beta(\pi)\,\lambda\pi}{4h\upsilon B}$$

Figure 4 S/N Over Depth of Field

The vortex detection experiment consisted of a CO_2 laser
Doppler velocimeter for remote vortex detection, a smoke bomb
elevated on a tower for vortex visualizations and C-47 and the
NASA-Gulf Stream 1 test aircraft for vortex generation.

Figure (5) illustrates the arrangement of trailing vortex
laser equipment and instrumentation.   In order to obtain a visualization of the vortex pattern generated, and to aid in correlating
between visual sightings and prior detection with the laser
Doppler system, a smoke bomb was elevated to the top of the tower.
A typical run consisted of the test aircraft, being directed by
radio contact, flying at 40 feet altitude up wind of the 31-foot
high tower.   The laser Doppler velocimeter was then focused 80
feet from the test van and approximately 10 feet up wind from
the tower.   With this test configuration the vortex would be
detected by the laser system prior to being visualized by the
smoke being entrained in the vortex.

The flight paths were chosen so that the vortex would drift
into the region of the smoke bomb near the top of the tower and,
therefore, the display on the spectrum analyzers was able to
indicate the arrival of a vortex many seconds before the smoke
tower indicated it.

The C-47 aircraft was flown in a "clean" configuration (flaps
and landing gear retracted) to produce a well-behaved vortex.
The air speed of the aircraft was in the vicinity of 90 knots
and the height above the ground about 40 feet so that the vortex
would pass the top of the tower.

A cross wind path was usually flown to allow westerly winds
to blow the vortex onto the tower.   On occasion, a northerly wind
made it necessary to fly a downwind path which then allowed the
vortex to settle on the tower from overhead.

Days with low, steady winds were best for the experiments.
High or gusty winds tended to dissipate.   The flight paths of
the plane were determined by observers outside the van.   Instruc-
tions for changes in the flight path were radioed to the RAA
flight tower and relayed from the tower to the plane.   On steady
days, the repeatability of the vortex position from run to run was
in the neighborhood of ten feet, so that there were many times
when parts of the vortex would be in the focal volume.

In the next Figure (6) we observe a system block diagram of
the trailing vortex laser equipment.   Basically a 20-watt CO_2
laser operating single mode and single frequency was coupled thru
the interferometer and optically focused by a 6-inch telescope.
Range at which the laser beam was focused was determined via a

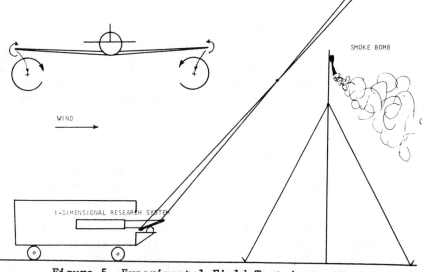

Figure 5 Experimental Field Test Arrangement

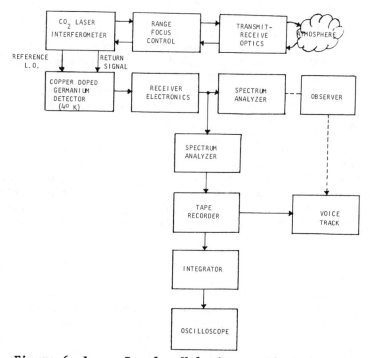

Figure 6 Laser Doppler Velocimeter Block Diagram

calibrated adjustment control on the telescope which effectively changed the focus point of the optical telescope. The laser radiation was therefore focused at a known point in space. The resultant electromagnetic field would then become incident on the natural aerosol particles suspended at this point in space within the near field of the optical system. The resultant backscattered energy was then collected by the same telescope and directed thru the interferometer to a copper doped germanium detector operating at 4° Kelvin.

After passing thru suitable receiver electronics a Doppler frequency corresponding to the motion of the natural aerosols would be observed in real time on a spectrum analyzer. Sixteen millimeter movies were made of the aircraft fly-by and the subsequent vortex roll-up in the smoke.

The remote measurement of wind speed and wind speed fluctuation was typically monitored by viewing the CRT display of a spectrum analyzer having a full scale frequency of 1.5MHz, corresponding to a velocity of 26 FPs. Additionally this signal was recorded on tape recorder along with voice track information for post flight data analysis.

In the next figure, the vortex core is well shown in the upper left, while the region near the smoke tower is changing rapidly with no clearly defined pattern. Shortly after this photo was taken, the upper left portion of the core dipped down to within a few feet of the ground and dissipated itself. The vortex shown in the next figure is an example of a well-behaved one which occurred on a day of low steady wind. Also indicated in these photographs is a fast axial flow in the direction of the aircraft. As a result, the vortex core visualization extends further in an axial direction than the outer regions do.

Demonstration of this effect is shown in the next figure. The wind here was to the right as evidenced by the smoke in the outer portions of the vortex. In spite of this, a considerable core flow has occurred against the wind. The flight direction in this photo was to the left. In the next figure, the aircraft direction has been reversed and the flow is to the right and out of the photograph.

The maximum core velocity for a C-47 can be calculated to be on the order of 19 to 25 feet per second. This would place the maximum detected Doppler frequency in the 1 to 1.5 MHz region which is consistent with the maximum frequencies observed. When measurements are extended to larger aircraft, the vortex velocities will be higher and the vortex signals will be more distinct through their greater separation from the ground wind signal.

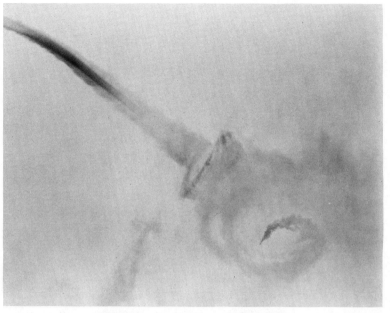

Figure 7   Vortex in Smoke

Figure 8   Vortex in Smoke

Figure 9 Vortex in Smoke

Figure 10 Vortex in Smoke

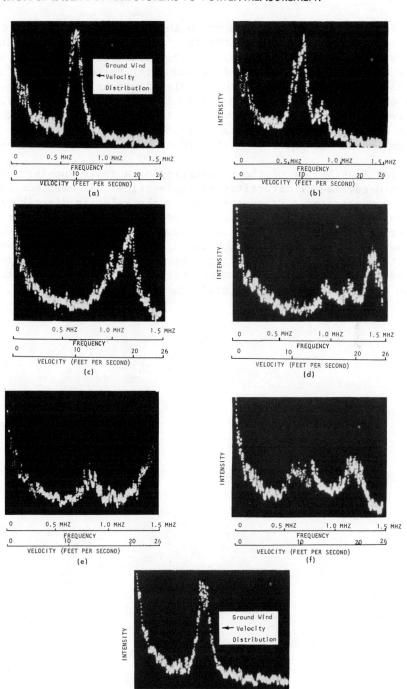

Figure 11   Time Sequence Electronic Vortex Diagram

The sequence of photographs in this next figure (11) represents the vortex velocity within the instrument sensitive volume at seven instants during the traverse of the vortex. Figure A and G represent conditions before and after vortex traversal and correspond to ground wind velocity distributions. As the laser beam enters the vortex the peak velocity increases (Figure B). Figures B, C, D, E, and F show the velocity distribution as the laser beam approaches and recedes from the core. After the vortex leaves the laser beam the ground wind signature returns.

The feasibility of the application of a laser-Doppler system to the detection of aircraft trailing vortices has been successfully demonstrated. Using this system the velocity distribution of the vortex generated by a C-47 aircraft was measured. For all measurements where the vortex was known to be in the sensitive volume, there was correlation between visual observation and prior detection using the laser system. Some measurements were performed without visual observation which emphasized that the technique does not require artificial tracers.

AN ANALYSIS OF FLIGHT MEASUREMENTS IN THE

WAKE OF A JET TRANSPORT AIRCRAFT

B. Caiger and D. G. Gould

National Aeronautical Establishment

National Research Council of Canada

ABSTRACT

Peak velocity vectors measured in the transverse plane of the wake of a Convair 880 aircraft were found to be four times as high as anticipated. The peak velocities are shown to occur in small intense vortices which are separate from the main tip vortices. These exist up to 30 seconds after generation. The main tip vortices are shown to have a core size which is roughly of the size anticipated but whose diameter does not increase significantly up to 30 seconds after generation.

1. INTRODUCTION

In November, 1967, the National Aeronautical Establishment of Canada participated in joint trials with the Federal Aviation Administration and the National Aeronautics and Space Administration to measure the wake characteristics behind an F.A.A. Convair 880 four jet transport aircraft (Figure 1). The aircraft has a wing span of 120 feet and a wing area of 2,000 square feet giving an aspect ratio of 7.1 with a sweepback of 35° at the 30% chord line. The aircraft was flown at an indicated airspeed of 220 knots at an altitude of 31,000 feet corresponding to a true airspeed of approximately 600 feet per second. At a mean all-up-weight of 135,000 lb, the lift coefficient was 0.44.

The N.A.E. provided an instrumented T33 aircraft to do the wake measurements (Figure 2). The aircraft was fitted with a nose boom supporting two free floating vanes, mounted on vertical and lateral axes, as well as an accurate static pressure probe. The

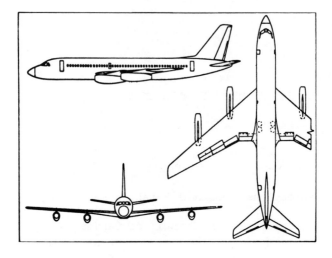

Fig. 1   The Convair 880 Transport Aircraft

Fig. 2   The N.A.E. T33 Turbulence Measuring Aircraft

vanes and filter were designed to have a natural frequency of 30 Hz
under the test conditions with a damping ratio of 0.7. The aircraft
was also equipped with accelerometers and rate gyros to measure the
aircraft motion about all three axes, as well as transducers for the
measurement of static and dynamic pressures, air temperature, small
variations in total head, and aileron angle. The data were recorded
on a 14-channel F.M. tape recorder.

2. TEST TECHNIQUE

The T33 was flown behind the 880 at approximately the same
speed and made slow lateral traverses through the wake at separation
distances varying from ½ to 3 nautical miles. The separation distance between the aircraft was obtained from tracking by the NASA
Wallops Island Radar Station. Location of the vertical position
of the wake was assisted by natural condensation trails and the
dumping of dyed fuel from the 880, but these were not adequate to
ensure interception of the wake on every traverse. When the first
tip vortex was encountered, the aircraft was rolled up to 20° in
spite of corrective action by the pilot and was usually pushed below
the second vortex by the downwash.

3. ANALYSIS OF THE MEASUREMENTS

In a brief report issued after the trials[1], it was concluded
that the peak vertical and horizontal velocities in a cross-
sectional plane of the wake were of the same order as indicated by
potential theory. However, further study showed that the peak
resultant velocities in that plane were in fact up to 135 feet per
second or four times those estimated using the best available data
for viscous vortices. It was therefore decided to make a more detailed analysis of the time histories of the measured velocities
in this transverse plane in order to obtain a better understanding
of this discrepancy. The velocity vectors were determined from the
measured vane deflections and the true aircraft forward speed with
full correction for the motion of the aircraft.

In order to convert the time histories of each traverse into
spatial distributions, the motion of the T33 across the transverse
plane was required. Unfortunately, the instrumentation on the T33
only permitted the determination of changes in the aircraft velocity
across the plane, and not its absolute value. The following procedure was therefore adopted.

From a wing loading distribution provided by Convair with
modification to allow for the lift induced on the body, it was estimated that the two trailing vortices each had a strength of 4,000
square feet per second. The distance between the vortices was

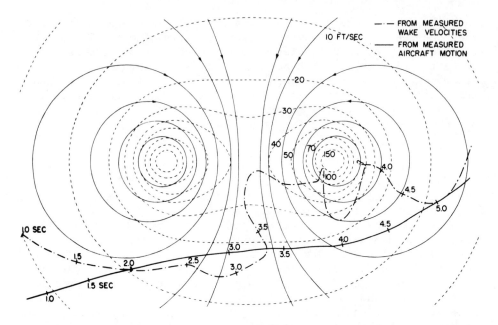

Fig. 3   The Derivation of a Lateral Path of the T33 Through the
         Wake

determined from preservation of the lift impulse[2] and the potential
flow solution for the induced flow derived (Figure 3). For one
particular traverse, that indicated a peak velocity of 135 feet per
second at a separation distance of nearly 3 nautical miles, the
apparent path of the T33 through the wake was determined by matching
the measured velocity vectors with the potential flow solution. The
result is illustrated in fig. 3. It can be seen that the path obtained was extremely erratic.

       Based on the measurements of the T33 motion, a more realistic
path was derived which was matched to the first path at two selected
points and this is also shown. The transverse velocity vectors are
plotted along this more realistic path in the upper traverse of
fig. 4.

       It can be seen that the velocity vector changes extremely
rapidly in the region of the peak value. A more detailed picture
of this region in which the spatial dimensions have been increased
by a factor of ten is also shown. The high velocities are obviously confined to a very small area. In this particular case, their
variation along the traverse suggests the presence of a secondary
vortex with a strength of approximately 700 square feet per second
and a core diameter of only 1.6 feet rotating in the same direction
as the main vortex and separated from it by a distance of 12 feet.

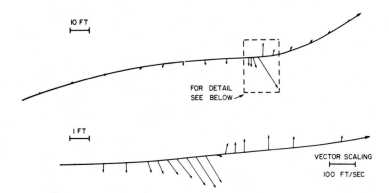

Fig. 4   Transverse Velocity Vectors Along a Lateral Path of
         the T33. Aircraft Separation 16,910 feet.

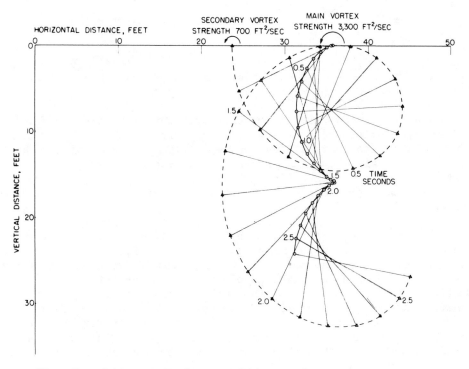

Fig. 5   Computed Motions of Main and Secondary Vortices

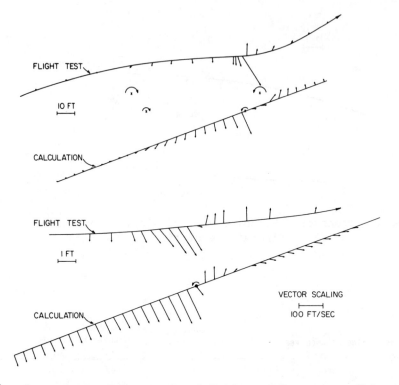

Fig. 6   Comparison of Measured and Estimated Transverse Velocity
          Vectors. Aircraft Separation 16,910 feet.

Due to the low strength of this apparent secondary vortex, its
effect is confined to a relatively small area. In thirty-six traverses of the Convair 880 wake at separation distances of less than
½ to 3 nautical miles, the presence of a secondary vortex of this
nature was apparent on sixteen occasions. In one traverse, secondary vortices were intercepted on both the left hand and right hand
sides of the wake.

Based on these results, a revised vortex wake model having two
main vortices, each with a reduced strength of 3,300 square feet
per second, and two secondary vortices, each of 700 square feet per
second, was considered. The core diameter of the main vortices was
assumed to be 18 feet and of the smaller vortices 1.6 feet. The
motion of such a combination of vortices with time has been calculated and that of the right hand pair is plotted in fig. 5. It
can be seen that the vortices descend under the influence of the
opposite pair, but that they also migrate around each other taking
about 1.7 seconds per revolution. The small vortex rotates about
a radius of the order of 10 feet and the large one about a radius
of 2 feet. The slow growth of the distance between them with time

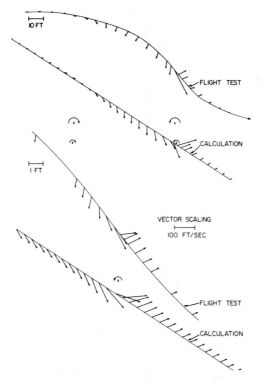

Fig. 7   Comparison of Measured and Estimated Transverse Velocity
          Vectors. Aircraft Separation 2,840 feet.

may be a result of approximations made in the calculations.

     In the flight traverses, the vortex motion is eliminated by
flying at an approximately constant distance behind the generating
aircraft, i.e. at a constant time on fig. 5. Since the aircraft
are flying at roughly 600 feet per second true air speed, the vortices complete one revolution about each other over a distance of
the order of 1,000 feet. As the separation distance between the
aircraft was changed for different traverses, the relative positions
of the large and small vortices would also change.

     For the traverses of fig. 4, the four vortex model was compared
with the flight tests by calculating the transverse velocity vectors
along a straight inclined traverse through the right-hand small
vortex with the most suitable relative positions of the large and
small vortices, corresponding to t = 0.2 seconds on fig. 5. The
comparison of the measured and calculated results is given in fig. 6
together with the corresponding vortex positions for the model.
Whilst the agreement is far from perfect, it can be seen that the
model does reproduce the fine scale variation in the velocity vector through the secondary vortex reasonably well.

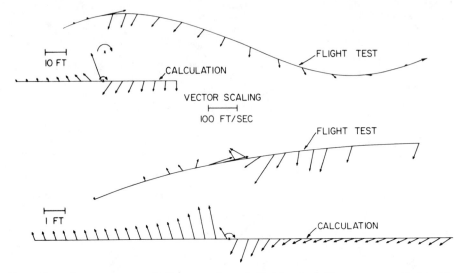

Fig. 8   Comparison of Measured and Estimated Transverse Velocity
          Vectors. Aircraft Separation 2,290 feet.

        A similar comparison for a traverse at a separation distance
of less than ½ nautical mile is shown in fig. 7. A peak transverse
velocity of 132 feet per second was measured in this case. The
same vortex model has been used, but with the relative positions
of the vortices corresponding to those of t = 0.4 seconds on fig. 5.
For the best agreement with the measurements, the inclined traverse
has been chosen to pass just below the right-hand secondary vortex.
Again, it can be seen that there is relatively good agreement between the model and the measured results.

        Figure 8 has been included to show that the agreement was not
always so good. This is a traverse closer to the generating aircraft in which a peak velocity of 120 feet per second was measured.
The fine detail of the measurements shows a very irregular variation of the transverse velocity vector and the best model comparison
given shows considerable differences which have not been adequately
explained.

        The strength of the secondary vortices corresponds approximate-
ly to the extra lift generated by the body. This indicates one
possible source of these vortices which could be shed from the region
of the wing-body junction or the change in taper near the wing root.
The interesting feature is that these small vortices have not been
dissipated or absorbed into the main vortices, but still remain discrete up to 3 nautical miles or 30 seconds after generation.

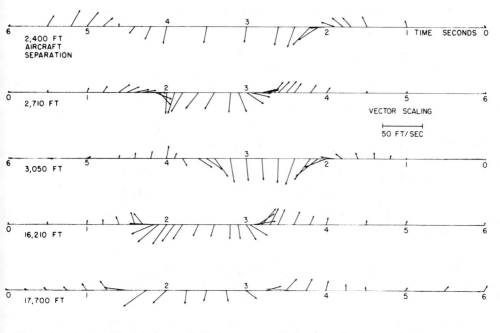

Fig. 9   Time Histories of Transverse Velocity Vectors from
           Traverses Avoiding the Secondary Vortices.

Consideration has been given to the possibility that the expected growth in their core size due to viscous dissipation may be
inhibited by the presence of high axial velocities in the core toward the generating aircraft. This could possibly result from the
generation of the vortices in regions of high loss in total head.
A sensitive pressure transducer was installed in the T33 for measurement of small variations in total head, but it did not have
sufficient high frequency response to detect such variations in the
small vortices.

If high axial velocities toward the generating aircraft do
exist in the core of these vortices, they might reduce the peak
transverse velocities derived from the flight measurements since
they are based on vane deflection angles and the relative airspeed.
However, the axial velocity is unlikely to be high at the radial
distance from the vortex centre at which the peak circumferential
velocity occurs, so that the error should not be too large.

One further point that should not be overlooked is that the
T33 is a poor probe, by wind-tunnel standards, with which to measure
even the main vortices, let alone the small ones.

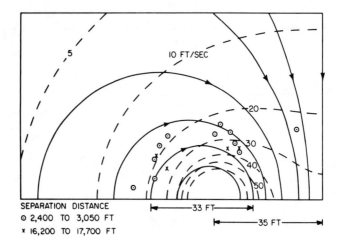

Fig. 10   Derivation of Main Vortex Core Size

Of course, it should be borne in mind that the small vortices are, to a large extent, only of academic interest in that they are too weak to upset any aircraft which might intercept the wake. However, they may induce severe local pressure loadings or cause difficulties with engine air intakes. Their presence does confirm that the structure of the wake can be very complex and not readily amenable to theoretical analysis.

Having discussed the secondary vortices at some length, we should now turn to the wing tip vortices since the main purpose of the flight tests was to determine their strength and rate of dissipation. Unfortunately, the presence of the small vortices did prevent an accurate assessment in many cases. However, a few traverses close to the main vortex cores in which the small vortices were not intercepted are shown in Figure 9.

In this figure, the induced velocity vectors have been plotted along a time scale, the direction of the scale being consistent with the direction in which the traverse was made. The peak velocity measured is 36 feet per second, and this occurs at separation distances of both ½ and 3 nautical miles. It would appear that only in the first and third traverses did the core of one vortex appear to be penetrated, the third one showing a rather irregular variation in the velocity vectors. In all other cases, the core centre was bypassed.

In an attempt to define an approximate core size from these measurements, Figure 10 shows one quadrant of the potential flow solution for a pair of vortices of 3,300 square feet per second circulation with a separation of 70 feet. Some of the peak velocities of the previous figure have been plotted according to their magnitude and direction. This has been done to allow for the variation in velocity occurring around the streamlines. The results tend to indicate a core size of approximately 33 feet diameter which does not vary for separation distances of ½ to 3 nautical miles. If allowance is made for viscous smoothing of the velocity gradient at the edge of the core, this could reduce to a diameter of 24 feet.

For comparison equating the kinetic energy of the vortices to the induced drag of the aircraft as in ref. 2 indicated an initial diameter of 20 feet. Using Squire's concept of an eddy viscosity[3] with $a = 2 \times 10^{-4}$ as in ref. 4, it was estimated that the core would grow to 21.7 feet diameter at ½ nautical mile separation and 29.5 feet diameter at 3 nautical miles. However, with the complex flow structure which recent full scale flow visualization has shown to exist in the core, these estimates cannot be considered too reliable.

In particular, flow visualization has suggested the presence of large axial flows in the core. Batchelor[5] has shown that such axial velocities can be generated in either direction. They will be towards the aircraft if the total head loss at the point of generation is large or away from the aircraft if the axial pressure gradient induced by centrifugal force as the vortex sheet rolls up is predominant. Some flow visualizations suggest that both of these conditions exist in the tip vortex core. This could be due to the spiral vortex sheet remaining in that form for considerable distances aft of the aircraft instead of diffusing into a homogeneous core as had previously been assumed. The total head losses and corresponding changes in the axial flow direction could then be confined to the narrow layers of the vortex sheet.

However, such fine detail could not be observed with the T33 instrumentation. Higher response flow direction and total head sensors would be necessary if this were to be done.

4.   CONCLUSIONS

Flight test measurements behind a Convair 880 aircraft have shown that the structure of aircraft wakes can be very complex. In particular, small vortices rotating in the same sense as the tip vortices appear to exist and remain discrete from the tip vortices up to 30 seconds after generation. Because of an exceptionally small core size, these vortices contain peak circumferential velocities up to four times those which normally occur in the tip vortices.

The complex nature of the wake is difficult to simulate in

model tests. It is therefore recommended that further flight test measurements be conducted with high frequency response vanes and/or hot film sensors to determine in more detail the core structure of both the main tip vortices and the smaller ones. Measurements should be made over the maximum possible range of aircraft separations.

Model tests are desirable to determine the origin of the smaller vortices and whether they exist on a wide range of aircraft configurations.

5.   ACKNOWLEDGEMENT

The authors would like to thank the Director of the National Aeronautical Establishment for permission to publish this paper.

6.   REFERENCES

1.   Treddenick, D.S.   Flight Measurements of the Vortex Wake
                                Behind a Convair 880.
                                National Aeronautical Establishment
                                Report LTR-FR-4, April 1968.

2.   Spreiter, J.R.   The Rolling Up of the Trailing Vortex
      Sacks, A.H.   Sheet and Its Effect on the Downwash
                                Behind Wings.
                                Journal of Aeronautical Sciences,
                                Vol. 18, p. 21, January 1951.

3.   Squire, H.B.   Growth of a Vortex in Turbulent Flow.
                                Aeronautical Research Council
                                FM 2053, 1954.

4.   Rose, R.   Aircraft Vortex Wakes and Their Effect
      Dee, F.W.   on Aircraft.
                                Ministry of Aviation, ARC CP No. 795,
                                December 1963.

5.   Batchelor, G.K.   Axial Flow in Trailing Line Vortices.
                                Journal of Fluid Mechanics, Vol. 20,
                                Part 4, p. 645, 1964.

VORTEX CONTROL

M. T. Landahl and S. E. Widnall

Massachusetts Institute of Technology

ABSTRACT

The fluid dynamical aspects of the control of free vortices
emanating from the wing or other parts of a flight vehicle are
reviewed. Simple calculations based on slender-body theory are
used to demonstrate the importance of controlling the strength,
position and core stability and thickness of such vortices. From
the equations describing vortex motion, conclusions are drawn as to
how vortex properties can be influenced. In particular, a one-
dimensional flow model for the vortex core flow is explored to
identify the factors important in the determination of core thickness and stability. Simple energy arguments are used to clarify
the phenomenon of vortex bursting and to draw conclusions as to
what can be done to prevent or induce bursting. Examples of
engineering applications of vortex control are given. In particular,
recent advanced control schemes employed for the Saab A37 "Viggen"
are described.

1.   INTRODUCTION

Flight vehicles of high span loading depend for their susten-
ance at low speeds on the shedding of a large amount of vorticity
in the surrounding air. This vorticity has a tendency to concentrate in more-or-less discrete vortices which can interfere with
the lifting and control surfaces to create strong aerodynamic
effects. It is therefore important to understand the fluid mechanics
of such vortices and to learn how they can be controlled. The
present paper is primarily concerned with the near effects as they
relate to the flight characteristics of the vehicles. However,

some distant effects, such as the persistance of the wake and the
prospects of decreasing vortex wake turbulence as well as a particular noise aspect will be touched upon.

2. EFFECTS OF VORTICES ON LIFTING SURFACES

A simple analysis based on slender-body theory can be used to
demonstrate the importance of strong free vortices on the aerodynamic
characteristics of a low-aspect-ratio wing. (See Fig. 1)

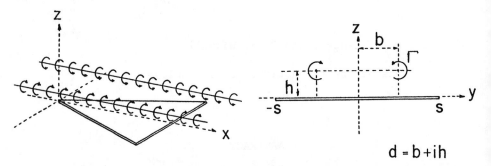

Figure 1. Vortex - wing interaction model.

A pair of vortices of equal but opposite strength are symmetrically located near the upper wing surface as shown. The vortices
may emanate from the wing leading edges, a forward control surface
(canard), the fuselage or the air intake. For simplicity, we shall
assume that they have been completely rolled up in such a manner
that their location and strength do not vary in the downstream
direction. Naturally, these are very unrealistic restrictions, but
the model problem still retains the most essential features that
we want to study. Employing the slender-body approximation of a
two-dimensional constant-density crossflow in the y,z-plane, one
finds from the application of a simple conformal mapping that the
crossflow complex potential $W = \Phi + i\Psi$ is given by

$$W = \Phi + i\Psi = - \frac{i\Gamma}{2\pi} \ln \frac{(X^2-s^2)^{1/2} - (d^2-s^2)^{1/2}}{(X^2-s^2)^{1/2} + (d_*^2-s^2)^{1/2}} \tag{1}$$

where
$$X = y + iz$$

$$d = b + ih$$

and star denotes complex conjugate. By using Bernoulli's equation
one finds the following integrals to evaluate for determining the
lift per unit chord:

$$\frac{dL}{dx} = \rho U \frac{d}{dx} \int_{-s}^{s} (\Phi_u - \Phi_\ell) dy + \frac{\rho}{2} \int_{-s}^{s} \left[\left(\frac{\partial\Phi}{\partial y}\right)_u^2 - \left(\frac{\partial\Phi}{\partial y}\right)_\ell^2 \right] dy \qquad (2)$$

index u and ℓ referring to upper and lower wing surfaces respectively.
Both of these integrals can be transformed to contour integrals of
the analytic function W and $(dW/dX)^2$ -- the second in the spirit
of Blasius' formula -- and hence easily evaluated using residue
calculus. It is instructive to separate the answer into two portions
corresponding to the linear and nonlinear terms respectively in
the equation for the pressure. The first term is the usual circulatory lift distribution associated with the downwash from the vortices.
The second, nonlinear, term may be thought of as resulting from the
suction created in the two-dimensional flow field because of the
vortices. The final result may therefore be written as follows:

$$L = L_d + L_s \qquad (3)$$

$$\frac{dL_d}{dx} = -2\rho U \Gamma . R.P. \left\{ \frac{s \frac{ds}{dx}}{(d^2-s^2)^{1/2}} \right\} \qquad (4)$$

$$\frac{dL_s}{dx} = - 2\rho \Gamma^2 R.P. \left\{ \frac{d^2}{d^2-s^2} \left[\frac{1}{(d^2-s^2)^{1/2}+(d_*^2-s^2)^{1/2}} \right. \right.$$

$$\left. \left. + \frac{2s^2}{d(d^2-s^2)} - \frac{s^2}{(d^2-s^2)^{1/2}} \right] \right\} \qquad (5)$$

where index d and s refer to "downwash lift" and "suction lift",
respectively. As the vortices approach the upper wing surface, i.e.
as h becomes very small (more precisely, as h << (s-b)), these
expressions simplify to

$$\frac{dL_d}{dx} \approx - 2\rho U \Gamma \frac{bhs \frac{ds}{dx}}{(s^2-b^2)^{3/2}} \qquad (6)$$

$$\frac{dL_s}{dx} \approx \rho \Gamma^2 \frac{b}{\pi h (s^2-b^2)^{1/2}} \qquad (7)$$

The suction lift becomes theoretically infinite as $h \to 0$. However,
the analysis has not taken into account the finite core size of the
vortex. This can be done in a very simple manner following an
idea by Widnall[1]. The downwash on the lifting surface due to a
vortex of core radius a can be approximated by that of a point vortex at an effective distance

$$h_{eff} = \sqrt{h^2 + a^2} \qquad\qquad (8)$$

One is then led to the following approximation for the total lift:

$$\frac{dL}{dx} \simeq \frac{\rho \Gamma b}{\sqrt{s^2 - b^2}} \left[\frac{\Gamma}{\pi \sqrt{h^2 + a^2}} - U \frac{\sqrt{h^2 + a^2}}{s^2 - b^2} s \frac{ds}{dx} \right] \qquad (9)$$

For large values of Γ, and small h and a, the first term, representing
the suction lift will dominate. A large increase in core radius,
as will take place at vortex bursting, will lead to a large loss
of suction lift. Also, as the vortices first begin to burst near
the wing trailing edge, a large pitch-up moment will be created.
For stability of the aircraft it is therefore important that vortex bursting be avoided or controlled.

Free vortices can also have a strong adverse effect on yaw
stability as noted by Behrbohm[2]. Again, a simple slender-body
analysis can be employed to illustrate this. Consider a slender
delta wing - vertical tail configuration (see Fig. 2) which is

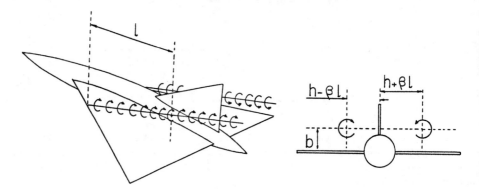

Figure 2.   Yaw destabilization due to free vortices.

influenced by a vortex pair, originating upstream in a symmetric
manner. It is sufficient for the present purpose to ignore the influence of the fuselage and to treat the wing as an infinite
reflecting plane. The results from the above wing calculations can
then be applied directly. Because the windward vortex will get
closer to the tail than the leeward one when the airplane is yawed,
there will be a differential suction force induced by the vortex
pair. Denoting the distance upstream from the section considered
to   the origin of the vortices by ℓ, one easily finds that the

yawing moment contribution per unit chord is

$$\frac{dN}{dx} \simeq \rho \beta x \left[\pi s \frac{ds}{dx} U^2 - \frac{\Gamma^2 b \ell}{\pi h^2 \sqrt{s^2 - b^2}} \right] \tag{10}$$

Here, s is the local tail height and the remaining symbols being
defined in the figure. For small h and large Γ the destabilizing
effect of the vortices given by the second term can become very
large, in fact even so as to cause the tail to loose its effectiveness completely. The formula indicates the possible cures: change
the vortex location (if feasible) so as to decrease b and ℓ and
increase h, or decrease the tail leading edge slope (increase ds/dx)
and thereby make the non-vortex contribution comparatively larger.
Also, an increase in vortex core size through, e.g., bursting,
would alleviate the problem.

There are also other situations in which it is desirable to
induce an increase in vortex core size. An obvious example is
presented by the vortex wake turbulence problem; an increase in
vortex core size would lead to a corresponding decrease of maximum
swirl velocity. Another example is given in Figure 3. Here is
shown measured and calculated noise due to a helicopter blade
interacting with a vortex shed by the previous blade passage. This
is the mechanism responsible for the blade slapping noise of a tan-
dem rotor helicopter. Theoretical results were obtained by Widnall[1]
using the approximation mentioned above to take the finite core
size into account; the experimental results were obtained by Schairer
of Boeing Vertol.[1] From the theoretical curves for doubled and
tripled core sizes it is evident that substantial reductions of
slapping noise could be achieved if one could find feasible
methods to increase the core size.

3. VORTEX DYNAMICS

Vortex motion is most conveniently analyzed by aid of the
following well-known equation (constant-density flow being considered)

$$\frac{D\bar{\zeta}}{Dt} = (\bar{\zeta} \cdot \nabla)\bar{Q} + \nu \nabla^2 \bar{\zeta} \tag{11}$$

Here $\bar{\zeta}$ is the vorticity vector ($\bar{\zeta} = \nabla \times \bar{Q}$), and $\bar{Q}$ the velocity vector.
The usual interpretation of this equation is that the rate of change
of vorticity for a fluid particle is given by the difference between
the gain due to vortex stretching and loss through viscous diffusion.
Since viscosity is normally very small in the problems of interest
here, the last term can only be large in regions of large shear
gradients as are to be found in the boundary layers. Elsewhere,

the flow is essentially inviscid, and the vorticity lines will then
follow the stream lines. Another consequence is that once a free
vortex has been created by boundary layer separation, it will take
a large time to decay, unless it breaks down through some
instability process.

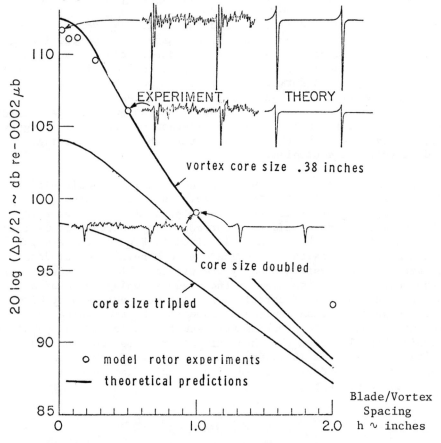

Figure 3.   Effect of tip vortex core size on helicopter
noise due to blade/vortex interaction.

A very important effect of viscosity being small is that the
vorticity usually becomes concentrated in thin sheets (except
possibly where separation is spread over a large wing surface area)
which have the tendency to roll up into more-or-less concentrated
vortices with small cores of rotational flow. In the calculation
of the location of the vortices, one may therefore approximate them
as concentrated line vortices. Each vortex has an induced velocity
field associated with it which will influence the streamlines and
hence the location of the vortex itself, leading to an often difficult nonlinear problem. Most of the calculations on wake rollup

presented in the literature to date have been based on reducing the number of space variables by one, either as in the vortex wake rollup problem through the replacement of the original three-dimensional problem by a two-dimensional initial value one with time replacing x/U_∞, or by the assumption of a conical flow as in the treatment of leading edge vortices from a slender delta wing. For the fully three-dimensional case one can in principle proceed by stepwise integration in time from an unsteady starting flow. This has been tried with some success for the indicial lift problem by Djodjodihardjo and Widnall.[3] Other methods that could be used are iteration techniques or step-by-step variation of some parameter in the problem like the angle of attack. It should be noted, however, that for a fully three-dimensional flow a fundamental difficulty will arise when replacing a vortex of finite core thickness by a concentrated line vortex. A curved vortex will produce an induced velocity on itself that has one part proportional to the logarithm of the ratio of vortex core and curvature radii which will become singular if the vortex is assumed to have zero thickness[4]. Three-dimensional vortex calculations based on concentrated vortices may thus be of limited value, but might be helpful for giving guidelines as to the effect of configuration changes on the vortex location.

Calculation of vortex strength is only feasible when the separation location is fixed by sharp edges (trailing or leading edges) so that the Kutta condition can be applied. For three-dimensional separation at non-sharp surfaces, theory in its present stage is not able to provide much guidance as to how the vortex strength could be controlled, so here one would have to rely primarily on wind tunnel experimentation.

A relation intimately connected with (11) is Kelvin's theorem

$$\frac{D\Gamma}{Dt} = 0 \tag{12}$$

stating that the circulation around a closed path comprising always the same particles in a nonviscous fluid will be independent of time. Thus, a free vortex conserves its circulation, even though vortex stretching may change the vorticity strength in the core. If the vortex is stretched, the core size will decrease by the corresponding amount, so that the circulation (which by Stoke's theorem is the integral of the vorticity through the core) will remain the same. Hence, it might seem that control of the vortex core size would be simply a matter of letting the vortex pass through a region of suitably accelerated or decelerated flow. However, as the subsequent analysis will show, this is a gross oversimplification as the flow in the core has a "life of its own" to a certain degree independent of the exterior flow.

Analysis of vortex core flow is usually based on the quasi-cylindrical assumption, i.e. that radial derivatives dominate over axial ones. A consequence of this assumption is that the pressure in the core is given to a good approximation by

$$p = p_\infty - \int_r^\infty \rho \frac{V^2(r_1)}{r_1} \, dr_1 \qquad (13)$$

where p_∞ is the undisturbed pressure and V the swirl velocity. Calculations of vortex core flows on this basis have been carried out by Hall[5], Batchelor[6] and others. However, such calculations are fairly complicated and the results are difficult to interpret physically, in particular as they relate to the phenomenon of vortex breakdown.

The vortex breakdown phenomenon has received much attention in recent years and a number of different explanations have been proposed of which the following is a partial list:

i) Instability of the swirling shear flow (Ludwieg[7], Jones[8])

ii) Phenomenon associated with the appearance of a standing wave (Squire[9] and others)

iii) Stagnation of core center (Hall[10], Bossel[11] and others

iv) Finite transition (Benjamin[12,13])

The instability mechanism is now generally regarded as not being the likely primary cause of breakdown, and the main controversy at present is between stagnation and finite transition with perhaps the latter holding the edge at the moment. In Benjamin's theory of finite transition, the breakdown is envisaged as a process similar to the hydraulic jump or a shock wave. This process can be analyzed qualitatively in a very simple manner by aid of a "one-dimensional" model in the spirit of the usual treatment of compressible flow in a stream tube or the hydraulic flow over a sloping bottom.

In this model, a Rankine type vortex is assumed, i.e. one having a core rotating as a solid body surrounded by potential flow. Also, the axial velocity, U, is assumed to be uniform across the core. Relations between the conditions upstream and downstream of the transition are obtained from the application of the conservation equations for mass, axial momentum and angular momentum.

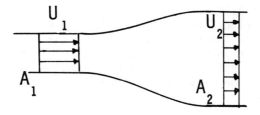

Figure 4.   One-dimensional model of finite transition.

Conservation of mass and angular momentum shows that the edge of the rotating core must be a streamline, and thus

$$U_1 A_1 = U_2 A_2 \qquad (14)$$

In the conservation of axial momentum it is necessary to take into account the suction force on the core boundary produced by the swirling motion (see eq. (13)). This force will introduce logarithmic terms in the final results. The details of the problem have been worked out by Barcilon[14]. The relation between upstream velocity and ratio of core areas downstream and upstream of the transition can be cast in the following form:

$$\frac{U_1}{U_*} = \frac{\ln\left(\frac{A_2}{A_1}\right)}{1-(A_1/A_2)} \qquad (15)$$

where

$$U_* = \frac{\Gamma^2}{8\pi A U} = \frac{\Gamma^2}{8\pi V} \qquad (16)$$

V = AU being the volume flux in the core.

In this model U_* plays the role of a critical velocity analogous to a_* in a compressible flow. When $U_1 = U_*$, $A_1 = A_2$ and U_* then gives the propagation velocity of a small disturbance preserving uniform axial velocity. For a given core, U_* is an invariant quantity.

In analogy with a hydraulic jump or shock wave one would expect the transition always to occur from a velocity higher than the critical velocity U_* to one that is lower. Formally, (15) has two solutions, the other one being obtained by interchanging U_1 and U_2. To confirm that the proper solution is the one that gives a change from a supercritical to a subcritical state, one should employ the energy equation to determine which of the solutions is associated with a loss of mechanical energy. The flux of mechanical energy is given by

$$E = 2\pi \int_0^a [p + \frac{\rho}{2}(U^2+V^2)]Urdr \qquad (17)$$

The pressure can be found from (13). It turns out that it is possible to determine, through the interchange of the order of integrations in (17) and (13), the integral

$$\int_0^a (p + \frac{\rho}{2} V^2) rdr$$

independently of the swirl velocity distribution $V(r)$. For the assumption of a uniform axial velocity the final result therefore takes on a simple form, independent of what one choses for $V(r)$. A convenient way to write the final result is as follows:

$$E - E_* = \frac{M}{2} (U-U_*)^2 \qquad (18)$$

where $M = \rho UA$ = mass flux
and the "critical energy" E_* is given by

$$E_* = \frac{p_\infty}{\rho} - \frac{1}{2} MU_*^2 \qquad (19)$$

An instructive way to present this result is given in Figs. 5 and 6. Figure 5 shows how the axial velocity varies with the core energy flux. The corresponding area ratios are presented in Fig. 6 obtained from the continuity equation (14) written in the form

$$U_*A_* = UA \qquad (20)$$

From these diagrams one can determine qualitatively the effects of core energy changes on axial velocity and core area. For a super-critical vortex (low swirl) an increase in core energy would cause an increase in axial velocity and decrease in cross-sectional area. For a subcritical vortex, on the other hand, the effect of an energy increase would be the opposite; its velocity would decrease and core area increase. Note that a decrease in free stream pressure makes the critical energy decrease and thus is equivalent to an increase in E.

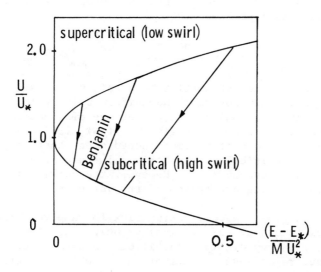

Figure 5.   Axial velocity as a function of core energy flux

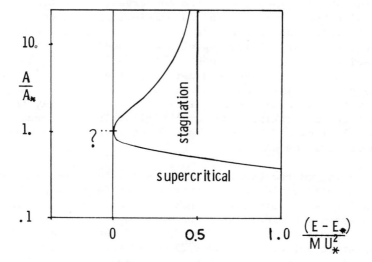

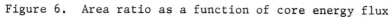

Figure 6.   Area ratio as a function of core energy flux

Thus, if the core goes through a pressure drop from p_{∞_1} to p_{∞_2}, the energy gain is approximately

$$\Delta E = \frac{p_{\infty_1} - p_{\infty_2}}{\rho}$$

In reality it is somewhat smaller than this value, since the read-justment to solid body rotation and uniform axial velocity at each section will entail a loss of energy through dissipation. The end states of the finite transition given by (14), (15) can be directly plotted in Figure 5 and it becomes indeed clear that only the transition from a supercritical to a subcritical state is possible as only this results in an energy loss. However, the curves also reveal other very interesting properties of the vortex core.

The lowest possible core energy flux for a given circulation and mass flux occurs at the critical velocity U_*. Thus, if the core has a net energy dissipation, it will always be driven to the critical state regardless of whether the flow is supercritical or subcritical to begin with. Once the core flow has reached the critical state, it cannot loose any more energy; further dissipation must thus force the flow to change radically to a new state with a lower total energy flux. Precisely what type of flow will result after breakdown is not yet clear; presumably there will be rapid turbulent mixing with the flow outside the core, but the one-dimensional equations seem to rule out the possibility that the axial flow then resembles one of uniform velocity. Possibly, a core structure consistent with the conservation equations would be one in which the vortex core has a multicell structure. In the absence of clear experimental data on this point, it is hazardous at present to construct meaningful simple models. Nevertheless, the one-dimensional model forces one to conclude that the breakdown of the vortex is a considerably more complicated and violent process than the simple transition model of Benjamin[12,13] would suggest, and the vortex breakdown should therefore more properly be described as a choking or blocking phenomenon similar to what appears in the compressible flow in a constant-area duct with heating when the Mach number reaches unity.

Depending on the conditions in the outside flow, the net energy flux in the core could be both increased or decreased, and the diagrams will give the associated changes in core velocity and area. The simplest way of increasing the core enery is to let it pass through a decreasing pressure as discussed above. Energy can also be transferred between the outer flow and the core through friction. A core of lower axial velocity than the free stream will then gain energy, whereas one with an excess velocity will lose energy. Complete stagnation can only occur for a subcritical vortex subjected to an energy addition. Such a vortex will have a finite length

with an infinite final cross-sectional area. Calculation of the
shape of a subcritical vortex in a constant negative axial pressure
gradient (modeling the atmospheric pressure distribution) gives one
that is highly suggestive of the stable "elephant trunk" shape
observed for strong tornadoes. The observations[15] seem to indicate
flow conditions typical of a subcritical vortex, there is also evidence of axial stagnation at the top, but what little there is
known about the core points to a considerably more complicated
structure (with possibly a multicell structure) than could adequately
be described by the one-dimensional model.

Comparisons with more detailed vortex calculations such as
those by Hall[5] and by Gartshore[16] confirm that the energy diagram
can explain qualitatively the behavior of the vortex core under
different conditions of the exterior flow. The simple diagram cannot of course replace accurate calculations based on the flow field
equations, but one can draw certain conclusions from it as to how
one can influence the core flow. To prevent bursting one must
remove the vortex from the criticality neighborhood by energizing
the core. This can be done by passing the vortex through a region
of falling pressure, or by blowing downstream in the core. The
increase in sweep of a low-aspect-ratio wing will have the same
qualitative effect as blowing since the rollup of the vortex sheet
from the wing leading edge is such as to produce an increase in
downstream axial velocity.

An important conclusion is that, unless the core energy is
constantly replenished, every vortex will eventually reach a
critical state through dissipation, and thus burst sooner or later.
The trailing vortices from an airplane of high aspect ratio are
normally highly supercritical at their inception and will therefore
take a long time to reach the bursting state, particularly if the
core velocity is slightly less than the free stream velocity. On
the other hand, a separated leading-edge vortex from a low-aspect-
ratio wing at high lifting conditions may be close enough to the
critical state so that the increase in pressure it experiences when
passing through the wing flow field towards the trailing edge may
be sufficient to cause it to burst.

4.   EXAMPLES OF VORTEX CONTROL

Traditional approaches to vortex control have centered almost
exclusively on the control of flow separation. For reasons of
stability, it is essential that the separated region is not allowed
to wander freely over the wing surface since otherwise severe
pitching or rolling disturbances will be experienced. The most
commonly used devices to control separation are vortex generators,
boundary layer fences, and wing leading edge notches or extensions.

(See Fig. 7)   The principal mechanism of operation common to all
three seems to be the creation of a weak secondary vortex of opposite

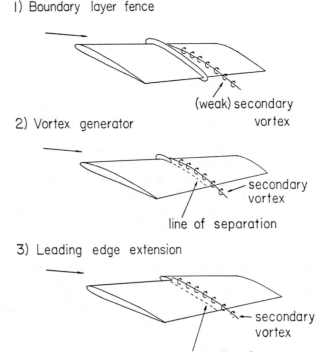

1) Boundary layer fence

(weak) secondary
vortex

2) Vortex generator

secondary
vortex

line of separation

3) Leading edge extension

secondary
vortex

separation line

Figure 7.   Control of position of flow separation
by vortex generators.

sign from the main wing vortex which will tend to fix the location
of the separation line.   Blowing and suction are also efficient to
control separation, but design complications have severly restricted
the practial use of such methods.   For a low-aspect-ratio wing it
is also possible to tailor the leading edge separation and hence
the vortex strength by carefully shaping and cambering the leading-
edge area[17].

More novel and radical approaches to vortex control have been
tried with considerably success in the development of the Saab A37
Viggen (Fig. 8).   This is a trisonic attack fighter design with a

Figure 8.   Saab A37 "Viggen".

low aspect ratio canard configuration chosen so as to achieve good
low speed STOL characteristics. The aerodynamic philosophy underlying the design and a summary of the experimental work carried out
in the course of development of this configuration are described
in the report by Behrbohm[2].

   The Viggen configuration is a so-called short-coupled canard delta which has the inherent advantage of being able to produce a high
trimmed lift coefficient. A difficulty with such a configuration,
however, is that the canard tends to stall earlier than the main
wing as the former is located in the upwash field from the latter.
The stalling is primarily manifested as vortex bursting over the
rear of the canard surface. By carefully positioning the canard
surface slightly above the plane of the main wing so that the leading-
edge vortices from the canard will pass through the pressure drop
caused by the main wing flow field and thus be energized, it was
found possible to delay the bursting to much higher angles of attack.
Through this simple stratagem, the designers of Viggen were able to
realize an effective angle of attack up to an incredible 50° before
the canard would stall (Fig. 9). In addition, the mutual interference between the vortex systems from the canard and from the
main wing helps stabilizing the wing vortex separation point.

   The vortices from the canard surface tend to pass close to the
vertical tail and therefore will have the adverse effect on yaw
stability discussed in Section 2. Figure 10 shows measurements of

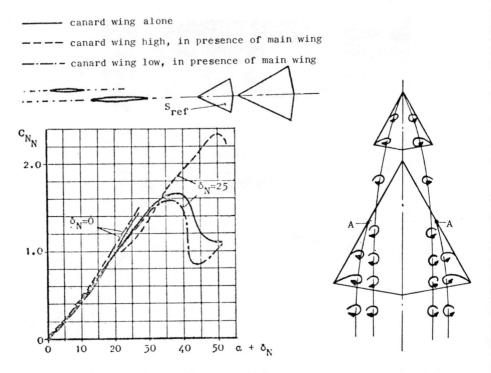

Figure 9.   Coefficient of normal force as a function
of angle of attack with and without main wing[2].

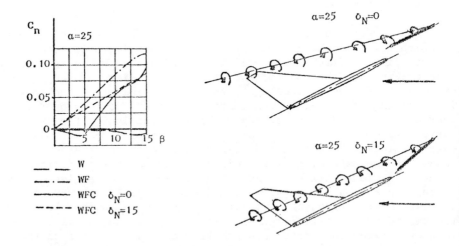

Figure 10.   Yawing moment coefficient for a simple
canard configuration with a swept vertical tail[2].

yawing moment on one of the early idealized configurations tested
having a highly swept vertical tail. As seen, complete loss of
yawing stability occurs when the vortices are located high up on
the tail. The situation is much improved when the canard is given
a large angle of attack so as to bring the vortices closer to the
wing. On the final configuration selected, the remedies listed in
Section 2 were applied such as shortening the tail root chord
(decrease of the leading edge sweep). The final configuration has
stable characteristics even at low speeds and very steep descent
paths. Thus, by application of vortex control the designers were
able to achieve good STOL characteristics without the use of
"active" devices such as boundary layer blowing.

5.   CONCLUSIONS

The survey presented on the need for and applicability of vortex control points to the potentialities for substantial improvement
of aerodynamic performance in low speed flight that can be realized
through control of vortex strength, position and stability. Particularly interesting approaches to STOL flight might involve unconven-
tional aerodynamic configurations based on advanced vortex control
principles, and here is clearly a fruitful area for further work.

The simple analysis of the energy flux in the core given in
the paper suggests that any vortex core could be broken up by dissipation of enough energy to make the core reach the critical state.
This could perhaps be accomplished by slowing the fluid down in the
core through the increase of the boundary layer or profile drag,
for example, by equipping the wing with properly located air brakes.
This may or may not be a feasible solution to the problem of dissipation of the strong vortex wake behind a heavy transport, however,
as the added drag may be too high a penalty to pay. In fact, simple
estimates show that the drag addition required to slow the core flow
down to critical could constitute a substantial portion of the total
drag of the airplane. It is highly unlikely that such a method would
be economical for thickening the vortex core coming off a helicopter
blade so as to decrease blade slapping noise as the added drag would
cause too much loss of power. Increase of the mass flux in the core
through blowing, as has been suggested[18], may instead be more practical.

5.   ACKNOWLEDGEMENTS

This work was supported by the Air Force Office of Scientific
Research (OAR) under contract numbers AF 49(638)-1622 and
F44620-69-C-0900.

1. Widnall, S. E., "Helicopter Noise Due to Blade/Vortex Interactio⌐
    to be published in Journal of the Acoustical Society of Americ⌐

2. Behrbohm, H., "Basic Low Speed Aerodynamics of the Short-Coupled⌐
    Canard Configuration of Small Aspect Ratio", Saab Technical
    Notes,Saab TN 60, July 1965.

3. Djojodihardjo, H. and Widnall, S. E., "A Numerical Method for
    the Calculation of Nonlinear Unsteady Lifting Potential Flow
    Problems", presented at the AIAA 7th Aerospace Sciences
    Meeting, New York, January 1969; AIAA Journal, 7, no. 10, 1969⌐

4. Betchov, R., "On the Curvature and Torsion of an Isolated Vortex⌐
    Filament", J. Fluid Mechanics, 22, 3, 471-479, July 1965.

5. Hall, M. G., "A New Approach to Vortex Breakdown", Proceedings
    of the 1967 Heat Transfer and Fluid Mechanics Institute,
    edited by P. A. Libby, et al., Stanford University Press, 1967⌐

6. Batchelor, G. K.,"Axial Flow in Trailing Line Vortices", J.
    Fluid Mechanics, 20, p. 645, 1964.

7. Ludwieg, H., "Zur Erklärung der Instabilität der über angestellte⌐
    Deltaflügeln auftretenden freien Wirbelkerne", Z. Flugw., 10,
    242, 1962.

8. Jones, J. P., "The Breakdown of Vortices in Separated Flow",
    Dept. of Aero & Astro, Univ. of Southampton, Rep. No. 140, 196⌐

9. Squire, H. B., "Analysis of the Vortex Breakdown Phenomenon",
    Part 1, Aero Dept. Imperial College Rep. No. 102, 1960.

10. Hall, M. G., "The Structure of Concentrated Vortex Cores",
    Progress in Aeronautical Sciences, 7, p. 53, edited by
    D. Kuchemann, et al., Pergamon Press, 1966.

11. Bossel,H. H., "Vortex Breakdown Flowfield", Physics of Fluids,
    12, 3, 1969.

12. Benjamin, T. B., "Theory of the Vortex Breakdown Phenomenon",
    J. Fluid Mechanics, 14, p. 595, 1962.

13. Benjamin, T. B., "Some Developments in the Theory of Vortex
    Breakdown", J. Fluid Mechanics, 28, part 1, 65-84, 1967.

14. Barcilon, A., "Vortex Decay above a Stationary Boundary", J.
    Fluid Mechanics, 27, pt. 1, 155-175, 1967.

15. Morton, B. R., "Geophysical Vortices", Progress in the Aero-
     nautical Sciences, 7, p. 145-194, edited by D. Kuchemann et al.
     Pergamon Press, 1966.

16. Gartshore, I. S., "Some Numerical Solutions for the Viscous
     Core of an Irrotational Vortex", National Research Council
     of Canada, Aeronautical Report LR-378, 1963.

17. Poisson-Quinton, P. and Ehrlich, E., "Hypersustentation et
     Equilibrage des Ailes Elanchées", Paper presented at the
     ONERA Colloquim of Applied Aerodynamics, Paris, Nov. 1964;
     also available as NASA TT F-9523, "Hyperlift and Balancing
     of Slender Wings".

18. Rinehart, S. A., "Study of Modification of Rotor Tip Vortex by
     Aerodynamic Means", Rochester Applied Science Associates, Inc.
     RASA Report 70-02, 1970.

THE EFFECT OF A DROOPED WING TIP ON ITS TRAILING VORTEX SYSTEM[1]

Barnes W. McCormick and

Raghuveera Padakannaya

The Pennsylvania State University

ABSTRACT

The effect of wing tip droop on the structure and position of its trailing vortex is studied. Spanwise load distributions determined by vortex lattice theory show that the stronger vortex moves from the tip of the wing to the hinge of the drooped tip as the droop angle increases. Experimental results on model wings are given which present the strength and the induced velocity profiles of the rolled-up vortex as a function of tip geometry. These results confirm at least qualitatively, the analytical prediction. It is concluded that a droop angle of approximately 90° is optimum and results in a maximum induced velocity which is half of that produced by a plane wing.

NOMENCLATURE

a   = core radius where $v(r)$ is a maximum

A   = aspect ratio = b^2/S

b   = wing span

c   = local wing chord

C_ℓ = section lift coefficient

L   = lift per unit span

r   = radial coordinate

---

[1] by NASA (Langley) under Research grant number NGR 39-009-111

R   = radius

s   = semispan

S   = wing area

$v(r)$ = tangential velocity at radius r

v_{max} = maximum value of $v(r)$ = $v(a)$

V   = free stream velocity

W   = downwash velocity

x,y,z   = coordinates of the control point

y'   = spanwise coordinate of the wing

$X=x/y_v$

$Y=y/y_v$ = nondimensional coordinates of the control point

$Z=z/y_v$

Z_1 = distance downstream of wing trailing edge

y_v = semiwidth of the horseshoe vortex

θ = droop angle

α = angle of attack

$\Gamma(\eta)$ = circulation at the section η

η = $y'/(b/2)$ nondimensional spanwise coordinate

ζ = vorticity at any radius r

ζ_{max} = center value of ζ $(r=0)$

Subscripts

i,j   = number of chordwise and spanwise locations of control
                 points

ℓ,m   = number of chordwise and spanwise locations of vortices

INTRODUCTION

Any lifting surface produces a trailing vortex sheet as a result of the circulation shed along the span. This vortex system
rolls up rapidly behind the lifting surface to form a pair of discrete vortices.

A helicopter rotor blade, which provided the motivation for
this study, can intersect the tip vortex generated by a preceding
blade under certain flight conditions. It is important to avoid
blade-vortex interaction as this is one of the mechanisms for

producing "blade slap", the term used by some to describe the sharp
cracking or popping sound associated with helicopter rotors.

The purpose of this investigation was to study the possibility
of displacing the rolled-up vortex vertically from the tip path
plane a sufficient distance so that a following blade will not intersect the vortex. Results of analytical and experimental studies
utilizing fixed-wing models having different tip shapes are presented. This paper is based primarily on the results of reference 1.

ANALYTICAL STUDY

The spanwise airload distributions on the drooped-tip models
were calculated in order to determine qualitatively the effect of
the droop on the strength and location of the rolled-up vortex.
A lifting surface model, used for the analysis, was applied to the
configuration shown in figure 1. For the analysis the fluid was
assumed to be inviscid and incompressible, finite thickness of the
wing was neglected, and the trailing vortex of the system was
assumed to be aligned with the direction of flight.

Following the lead of Reference 2, the continuous distribution
of circulation along the span of the wing was approximated by a
number of horseshoe vortices, each extending over a small fraction
of the span and placed so as to form a rectangular lattice. Figure
2 shows a typical example of a rectangular wing. In the vortex
lattice theory both the spanwise and chordwise loadings are made
stepwise discontinuous. The boundary condition which determines
the strength of each horseshoe vortex is that the resultant of the
free stream velocity and that induced by the horseshoe vortices
must be tangent to the wing at each control point. The number of
control points chosen is equal to the number of unknown vortex

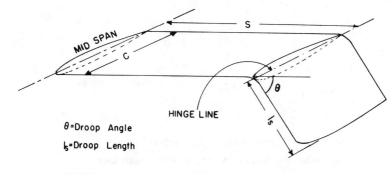

θ=Droop Angle
l_s=Droop Length

FIGURE I   DROOPED   WING   MODEL

strengths.   In order to satisfy the Kutta condition, the most aft
spanwise row of control points must lie downstream of any spanwise
bound vortex lines.   Referring to figure 3, the complete downwash,
$W_{\ell m}$, at a point $P(x,y,z)$ due to a rectangular vortex of strength
$\Gamma_{\ell m}$ is

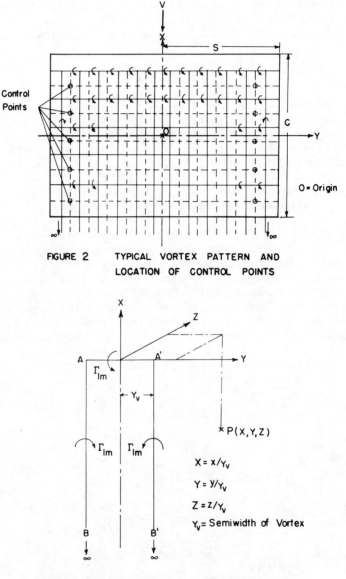

FIGURE 2   TYPICAL VORTEX PATTERN AND
LOCATION OF CONTROL POINTS

$X = x/Y_v$

$Y = y/Y_v$

$Z = z/Y_v$

Y_v = Semiwidth of Vortex

FIGURE 3   GEOMETRIC RELATIONSHIP DEFINING
THE INDUCED VELOCITY BY A RECTANGULAR
VORTEX ELEMENT

$$W_{\ell m} = \frac{\Gamma_{\ell m}}{4\pi y_v} \left[-\frac{X}{(X^2 + Z^2)} \frac{(Y + 1)}{\sqrt{X^2 + (Y + 1)^2 + Z^2}} - \right.$$

$$\frac{(Y - 1)}{\sqrt{X^2 + (Y - 1)^2 + Z^2}} - \frac{(Y - 1)}{(Y - 1)^2 + Z^2} \left[1 - \frac{X}{\sqrt{X^2 + (Y - 1)^2 + Z^2}} \right]$$

$$\left. + \frac{(Y + 1)}{(Y + 1)^2 + Z^2} \left[1 - \frac{X}{\sqrt{X^2 + (Y + 1)^2 + Z^2}} \right] \right] = \frac{\Gamma_{\ell m}}{2y_v} \frac{H(\ell,m)}{2\pi} \quad (1)$$

where X,Y,Z are the (nondimensional) coordinates of the point P.

y_v = semiwidth of the horseshoe vortex.

When the point P lies in the plane of the wing (Z = 0) the expression for the downwash reduces to the following simple form.

$$W_{\ell m} = \frac{\Gamma_{\ell m}}{4\pi y_v} \left[\frac{(Y + 1)\sqrt{X^2 + (Y - 1)^2} - 2X - (Y - 1)\sqrt{X^2 + (Y + 1)^2}}{X(Y - 1)(Y + 1)} \right]$$

$$(2)$$

The total downwash W at a chosen control point is obtained by summing the downwashes due to individual vortices. The boundary condition at the control point can be written as follows:

$$\frac{W}{V} = \frac{1}{V} \sum_{\ell,m} W_{\ell m} = \tan \alpha \simeq \alpha \text{ (for small } \alpha)$$

$$= \frac{1}{2Vy_v} \sum_{\ell,m} \Gamma_{\ell m} \frac{H(\ell,m)}{2\pi} \quad (3)$$

where ℓ and m are the number of vortices in the chordwise and spanwise directions respectively. The summation is taken over all the horseshoe vortices. The control points are chosen midway between two successive vortices. By satisfying the boundary condition at all control points a system of linear equations for the unknown vortex strengths is obtained which can be written as

$$\frac{1}{2Vy_v} \sum_{\ell,m} \Gamma_{\ell m} \frac{H_{ij}(\ell,m)}{2\pi} = \alpha_{ij} \quad (4)$$

where α_{ij} = α = Angle of attack of the plain wing

$\alpha_{ij} = \alpha \cos \theta$ on the drooped portion of the wing, θ being the droop angle.

The above system of equations was solved by the Gauss-Jordan elimination method using an IBM 360/67 computer.

Knowing the circulation of each section of the wing, the section lift coefficient, C_ℓ, is determined from,

$$C_\ell = \frac{2\Gamma(\eta)}{Vc} \text{ where } \Gamma(\eta) = \sum_\ell \Gamma_{\ell m},$$

m corresponding to the section considered.

EXPERIMENTAL STUDY

The experimental program was designed to provide data on the strength, displacement and rolling-up with distance downstream of the trailing vortex for a number of drooped tip wings. Model testing was performed in the Pennsylvania State University's 3' x 3' subsonic wind tunnel at a speed of 65 mph. Semi-span wing models were employed with droop angles of 0°, 70°, 80°, 90° and 110° and drooped lengths of about 45% of the semispan. Surveys at various downstream transverse planes were made using a vortex meter capable of measuring the rotational velocity of the fluid at a point. A description of this meter can be found in reference 3.

DISCUSSION OF RESULTS

Analytical Results

The strength of the circulation obtained by the vortex lattice method depends on the location and the number of vortices distributed along the span and chord of the wing. Calculations performed with different locations of vortices and control points in the case of plain rectangular wings showed that 19 vortices of 0.1s width each, along the span and 5 vortices along the chord, placed at 0.05c, 0.25 0.45c, 0.65c, 0.85c are sufficient by comparison with the earlier results of Reference 4. Using this number and location figure 4 was obtained which presents the calculated spanwise variation of the bound circulation for the drooped tip wings.

Any change of circulation along the wing span must be accompani by the shedding of vorticity from the wing. It is clear from figure 4 that for droop angles less than approximately 45° a stronger vortex will be located near the tip and a weaker secondary vortex at the hinge line. As the droop angle increases beyond 45°, the loadin

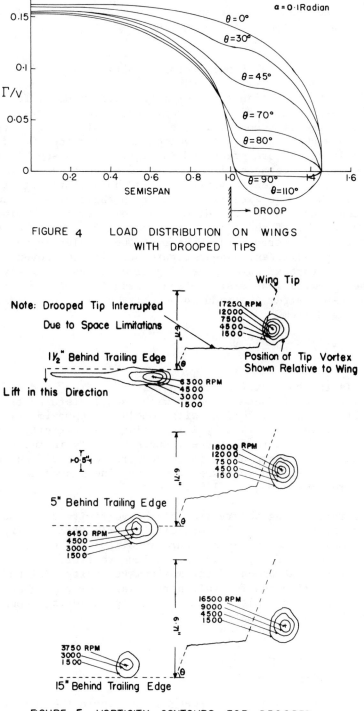

FIGURE 4   LOAD DISTRIBUTION ON WINGS
WITH DROOPED TIPS

FIGURE 5   VORTICITY CONTOURS FOR DROOPED
TIP WING, $\theta = 70°$

becomes weaker near the tip and the stronger vortex would be
expected to move toward the hingeline.

Experimental Results

Measurements, using a vortex meter, were made at four transverse planes behind each model wing. Each wing was tested at approximately the same lift coefficient of 0.4. Typical vorticity contour
measurements obtained with the vortex meter for the wing with a
droop angle of 70° are presented in Figure 5. As seen in this figure
at a downstream distance of about 75% of a chord behind the trailing
edge completely rolled-up vortices exist both at the tip and at the
hingeline. With the exception of a droop angle of 110°, two vortices
one at the hingeline of the droop and the other at the wing tip were
observed for all of the wings. Figures 6 and 7 show the vorticity
distribution, ζ, through the center of the completely rolled up tip
and hingeline vortices. These figures confirm the movement of the
stronger tip vortex towards the hingeline as qualitatively predicted
from the spanwise load distribution. Assuming axisymmetric flow in
the rolled-up vortex the induced velocity v(R) at any radius R can
be calculated from

$$v(R) = \frac{1}{R} \int_o^R r\zeta \, dr \tag{5}$$

where ζ is the vorticity at any radius "r". Figures 8 and 9 present
the variation of induced velocity with radial distance from the
center of the vortex, for the wing models at approximately the same
lift. These were obtained by integrating the vorticity shown in
figures 6 and 7. Note that the induced velocity in the tip vortex
decreases as the droop angle increases from 0° to 90° and the induced velocity in the vortex at the hingeline increases as the droop
angle increases from 0° to 110°.

One interesting observation of the vorticity distribution is
shown in figure 10. If the vorticity distributions of figures 6 and
7 are normalized as shown in figure 10 all of the curves collapse
into one. Thus, if one knows the maximum vorticity in a vortex and
the core radius at one-half the maximum vorticity amplitude, the
vorticity distribution for this particular operating condition can
be easily obtained. Also, included in figure 10 is a comparison
with an exponential equation

$$\frac{\zeta}{\zeta_{max}} = \exp \quad -\left(\frac{r}{w}\right)^2 \ln 2 \tag{6}$$

Utilized in Reference 5. As seen in the figure the two distributions
agree fairly well.

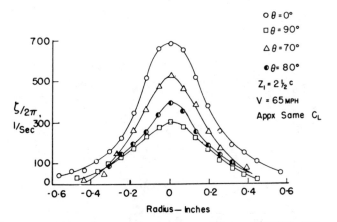

FIGURE 6   VORTICITY DISTRIBUTION THROUGH THE
ROLLED-UP TIP VORTEX OF MODEL WINGS

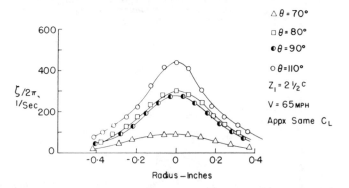

FIGURE 7   VORTICITY DISTRIBUTION THROUGH THE
ROLLED - UP VORTEX AT THE HINGE LINE OF THE
DROOP OF MODEL WINGS

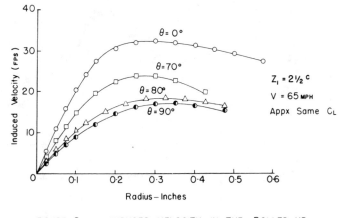

FIGURE 8   INDUCED VELOCITY IN THE ROLLED-UP
TIP VORTEX OF MODEL WINGS

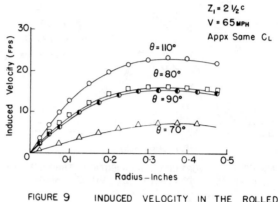

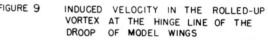

FIGURE 9   INDUCED VELOCITY IN THE ROLLED-UP
            VORTEX AT THE HINGE LINE OF THE
            DROOP OF MODEL WINGS

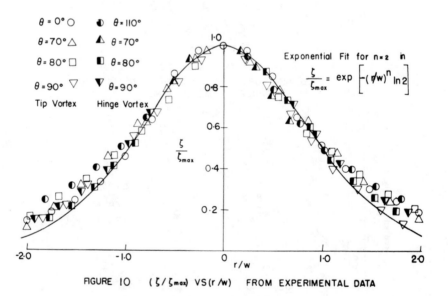

FIGURE 10 (ζ / ζ_{max}) VS (r/w) FROM EXPERIMENTAL DATA

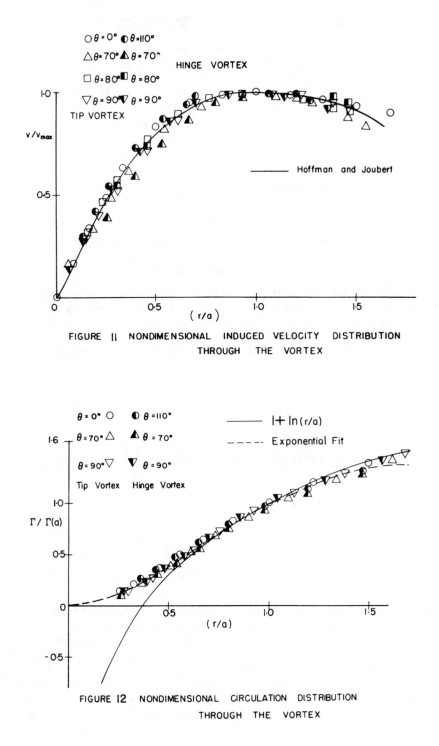

FIGURE II   NONDIMENSIONAL   INDUCED   VELOCITY   DISTRIBUTION
THROUGH   THE   VORTEX

FIGURE 12   NONDIMENSIONAL   CIRCULATION   DISTRIBUTION
THROUGH   THE   VORTEX

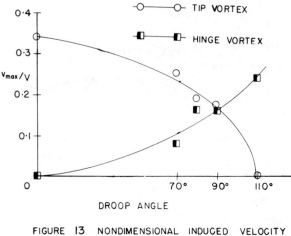

FIGURE 13   NONDIMENSIONAL INDUCED   VELOCITY
VS
DROOP ANGLE

Figure 11 presents the nondimensional induced velocity distribution through the rolled up tip and hingeline vortices of model
wings. Also shown in this figure are the results of Hoffman and
Joubert, reference 6, which appear to confirm the present results.
According to Reference 6, the circulation Γ for a turbulent vortex
can be written in the form

$$\frac{\Gamma}{\Gamma_{(a)}} = 1 + \ln\left(\frac{r}{a}\right) \tag{7}$$

where $\Gamma(a)$ is the value of circulation at the radius for maximum v.
For a laminar vortex the exponential variation of vorticity ζ, would
be expected to hold

$$\text{Using } \Gamma(R) = \int_{o}^{R} 2\pi r\zeta \, dr \tag{8}$$

the circulation, Γ, becomes;

$$\frac{\Gamma}{\Gamma(a)} = 1.392 \quad 1 - \exp\left[-1.262\left(\frac{r}{a}\right)^{2}\right] \tag{9}$$

The nondimensional circulation obtained from the measured vorticity
ζ of the model wings is compared with equations (7) and (9) in
figure 12. From this figure it appears that the exponential equation
is applicable for Γ for r/a values less than approximately 0.8. The
nondimensional maximum tangential velocities of the tip and hingeline vortices for the wings are plotted against the droop angle in
figure 13. As seen in this figure a droop angle of 90° is optimum
and results in a maximum induced velocity which is half of that
produced by a plane wing.

CONCLUSIONS

From the results of this study the following conclusions are drawn.

1. The spanwise load distribution, qualitatively, is indicative of the location and the strength of the rolled-up vortex.

2. For droop angles of 90° or less, two distinct vortices are generated, one at the tip and the other at the hingeline.

3. The maximum vorticity in the rolled-up tip vortex decreases as the droop angle increases from 0° to 110°.

4. The maximum vorticity in the rolled-up vortex at the hinge-line increases as the droop angle increases from 70° to 110°.

5. A droop angle of approximately 90° appears to result in the minimum value of induced velocity in the total combined vortex system.

6. The circulation distributions in both vortices are similar, and reduce to a universal normalized function. This function is fitted closely by predictions based on the Navier-Stokes equations for radii less than the core radius. For larger radii it agrees with a predicted logarithmic variation obtained from an analysis of a turbulent vortex.

REFERENCES

1. Padakannaya, R., "Effect of Wing Tip Configuration on the Strength and Position of a Rolled-up Vortex," M. S. Thesis, 1970, The Pennsylvania State University, University Park, Pa.

2. Falkner, V. M., "The Calculations of Aerodynamic Loading on Surfaces of any Shape", ARC, R & M No. 1910 (1941).

3. May, D. M., "The Development of a Vortex Meter," M. S. Thesis, 1964, The Pennsylvania State University, University Park, Pa.

4. Falkner, V. M., "Calculated Loadings Due to Incidence of a Number of Straight and Swept-pack Wings", ARC, R & M No. 2596 (1952).

5. McCormick, B. W., and Tangler, J. M., and Sherrieb, H. E., "Structure of Trailing Vortices," Journal of Aircraft, May-June 1968, pp. 260-267.

6. Hoffman, E. R. and Joubert, P. N., "Turbulent Line Vortices", Journal of Fluid Mechanics," Vol. 16, Pt. 3, July 63, pp. 395-411.

SOME WORK AT THE ROYAL AIRCRAFT ESTABLISHMENT ON THE BEHAVIOUR

OF VORTEX WAKES

P. L. Bisgood, R. L. Maltby, and F. W. Dee

Royal Aircraft Establishment

ABSTRACT

The paper reviews the work done at the Royal Aircraft
Establishment on the behaviour of vortex wakes. Measurements of
the development of wakes behind straight and swept wings are
described. The results of observations on the formation of loops
in a vortex wake, on the behaviour of the wakes close to the ground
and on the development of the wake from a slender wing are given in
some detail.

1. INTRODUCTION

A number of studies of the behaviour of trailing vortices and
their effects on following aircraft have been made at the Royal
Aircraft Establishment since 1954. These have provided a useful
working understanding of the problems and have made it possible to
give some guidance to air traffic control authorities on the
separation distances required between aircraft in a traffic
pattern. This paper reviews the principal results obtained and
describes in more detail the work on the motion of vortex trails
close to the ground, the formation of loops in the trails and the
work currently in progress to examine the wake from a slender wing.

Much of the experimental work concerned was based on
theoretical work by Squire[1] in which he suggested that the development of a simple trailing vortex from a wing depends on an eddy
viscosity, ϵ, which can be taken as proportional to the circulation.
Thus the circumferential velocity at radius r and time t is given

by
$$w = \frac{K}{2\pi r} \left[1 - \exp \frac{-r^2}{4(\nu + aK)t} \right] \tag{1}$$

where K = the circulation of the line vortex
$\quad \nu$ = kinematic viscosity
$\quad a$ = eddy viscosity factor = $\frac{\varepsilon}{K}$.

In the conditions considered, ν is small compared with a K and is neglected.

Substituting $K = \frac{4}{\pi \rho \nu} \left(\frac{W}{b} \right)$, which is appropriate for an elliptic wing, it can be shown that the maximum circumferential velocity in the vortex, $\hat{w}$, and the radius at which it occurs, r_1, are given by

$$\hat{w}^2 = \frac{0.0033\ W}{a\ \rho\ V\ b\ t} \tag{2}$$

and

$$r_1^2 = \frac{6.37\ a\ W\ t}{\rho\ V\ b} . \tag{3}$$

The velocity at which the vortex system from an elliptically loaded wing is propelled vertically downwards is given in Ref.2 as

$$\dot{z} = \frac{0.258\ W}{\rho\ V\ A\ S} . \tag{4}$$

Thus, for an elliptically loaded wing, the development of the vortex system can be calculated provided an appropriate value for the eddy viscosity factor, a, is chosen. However, in the case of a marked discontinuity in the span loading, for example a wing with part-span flaps deflected, this simple approach is inadequate.

The initial aims of the experimental work at the Royal Aircraft Establishment were to assess the importance of the effects of trailing vortices on following aircraft and to determine the limitations of Squire's theory in predicting these effects.

The first series of tests[3] examined the wake from a Lincoln aircraft by flying a chase aircraft along the axis of one of the vortices. The strength of the vortex was assessed from the aileron deflection required to hold the induced rolling moment.

The Lincoln aircraft (Figure 1) was a comparatively lightly loaded aircraft compared with modern jet transports, having a mass of only 30000 kg and a span of 36.6 m, but the wake was found to have a marked effect on the Devon chase aircraft (Figure 2) which

Fig.1 Avro Lincoln

Fig.2 Hawker Siddeley Devon

had a mass of 3600 kg and a span of 17.3 m. Briefly the results
showed that full aileron was required to control the chase aircraft
up to 150 sec behind the Lincoln flying in the clean configuration
at 130 kt. The induced rolling moment appeared to fall rapidly at
greater separations. With flaps deflected the disturbance was less
and it began to decay rapidly at a separation of only 105 sec.

The results (Figure 3) indicated that Squire's theory
predicted the induced rolling moment up to the point of rapid decay
reasonably well and that a very simple treatment of the effect of
part-span flaps also gave a reasonable prediction. A value of
0.0004 for the eddy viscosity factor was recommended.

These tests, therefore, gave a useful indication of the
seriousness of the problem but it was recognised that the method
used for measuring the vortex strength was not sufficiently precise
for a detailed study.

The second experiment[4] was designed to measure the vortex
structure in more detail. A Comet aircraft (Figure 4) (45000 kg and
33 m span) and a Vulcan (Figure 5) aircraft (50000 kg and 30 m span)
were used to generate the wakes. A small, instrumented fighter
aircraft was flown through the wake at right angles to its axis at
several time intervals behind the wake-laying aircraft so that the
vertical component of velocity could be measured. The maximum
vertical component of velocity from each run was then plotted
against time interval. The upper bound of these points was taken
to identify the interceptions through the centres of the vortex
cores so that the velocity distributions through the cores could
be obtained.

The results (Figure 6) suggested that Squire's theory gave a
reasonable prediction of the variation in vortex structure with
time although a rather lower value of the eddy viscosity factor
(0.0002) was indicated.

The variation of vertical position of the wake with time was
also measured for the two aircraft over a wide range of conditions
and was found to correspond well with the predictions of Reference 2.

It was in this experiment that the formation of loops between
the trailing vortices a long way downstream of the aircraft was
first noticed. A discussion of this phenomenon is treated
separately later in the paper.

The third experiment[5] concerned the measurement of the motion
of the wake in the neighbourhood of the ground. This work is
treated in detail in the next section.

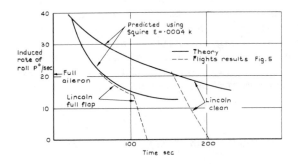

Fig.3 Comparison of flight test results and predictions
using the Squire theory

Fig.4 Comet 3B XP 915   Fig.5 Hawker Siddeley Vulcan B 1

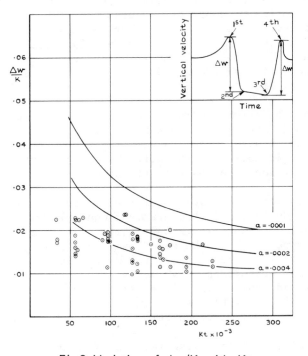

Fig.6 Variation of $\Delta w/K$ with Kt

The fourth experiment, which is in progress at the time of writing, is being undertaken to study any possible difference between the behaviour of the wake from a slender wing aircraft and that from a conventional aircraft. The object is to discover whether the wakes from typical supersonic transport aircraft have to be treated differently from those from current aircraft in the air traffic environment. The present state of this work is discussed in section 5.

2.   THE MOTION OF THE VORTEX WAKE CLOSE TO THE GROUND

Since air traffic is naturally most concentrated in the immediate neighbourhood of the runways at an airport, it is important to study the effect of the proximity of the ground on the behaviour of vortex wakes not only to establish safe operating limits but also to assist in the design of the layout of new airports.

Simple potential flow theory suggests that as the trailing vortices descend towards the ground their vertical velocity decreases and they begin to travel horizontally over the ground away from each other. In practice several effects may be expected to modify the behaviour of the vortices close to the ground. Because wind varies with height, so does its effect on the motion of the vortices. Compared with the earlier measurements at 10000 ft and above, atmospheric turbulence increases at lower altitudes and this, together with ground friction effects, should dissipate the vortices more rapidly than in the case of calmer air conditions away from the ground.

To obtain a better understanding of these effects some flight tests[5] were made at RAE using a Hunter aircraft with different coloured smoke injected into its two wing-tip vortices. In the tests the overall motion of the vortices was observed, but measurement of local velocities was not attempted.

The tests were made over an airfield situated in flat country-side with few buildings in the area and none within half a mile of the test area. The Hunter aircraft (Figure 7) which has a span of 10.75 m and a mass of 7450 kg was flown along the runway at a height of about 10 m and a speed of 170 kt. The smoke in the vortex wake was photographed by one camera stationed to one side of the track and from another in a helicopter hovering 150 m above the track. Thirty-eight runs were made with wind speeds varying from 0 to 15 kt and directions between 0° and 90° to the track.

A limited selection of the results is shown in Figure 8, which presents the movements of the vortices in the measuring plane

Fig.7 Hunter 6

for wind conditions from about 2 kt up to 16 kt. Figures 8a and 8b correspond to the calmest conditions, and it will be seen that the upwind vortex descended almost vertically and remained very near the runway centreline. The very light cross-wind was sufficient to neutralise the upwind movement due to the vortex-induced velocities. This could present a hazard to following aircraft using the same runway. For the downwind vortex the two effects were additive. Figure 8c shows the vortex movement with a 6½ kt cross-wind component, both the vortices being blown clear of the runway centreline quite rapidly. Again, in Figure 8d, with an 8 kt cross-wind component, the lateral vortex motion was very rapid, and the vortices would have presented no hazard to following aircraft using the same runway but could present a possible hazard on nearby parallel runways.

Figures 9-12 present the results shown in Figure 8, together with theoretical predictions based on the initial conditions in each case. The expected paths of the vortices in still air were calculated from the expressions derived in Reference 5 and an allowance made for the effects of cross-winds.

The actual cross-wind component during the first 3 sec of recording was derived from the mean horizontal velocity of the vortices during this period. This cross-wind was then applied to the calculated still air motion, assuming a 1/7 power law for the variation of wind strength with height in the earth's boundary layer and using a step-by-step integration process.

It will be seen that the theoretically predicted paths are in reasonable agreement with the measured results, suggesting that the motion derived in Reference 5, together with a 1/7 power law for wind velocity, provides an adequate prediction for vortex motion near the ground.

Smoke dissipation or vortex mutual interaction limited the period of measurement in the present tests to less than 20 sec. For large conventional aircraft the mutual interaction time appears to be considerably greater, so it is possible that their vortices would persist near the ground for considerable periods, as they do away from the ground. Thus, any information on vortex decay near the ground would be valuable.

No knowledge was gained of the velocity distributions within the vortex near the ground so it is not possible to predict how rapidly the peak velocities within the vortices will decay or to assess whether the expression, given by Squire[1], for velocities within the vortex still holds under these conditions.

It was concluded that, for a period of up to 20 sec after the passage of the aircraft, the predictions of existing theory are in

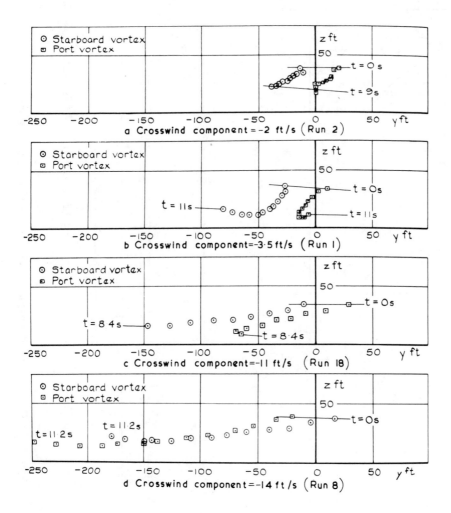

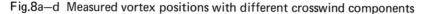

Fig.8a—d Measured vortex positions with different crosswind components

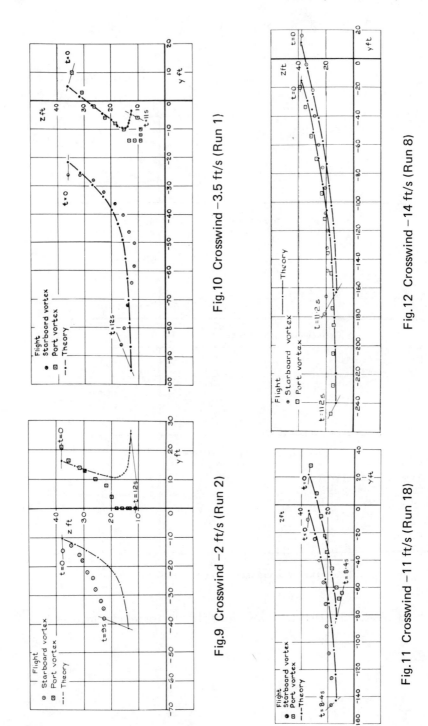

Fig.9   Crosswind −2 ft/s (Run 2)

Fig.10   Crosswind −3.5 ft/s (Run 1)

Fig.11   Crosswind −11 ft/s (Run 18)

Fig.12   Crosswind −14 ft/s (Run 8)

Figs.9—12   Comparison of measured and theoretical vortex positions

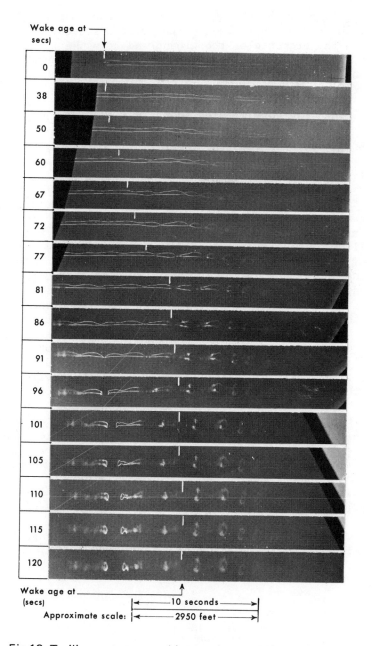

Fig.13 Trailing vortex mutual interaction away from the ground.
Comet aircraft. 150 km eas at 10000 ft

good general agreement with observed vortex motions. However,
significant differences did occur, the two vortices sometimes
descending to a minimum height and then rising again. There was
no clear indication as to whether the vortices decayed more
rapidly in the presence of the ground and atmospheric turbulence
than would have been expected from earlier measurements away from
the ground in calm air.

3. THE FORMATION OF LOOPS IN THE VORTEX TRAIL

During the tests with the Comet aircraft mentioned above, it
was observed[4] that there was a mutual interaction between the pair
of trailing vortices a long way behind the generating aircraft.
Four stages in the process of interaction are shown in Fig.13.
After a time interval of the order of 40 sec a sinusoidal distortion began to appear in the vortex trails, apparently in the
horizontal plane. The amplitude increased with time until,
eventually the two trails appeared to touch. At this point, some
90 sec after generation, the trails began to join up to form loops
having the appearance of smoke rings, which persisted for at least
a further 30 sec. Subsequent tests[5] with the Hunter in free air
showed a similar phenomenon although, in this case, the loops were
found within about 12 sec of the passage of the aircraft.

Maskell has suggested that a possible explanation of the
shorter interaction time of the Hunter may be derived from
dimensional analysis of a simplified model of the developing wake.
Suppose the vortex system to be characterised in a given
neighbourhood at time t = 0 by a vortex strength, K, and the two
lateral length scales s' and d, where s' is representative of the
large-scale distribution of vorticity (e.g. half the separation
distance of the vortex cores) and d of the smaller-scale properties
of the cores themselves. We might then expect the subsequent
development of the field in the given neighbourhood - non-
dimensionalised with respect to K and s' - to depend solely on
the non-dimensional time $\tau = K \, t/4s'^2$ and the ratio d/s'.

It follows that any well-defined qualitative change in the
wake development - like the appearance of loops - will occur at a
critical value τ_c of the non-dimensional time, such that

$$\tau_c = f\left(\frac{d}{s'}\right) .$$

However, the precise value to be assigned to d/s' is linked with
the choice of time origin. So provided that the possible variations in d/s' are small, we might expect to be able to account for
them by correspondingly small adjustments to the origin of time,

whence $\qquad \tau_c = \text{constant} = C = \dfrac{K\,t}{4s'^2}$. $\qquad$ (5)

It should be noted that the foregoing conclusion depends primarily on the adequacy of the description of the field at time $t = 0$, at which instant the distribution of vorticity in the given neighbourhood is supposed universal in form, and the velocity and length scales of the turbulence in the wake core supposed proportional to K/d and d respectively. But the argument implies no further restriction on the development of this field - neither on the magnitude of any disturbance to it that may develop subsequently, nor on the nature and magnitude of the subsequent development of the turbulence properties. Thus, within the implied restrictions on the initial state of the field, the conclusion would appear to have considerable generality.

It seems plausible to expect that our initial restrictions on the field are essentially satisfied by wings that are approximately elliptically loaded and from which the trailing vortex sheet emanates only from the trailing edge, in the classical manner. For this family of wings we write:

$$2s' = \frac{\pi}{4}\,b$$

where b is the geometric wing span, and

$$K = \frac{4L}{\pi\,\rho\,V\,b} \, ,$$

and if some significant interaction - like looping, for example, - is supposed to occur at time $t = T$, we infer that

$$T = C'\,\frac{\rho\,V\,b^3}{L} \qquad \text{or} \qquad T = C'\,\frac{2b\,A}{C_L\,V} \, , \qquad (6)$$

where $C' = (\pi/4)^3\,C$.

It is difficult to assign a precise value to C' from the Comet and Hunter tests, but both sets of observations are consistent with a value of roughly 10. A subsequent observation on a B 47 gives further support to this value.

In his study of the cause of the instability which leads to looping, Crow[6] makes initial assumptions that are inevitably more restrictive than these adopted here, and are therefore contained within them. In consequence, he also finds a critical time - in his case, to the development of a given non-dimensional amplitude of disturbance - such that

$$\tau_c = \text{constant} \quad,$$

and calculates the constant C' (corresponding to growth of his instability by a factor e) as 4.7. His analysis is therefore not inconsistent with the present conclusions for essentially elliptic wings.

It is interesting to note that when the Hunter was flying close to the ground mutual interaction between the vortices was never observed. However, it was seen that occasionally one or both of the vortices broke into a series of roughly semi-circular arches with both ends perpendicular to the ground (Figure 14). This suggests that each vortex can be considered to be interacting with its image in the ground in a manner resembling the mutual interactions in free air.

Having examined the occurrence of loops in terms of the wake age, it is necessary to consider whether loops always form at the expected time and whether the rates of decay of the 'looped' vortices are more or less rapid than that for a pair of line vortices. On the one hand the photographs of the smoke trail from the Comet[4] show visible rings for about 30 sec after the formation of loops, which may indicate that significant disturbances persist for at least this time. This is supported by reported observations of substantial induced rolling moments occurring much later than the predicted formation of the loops.

On the other hand, the results of the Lincoln trials[3] showed a rapid decrease in induced rolling moments at a time when "... a considerable amount of dispersion of the smoke has taken place: this usually takes the form of a vertical separation between sections of the smoke trail..."; these 'vertical separations' might well have been loops but there is, unfortunately, no positive confirmation of this. Putting C' = 10 in equation (6) suggests that loops would have formed at a wake age of about 125 sec, which compares quite closely with the observed age (150 sec) at the onset of the more rapid decay in rolling moments and suggests that the latter might have been associated in some way with the formation of loops. There have been many other reports which suggest that the rolling moments induced on probe aircraft decay much more rapidly once the wake exceeds a certain age.

The evidence, therefore, presents a rather conflicting and incomplete picture, and it must be concluded that the formation and stability of the loops, and their influence on the decay of the velocity field must be studied in greater detail before regulations can be framed to ensure a very low probability of accident.

4. THE WAKE FROM SLENDER WINGS

The preceding comments apply to straight and swept wings from which vorticity is shed predominantly from the trailing edge. This vorticity rolls up into simple vortex coils behind the wing tips and at the discontinuities at the ends of the flaps. The situation on a slender wing with leading-edge separation is, however, quite different, for vorticity is shed from both leading edges as well as from the trailing edge.

The leading-edge vortex contains complicated distributions of velocity both axially and circumferentially because it is fed continuously along the leading edge which can vary in sweep along its length. The rotation, however, is in the same direction as an ordinary tip vortex which, for the purposes of this description, will be taken as positive.

The vorticity shed from the outer parts of the trailing edge is negative and it rolls up in the negative direction to form a double-cored system with the leading-edge vortex on each side of the wing. This double core spirals in the positive direction as it proceeds downstream, for at least one chord behind the trailing edge. The nature of the flow close to the wing is illustrated in Figure 15.

Although the wake characteristics close to the wing are well understood there is no known information on the detailed structure in the far field where the vortices might affect following aircraft. On the one hand the large induced drag associated with slender wings suggests that the energy associated with the wake is large; on the other, the complex nature of the double cores, and the possible interactions between them, might well lead to a relatively rapid diffusion of vorticity and a correspondingly rapid redistribution of the energy. It is clear, at least, that the initial structure of the wake from a slender wing is sufficiently different from that of the conventional wake to require special consideration and, in an attempt to clarify the situation, some experiments have been made in flight.

It should be noted that, as a result of the special nature of the flow, a breakdown of the flow in the core of a leading-edge vortex occurs above or close to the wing at moderate to high angles of incidence. This so called vortex breakdown or vortex bursting should not be confused with the phenomenon of looping in the far field mentioned above.

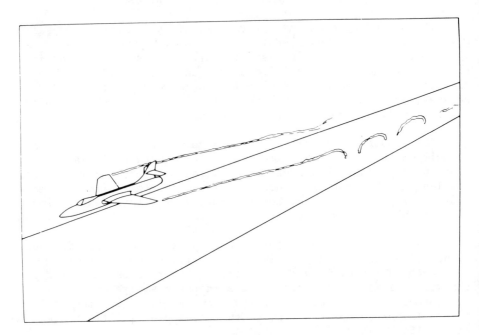

Fig.14 Sketch showing interaction sometimes observed between
individual vortices and the ground

Fig.15 Model of the flow past a slender
lifting wing

5. EXPERIMENTS WITH A SLENDER WING

5.1 Introduction

The object of the experiments described below was to study the velocity field downstream of a slender wing with particular reference to the maximum disturbance velocities liable to be encountered and the manner in which these diminished with increasing age of the wake. The work was still in progress at the time of writing but the results achieved so far seem sufficiently significant to warrant an interim statement. The technique consisted essentially in flying an instrumented probe aircraft through the wake generated by a slender-wing research aircraft, and recording the velocities normal to the probe's flight path.

5.2 Aircraft and Instrumentation

The Handley-Page HP 115 research aircraft was built to enable investigation of the in-flight characteristics of a slender-wing configuration at low airspeeds: the aircraft has been described in detail elsewhere, its leading particulars are presented in Table 1 and it is illustrated in Figure 16. The flow field around the aircraft is, broadly speaking, representative of the class of slender-wing flows which have just been described, so that the aircraft was a suitable vehicle for this experiment.

Since the aircraft was already comprehensively instrumented the only additional requirement was some means of making the vortex cores visible to the pilot of the probe aircraft. Ideally we would have chosen to inject smoke directly into the vortex cores, but it proved impossible to find a generator compatible with the aircraft and capable of producing a sufficiently long-lasting trail. The solution adopted was to fit a means of injecting a fluid (principally DERV) into the jet efflux aft of the final nozzle: this produced a dense plume of smoke which quickly became entrained by the vortex system and provided an acceptable target for the probe aircraft.

The probe aircraft employed was a Morane-Saulnier MS 760 'Paris' (Figure 17) owned by the Cranfield Institute of Technology and operated by them throughout these tests. The principal dimensions are given in Table 2. The aircraft had been modified by the addition of a nose-boom carrying a wind vane, measuring incidence, and a normal accelerometer; these instruments were situated approximately 6 m ahead of the aircraft's CG. An additional normal accelerometer and a pitch rate gyroscope were mounted close to the aircraft's CG. The output of all four transducers was recorded in analogue form on multi-track magnetic

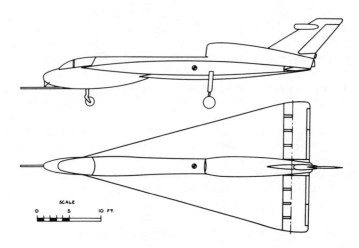

Fig.16 General arrangement of Handley Page HP 115

Fig.17 Morane-Saulnier MS 760 'Paris'

tape. A speech channel was used to record quasi-static quantities such as airspeed, altitude and fuel state, together with any comments made by the pilot. The tape recorder was run continuously throughout each flight.

It is worth noting here that wind vanes are not entirely satisfactory as sensors for an experiment of this kind because

(a)   they do not sense directly the quantity of interest (w, the velocity normal to the flight path) but instead sense α (where $\alpha = \tan^{-1} \frac{w}{U}$ and U is the airspeed along the flight path);

(b)   their natural frequencies are proportional to total airspeed;

(c)   their natural wave-lengths (typically in the order of a few feet) are comparable in size to important features of the vortex wake, and

(d)   they are naturally lightly damped.

It was recognised that all these features would give rise to difficulties in the analysis but no suitable alternative sensor was available to us.

5.3   Test Procedure

In the procedure most frequently adopted the two aircraft were flown, in calm air, at approximately the same height on mutually perpendicular tracks and were so positioned that the HP 115 would pass ahead of the Paris. Shortly before the HP 115 reached the track intersection the smoke and the recorders in it were switched ON, on command from the pilot of the Paris. Thereafter the pilot of the HP 115 concentrated on maintaining straight and level flight at constant airspeed with zero sideslip.

As the generating aircraft crossed the probe's track an event mark was inserted on the tape. The probe's pilot concentrated on penetrating the wake smoke-trail as near to its centreline as possible, making the minimum manoeuvres necessary to do this and maintaining a track perpendicular to the target's. The probe's airspeed, height and fuel state at wake penetration were read onto the tape record. Wake age was taken simply as the time interval between the event mark and the subsequent penetration of the vortex system (in fact, of course, the wake ages vary during a penetration but since this amounted, at most, to a few tenths of a second, the effect was ignored).

At the end of each run the recorders and smoke were switched off and the aircraft were re-positioned for a further run. The wake age at penetration was varied by changing the initial positioning of the aircraft.

The HP 115 was flown at one of three nominal speeds (80, 100 and 120 kt ias), the lowest being slightly above the minimum airspeed for level flight. No attempt was made to operate the probe aircraft at a pre-determined airspeed, though limitations were imposed for flight safety and structural reasons; in practice most penetrations were made at airspeeds between 120 and 180 kt ias. Height was not a critical factor in the experiment and was chosen, on the day, to avoid significant atmospheric disturbance; in practice nearly all tests were made between 3000 and 6000 ft (900-1800 m).

5.4 Results

One flight was devoted to developing the operating techniques and resolving the major problems. Thereafter, a total of fourteen flights yielded a total of 223 attempted penetrations, which were divided nearly equally between the three nominal airspeeds of the wake-generating aircraft. Of this total there were 35 cases in which no legible disturbance was recorded and in most of these the probe's pilot commented that he had missed the smoke trail; the proportion of misses rose as the wake age increased and the smoke trail became more diffuse and, hence, more difficult to track accurately.

The wake age at intersection ranged from about 2.5 sec, to about 40 sec, with most of the data being concentrated in the range 3.5-25 sec. Distances of the wake intersection point downstream of the generating aircraft ranged from about 100 m to 2300 m.

As was to be expected from the physical characteristics of the wind vane, its output was found to be heavily distorted at frequencies in the region of the vane's natural frequency. This distortion was greatly reduced by passing the signal through a network having a transfer function that was the inverse of the second-order approximation to the wind-vane's transfer function, the characteristics of the latter being determined experimentally for the mean conditions of each traverse: an example of the effect of this correction process is shown in Figure 18. It should be mentioned, however, that the velocity field associated with the

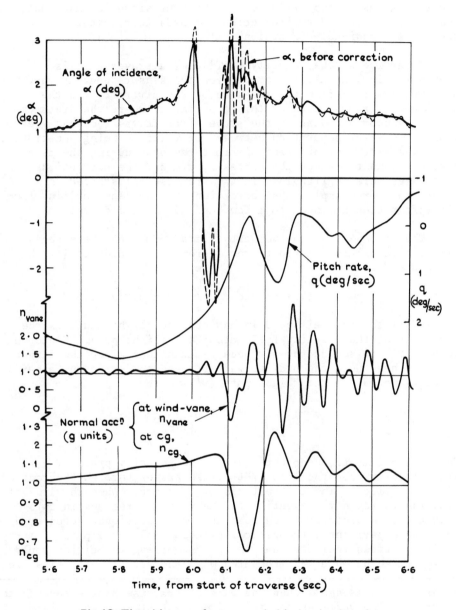

Fig.18 Time-history of a traverse behind a slender wing

vortices introduces a variation in airspeed at the vane* (and
hence in its natural frequency) during a traverse, so that the
use of a time-invariant transfer function appropriate to 'mean'
conditions was a simplification of the true situation and must, in
some cases, have led to 'corrected' records that contained
significant residual distortion.

To arrive at the velocity profile of a traverse it is
necessary, in general, to remove from the sensor's output that
part which arises from the response of the aircraft. In the
particular circumstances of this experiment, where the distance
between sensor and wing on the probe aircraft was somewhat greater
than the average separation between the vortices being penetrated,
there was little significant aircraft response during the time
that the vane traversed the vortex system and the corrections in
most cases were negligible: in the traverse illustrated in
Figure 18, for example, the corrections to incidence arising from
aircraft motion amounted to a few hundredths of a degree.

The component of the velocity field normal to the probe's
flight path (w) is related to the incremental incidence ($\delta\alpha$) by
the expression

$$w = (V + u) \tan \delta\alpha \approx (V + u) \delta\alpha \ ,$$

where V is the true airspeed in undisturbed flight and u is the
component of the velocity field along the flight path. Since we
were unable to record the airspeed fluctuations along the flight
path, we have used the undisturbed airspeed to derive an 'apparent'
velocity normal to the flight path (w') where

$$w' \approx V \delta\alpha = w \left(1 + \frac{u}{V}\right)^{-1} .$$

These simplifications do not distort the velocity profile
significantly when the line of traverse passes through the centres
of the vortices. The greatest distortions occur when the line of
traverse is roughly tangential to the vortex cores and in this
situation, combined with the most adverse conditions encountered
in these experiments, theoretical calculations for a pair of
Squire vortices indicate that the peak apparent velocities would
be increased by about 8% (compared with the actual velocities) in

---

* This effect is small when the axis of traverse passes through the
centres of the vortices and is at its greatest when the axis of
traverse is roughly tangential to the vortex cores. In the worst
case that might have been encountered in this experiment,
theoretical calculations indicated a maximum variation in airspeed
of about ±15% around the mean.

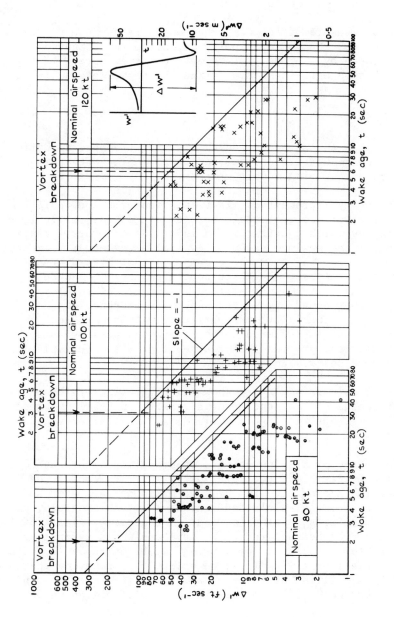

Fig.19 Peak change in apparent induced velocity near a vortex core, as a function of wake age

the vicinity of one vortex and reduced by a similar amount in the
vicinity of the other: however, even the exaggerated apparent
peak velocities are lower than the peaks encountered during a
traverse passing through the vortex centres.

A useful measure of the disturbance generated by a vortex is
given by the peak change in velocity normal to the flight path,
$\Delta w'$, experienced by a probe traversing the vortex core. If the
peak-to-peak change in incidence during the core traverse is $\Delta \alpha$,
then

$$\Delta w' \approx V \Delta \alpha .$$

This measure has the advantage that it avoids the need to define
a datum incidence and its use enabled us to employ data derived
from traverses involving modest manoeuvres.

The velocity increment $\Delta w'$ is a function of the circulation
(K) around the vortices, of their age (t) and of the vertical
distance (z) by which the traverse axis misses the vortex centres;
at a given age the maximum value of $\Delta w'$ for a vortex pair occurs
when z = 0. Proceeding from the assumption that miss distance, z,
would vary somewhat randomly throughout a series of penetrations,
it was argued that if $\Delta w'$ was plotted against t the envelope
forming the upper bound of the data should represent, with
reasonably high probability, the variation with age of the
velocity increment encountered during traverses passing through
the vortex centres, provided, of course, that a sufficiently large
number of traverses was considered.

Plots of $\Delta w'$ against t are shown in Figure 19 for the three
nominal airspeeds of the generating aircraft: not all the data
have been shown, but the figure includes all the larger values. of
$\Delta w'$ encountered.

It has often proved convenient, when dealing with measurements
of the velocity field induced by a vortex pair, to employ the
circulation (K) around the wing as a normalising parameter and to
present the data as $\Delta w'/K$ plotted against K t. This should, on
the basis of Squire's hypothesis, lead to a better collapse of the
data when K varies between runs. We have applied a variant of
this process to the present data, using the circulation (K_e) about
an equivalent elliptic wing (i.e. an elliptic wing having the same
span-loading and flying at the same airspeed) as a more convenient
normalising parameter in this case. The results are shown in
Figure 20.

If we define an 'apparent core diameter', d', as the distance
along the traverse that separates successive positive and negative

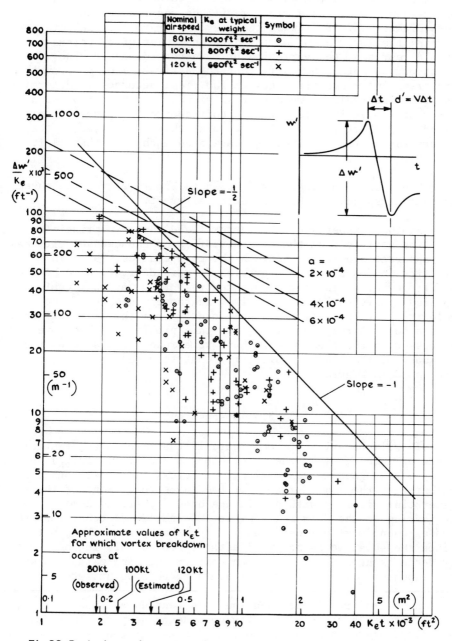

Fig.20 Peak change in apparent induced velocity near a vortex core as a function of wake age

peaks of w' in the vicinity of a vortex, then d' will be a function of K, t and z and its minimum value for a given t will occur when z = 0; this minimum will be regarded as the 'true' core diameter d. Following the previous argument, the envelope forming the lower bound of the data on a plot of d' versus t should represent the variation of true core diameter with wake age. It should be noted, however, that the derivation of d' from the traverse time – histories called for great accuracy in determining the time intervals between successive peaks in w' (or α) – an error of 1 msec in this interval corresponded, typically, to an error of 0.25 ft in d': it is thought that, in practice, the errors in d' are not likely to have been much less than ±1 ft (0.3 m).

Plots of d' against t are shown in Figure 21 for the three nominal airspeeds of the generating aircraft.

Another property of the vortex field that, in principle, could be derived from the time-histories of induced velocity was the apparent separation between the core centres, 2s'. In practice we found the main problem in this derivation to be the difficulty of defining the core centres* in a way that could be applied consistently, particularly in those cases where the velocity profiles differed significantly from the classical form. We are not entirely satisfied that this problem has been overcome successfully and therefore we have some reservations about the adequacy of the results which, for what they may be worth, are presented in Figure 22.

5.5 Discussion

It will be evident from Figure 19 that, for a substantial range of wake age (t), the upper bound of $\Delta w'$ (i.e. $\Delta w'_{max}$) varies as t^{-1}, to a good approximation; moreover, the upper bounds are essentially the same for each of the three conditions of the generating aircraft within this age band. There is some suggestion in the data relating to 100 and 120 kt conditions that the exponent may be lower than 1 for rather young wakes, but, since the envelope is not well established in this region, further evidence is needed before we can confirm this and establish what the exponent may be: we shall return to this point later.

---

* We have arbitrarily defined the core centres as those points, lying between successive positive and negative peaks in induced velocity, at which w' differed from the associated peak values by $\frac{1}{2}\Delta w'$. This definition is illustrated for an idealised case in Figure 22.

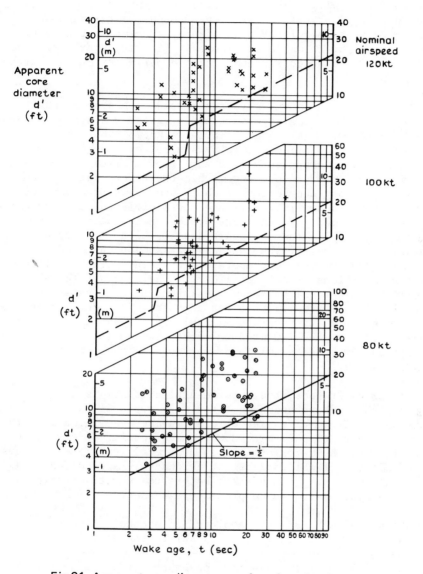

Fig.21 Apparent core diameter as a function of wake age

As mentioned earlier, theoretical studies[1,7] of the development of a vortex containing turbulence in its core have predicted that the peak change in velocity encountered during a (diametral) traverse (i.e. $\Delta w'_{max}$) would vary like $t^{-\frac{1}{2}}$, and these predictions are supported by a substantial body of experimental evidence[3,4,5] relating to vortices of the type shed by conventional aircraft (but prior to the formation of loops). Clearly, therefore, the unusual result obtained in the present experiments merits further discussion.

The phenomenon known as vortex breakdown has been observed to occur in the leading-edge vortices of slender wings at moderate to high incidence. Relatively little is known about the phenomenon, which manifests itself as a rather sudden increase in core diameter, accompanied by a change from a highly-organised flow to a more disorganised one within the core: for a given wing the position along the vortex at which breakdown occurs moves upstream as incidence increases and moves differentially when the wing is sideslipped. Vortex core breakdown has been observed in flight on the HP 115 and the positions at which it occurred have been measured, albeit rather crudely, at speeds up to 80 kt*; these results have been extrapolated to yield estimates of the breakdown positions at 100 and 120 kt. The approximate wake ages corresponding to these breakdown positions are indicated in Figure 19, whence it will be noted that they fall quite close to the regions where it appears reasonable to suggest that the upper bounds of the 100 and 120 kt data might change slope. Note also that all the 80 kt data relate to vortices in their post-breakdown state.

So far as it goes, the experimental evidence is not inconsistent with the hypothesis that the more rapid diffusion observed in the vortex field of a slender wing may be associated with the prior occurrence of core breakdown in the leading-edge vortices, indeed it would be surprising if so apparently vigorous a process had no effect, but more and better evidence is needed regarding the evolution of very young wakes before the merits of this rather attractive hypothesis can be properly assessed.

The normalising process employed in producing Figure 20 (i.e. plotting $\Delta w'/K_\epsilon$ against $K_\epsilon t$) shifts the data point of

---

* By observing and photographing the behaviour of smoke introduced into the vortex cores near the apex of the wing and noting the position at which the smoke tube changed in appearance and increased in diameter. At speeds above 80 kt, the smoke trails had become too diffuse for proper observation before the breakdown occurred.

Figure 19 along lines having a slope of -1 and therefore has no effect on the part of the envelope which is fitted by a line of the same slope, though the presentation of all the data on a single plot appears to improve the definition of this upper bound. At lower values of $K_\varepsilon t$ (less than 4000, say) the data suggest that $\Delta w'/K_\varepsilon$ varies like $(K_\varepsilon t)^{-\frac{1}{2}}$, which agrees with theoretical predictions for a single vortex, and suggests also that a reasonable value for the coefficient of eddy viscosity (a) would be 4×10^{-4}, which is in broad agreement with the findings of earlier experiments: it should be reiterated, therefore, that some of the data in this group relate to vortices prior to the occurrence of breakdown while others relate to the post-breakdown state - it seems unlikely that the same value of 'a' would apply to both situations and for the present, therefore, we must suspect that the data in this region may be incomplete and, perhaps, misleading.

The relationship suggested by Maskell for the interaction time of a vortex pair has been discussed earlier. Its derivation involves no assumption of the nature of the interaction and a similar form of relation might be expected to hold for the more complex flow field behind a slender wing provided that sufficient additional characteristics were introduced to describe the more complex flow field. The formation of loops may occur just as with the straight or swept wings but such observations of smoke trails that have been possible do not indicate that looping does in fact occur before the more rapid decay law begins.

Although the apparent core diameter could not be determined very accurately, the results shown in Figure 21 probably suffice to establish the gross trends with age. It can be said, therefore, that the lower bound of d' (i.e. d'_{min}, the 'true' core diameter) rises with increasing wake age (t) and is reasonably consistent with a core diameter that grows like $t^{\frac{1}{2}}$, at least for the data relating to 80 kt. A variation of this type is predicted by Squire's theory for a simple vortex in turbulent flow.

Possible interpretations of the 100 and 120 kt data, taking into account the wake ages estimated for the occurrence of vortex breakdown, are shown in the upper diagrams of Figure 21. Within the experimental accuracy, the vortex diameter in the post-breakdown phase appears to vary like $t^{\frac{1}{2}}$ in these cases also, and is essentially the same, at a given t, for all three airspeeds of the generating aircraft. The variations shown in Figure 21 for the pre-breakdown phase are speculative.

From photographic measurements made during the study of vortex breakdown, the diameters of the smoke-tubes associated with the leading-edge vortices were found to be about 1 ft (0.3 m) and

1.5 ft (0.45 m) when the wake was 1 sec old and the aircraft was flying at 100 kt and 80 kt, respectively: vortex breakdown at 80 kt was accompanied by an increase in diameter to about 4 ft (1.2 m). These observations are in broad agreement with the measurements made in the present experiments.

If the structure of the vortex core is roughly axisymmetric, or, more generally, if it retains a roughly universal form for all time, then a measure of the core strength (K', say) is given by

$$K' = \frac{\pi}{2} d' \Delta w' \quad .$$

On the basis of this hypothesis the present measurements of d' and $\Delta w'$ imply that K' would vary like $t^{-\frac{1}{2}}$ for the range over which $\Delta w'$ varies like t^{-1}, i.e. the core strength would diminish with time. This result is not entirely out of the question since it requires only a direct interaction between the expanding vortices, leading to mutual cancellation of vorticity in the two main regions of vorticity. Alternatively, this result might lead us to infer that the initial hypothesis is incorrect and that the vortex structure changes steadily in form while the overall strength remains constant, or that combined changes occur in both strength and structural form.

The downward momentum transported across any transverse plane is necessarily constant and equal to the lift on the wing. This constant value can be maintained with vortices of diminishing strength only if their lateral separation increases. With this in mind we turn now to the variation of core separation ($2s'$) with time shown in Figure 22, whence we can discern a fairly consistent trend for the core separation to increase with increasing wake age. Bearing in mind the reservations expressed earlier regarding the adequacy of this data, we will remark only that this result is not inconsistent with our earlier inference that the vortex strengths may diminish with age in the case of this slender wing.

In concluding this discussion it is appropriate to offer some observations on the structure of the velocity field encountered in these experiments. In the majority of cases the time-histories of incidence during a wake encounter exhibited approximately the classical W-shaped waveform, of which the trace reproduced in Fig.18 is a fair example, several instances were observed where this wave-form was distorted in that the incidence excursions differed substantially between the two vortices; this may have arisen from any combination of three main causes

(a) the variation in the component of induced velocity along the flight path (u), discussed in section 5.4;

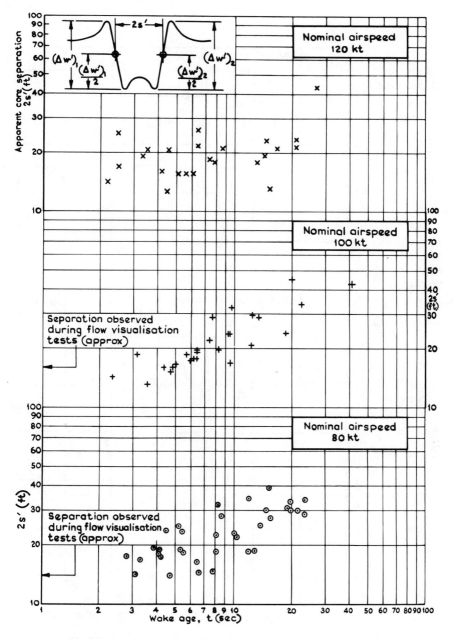

Fig.22 Apparent core separation as a function of wake age

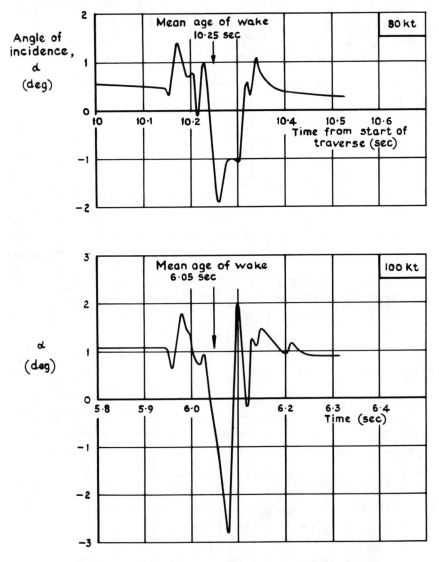

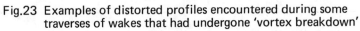

Fig.23 Examples of distorted profiles encountered during some
traverses of wakes that had undergone 'vortex breakdown'

(b)   an inclination, in the vertical plane, between the axis of traverse and the line joining the vortex centres, such that the probe passed closer to one vortex;

(c)   inadvertent sideslip of the generating aircraft.

Cases were observed which did not exhibit a reduction in downwash between the vortex cores: these cases usually were associated with old wakes or with relatively large miss distances from the vortex centres.

A considerable number of velocity profiles showed highly unusual features, some of which are illustrated in the examples shown in Figure 23.   About the only common property of these profiles was their rather disorganised appearance, with numerous peaks* in velocity tending perhaps to suggest that one or both of the main vortices had broken into several smaller ones: in this connection it seems significant that these highly distorted velocity profiles were observed only in those cases where the vortices had undergone breakdown.   Distorted profiles usually were associated with relatively large miss distances (inferred from the fact that $\Delta w'$ was lower than normal for the relevant age group), while traverses which, by inference, had been made closer to the vortex centres yielded velocity profiles that showed relatively little distortion: the significance of this is not clear at the present time.

5.6   Conclusions

These experiments have established that the wake shed by a slender wing develops in such a way that, beyond a certain age, the peak velocities induced by it vary like t^{-1} (approximately) and, for the limited range of circulation investigated, are independent of circulation at a given age.   For the tests made at 80 kt it has been established that the wakes to which these results apply have undergone vortex breakdown, and extrapolating the measurements of breakdown position suggests that the same may also be true of the tests made at 100 and 120 kt;   however, the data are insufficient to establish either that the results apply exclusively to post-breakdown wakes or, if this is the case, the laws that govern the pre-breakdown state.   There is a clear need for further data to clarify these points.

---

* It is likely that some of the minor peaks represent inadequacies in the wind-vane or in the rectification of its output, but it is clear that the major features cannot be dismissed in this way.

The above results do not agree with theoretical predictions for the growth of a simple turbulent vortex, nor do they agree with the findings of earlier experiments on conventional swept wings.

Vortex core diameter derived from the velocity profiles appeared to vary like $t^{\frac{1}{2}}$, and compared reasonably with photographic measurements in the few cases where these offered a check.

The variations in peak induced velocity and in core diameter with wake age are consistent with a vortex field in which changes in structural form or in vortex strength occur. The measurements of core separation distance showed a trend towards increasing separations with increasing age, which is not inconsistent with a vortex system that weakens with increasing age.

So far as we are aware there is no theory which explains the findings of these experiments, which are themselves incomplete. Thus, there is no means by which these results could be read across to other slender-wing aircraft, although they give grounds for hoping that the wakes from this class of aircraft will present a less serious problem than those from conventional aircraft. We hope to be able to study the wake behind a larger slender wing and thus to increase our understanding of the ways in which these wakes develop.

Acknowledgements

The authors wish to acknowledge the valuable assistance given by the Cranfield Institute of Technology in the experiment on the HP 115 aircraft and the helpful advice in the preparation of this paper given by Mr. E. C. Maskell of Aerodynamics Department, RAE.

Table 1

HP 115 PRINCIPAL DIMENSIONS

Overall length	15.3 m
Gross wing area	40.2 m^2
Span	6.1 m
Centreline chord	12.2 m
Leading-edge sweep	74° 42'
Training-edge sweep	0°
Dihedral	0°
Mass	2280 kg

Table 2

MS 760 'PARIS' AIRCRAFT PRINCIPAL DIMENSIONS

Overall length	10 m
Span	9.7 m
Gross wing area	18.4 m^2
Standard mean chord	1.9 m
Taper ratio	0.71
Leading-edge sweep back	$5\frac{1}{2}^\circ$
Dihedral	8°
Aspect ratio	5.1
Mass	3400 kg

SYMBOLS

A	aspect ratio
a	eddy viscosity factor (= ε/K)
b	aircraft span
d	typical diameter
K	circulation of line vortex
L	lift
r	radius
r_1	radius for maximum circumferential velocity
S	wing area
2s'	distance between vortex cores
t	time
V	free stream velocity
W	aircraft weight
w	circumferential velocity
$\hat{w}$	maximum circumferential velocity
$\dot{z}$	vertical velocity
α	angle of incidence
ε	eddy viscosity
ν	kinematic viscosity
ρ	air density
τ	non-dimensional time $\left(= \dfrac{K\,t}{4s'^2}\right)$

REFERENCES

No.	Author	Title, etc.
1	H. B. Squire	The growth of a vortex in turbulent flow. (ARC 16666) (1954)
2	J. W. Westmore J. P. Reeder	Aircraft vortex wakes in relation to terminal operations. NASA TN D1777 (ARC 24851) (1963)
3	T. H. Kerr F. W. Dee	A flight investigation into the persistence of trailing vortices behind large aircraft. RAE Technical Note Aero 2649 (1957)
4	R. Rose F. W. Dee	Aircraft vortex wakes and their effects on aircraft. RAE Technical Note Aero 2934 (1963)
5	F. W. Dee O. P. Nicholas	Flight measurements of wing tip vortex motion near the ground. (ARC CP 1065) (1968)
6	S. Crow	Stability theory for a pair of trailing vortices. Boeing S : R.L. D1-82-0918 (1969)
7	P. R. Owen	The decay of a turbulent trailing vortex. (ARC 25818)
8	J. P. Jones	The calculation of the paths of vortices from a series of vortex generators, and a comparison with experiment. (ARC CP 361) (1955)

SPAN LOADING AND FORMATION OF WAKE

Peter F. Jordan

RIAS, Martin Marietta Corp.

ABSTRACT

Classical analyses of aircraft wake formation assume that the wing span loading is (essentially) elliptic, and that in consequence the wake starts out being (essentially) flat. This assumption is incorrect: actual span loadings contain a logarithmic term, and in consequence there is an infinite <u>upwash</u> directly behind the wing just inside the wing tips. This explains why the aircraft wake rolls up faster than the classical analyses predict.

1. INTRODUCTION

In this paper we are concerned with certain aspects of the theory of thin lifting airfoils ("wings") of finite aspect ratio in linear subsonic potential flow. In practical calculations made to determine the spanwise lift distribution over the wing (the "span loading"), approximate ("collocation") methods are being used widely, and with some confidence which is based on experience. However, such approximate results are inadequate if the problem is to calculate the role of the downwash field, behind the wing, in the mechanism by which the vortex trail rolls up.

Indeed, classical analyses of the rolling up process[1,2] make simplifying assumptions: the wake is assumed to start out as a flat vortex sheet, tangential to the wing at the wing trailing edge, and the spanwise distribution of the vortex strength is assumed to correspond to an elliptic span loading.

Recently, a reliable solution has become available for one

particular airfoil problem: the pressure distribution over a lifting circular wing in incompressible flow.   Thereby, it became
possible to calculate the downwash field behind the circular wing.
The results of this calculation are presented in the present paper.
They provide new, and somewhat unexpected insight into the mechanism
of wake formation.   This insight is not only of interest for the
rather specific case of a circular wing, but it is relevant for
wings of any planform in subsonic flow.

    We will show that the vortex trail has an infinite curvature
at the trailing edge, and that it thus does not start out with a wel
defined tangent.   (This cognizance concerns recent arguments regarding the starting slope of the wake behind wings of finite profile
thickness.)   More important is that the downwash is not constant
spanwise.   Namely, a spanwise constant downwash is implied in the
classical assumption of an elliptic span loading; the actual downwash, however, changes sign, and it becomes an infinite upwash inside the wing span as the wing tip is approached.

    If the downwash would be constant spanwise, then the rolling
up of the vortex trail would have to be started by the relatively
slow process of viscous interaction with the upwash outside the
wing span, or by whatever chance disturbances might happen to be
present.   In fact, however, since the outer parts of the vortex
trail are directly engulfed in a field of intensive upwash, the
rolling up process is started in a positive and accelerated manner.

    For our analysis of the downwash field, the span loading of the
circular wing is the necessary starting point.   We first briefly describe the solution of this problem*.

2.   SPAN LOADING

2a.   Notation and Formulation

    The exact formulation of the circular wing problem was given
by Kinner[4] in 1937.   (This formulation, since it uses the acceleration potential, does not involve a priori assumptions about the
paths of the trailing vortices.)   We first describe this formulation
Our notation, Fig. 1, differs somewhat from that of Kinner.   Also,
in order to eliminate the inconvenient (and trivial) pressure singularities of orders $\pm 1/2$ at the leading edge (l.e.) and the trailing
edge (t.e.), we define a nondimensional pressure function $\bar{p}$ as
follows (see Fig. 1a):

$$\bar{p} = (1-\xi^2)^{\frac{1}{2}} p/q \qquad\qquad (1)$$

* The exact solution of the circular wing problem is described in
more detail in a forthcoming paper[3].

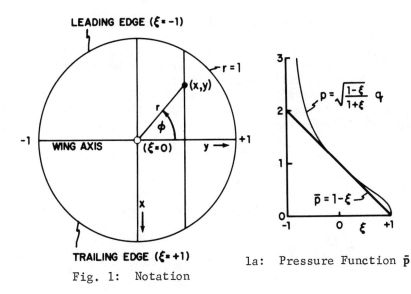

LEADING EDGE (ξ = -1)

$r = 1$

(x,y)

r

ϕ

-1   WING AXIS   (ξ=0) $y \rightarrow$ +1

x

TRAILING EDGE (ξ = +1)

Fig. 1: Notation

$p = \sqrt{\dfrac{1-\xi}{1+\xi}}\; q$

$\bar{p} = 1-\xi$

1a: Pressure Function $\bar{p}$

Here q is the dynamic pressure.

We do not lose any essential aspect of the problem when, as we do here, we confine the discussion to cases where the given downwash at the wing is symmetrical, and constant chordwise:

$$w(x,y) \equiv \tan\alpha(x,y) = w(y) = w(-y) \qquad \text{(at the wing)} \qquad (2)$$

The pressure function $\bar{p}$ is then composed of elementary solutions n, having unknown amplitudes C_n, as follows:

$$\bar{p} = \frac{4}{(1-y^2)^{\frac{1}{2}}}\left[\sum_{\varkappa=0}^{\infty} C_{2\varkappa} r^{2\varkappa}\cos 2\varkappa\phi + \sum_{\lambda=0}^{\infty} C_{2\lambda+1} r^{2\lambda+1}\sin(2\lambda+1)\phi \right] \qquad (3)$$

The span loading is

$$\ell(y) = 2qR(1-y^2)^{\frac{1}{2}}C_\ell(y) = 4\pi qR \sum_{o}^{\infty} C_{2\varkappa}P_{2\varkappa}(y) \qquad (4)$$

with R the wing radius, and the $P_{2\varkappa}$ the Legendre polynomials. (The local lift coefficient is denoted by $C_\ell(y)$. If the span loading would be elliptic, we would have $C_\ell(y) \equiv$ const.)

Individually, the elementary solutions in Eq. (3) are not physically meaningful. For their sum $\bar{p}$ to be meaningful, the amplitudes C_n have to fulfill certain conditions. There is first the Kutta t.e. condition $\bar{p}(\xi=1) \equiv 0$. From it follows that the two sums in Eq. (3) must be equal for all points on the l.e., that is, when $r=1$ and $0<\emptyset<\pi$. Therefore,

$$C_{2\lambda+1} = \frac{4}{\pi} \sum_{\kappa=0}^{\infty} \frac{(2\lambda+1)C_{2\kappa}}{(2\lambda+1)^2-(2\kappa)^2} \tag{5a}$$

Writing the same relation in reversed form:

$$C_0 = \frac{2}{\pi} \sum_{\lambda=0}^{\infty} \frac{C_{2\lambda+1}}{2\lambda+1} \; ; \; C_{2\kappa} = \frac{4}{\pi} \sum_{\lambda=0}^{\infty} \frac{(2\lambda+1)C_{2\lambda+1}}{(2\lambda+1)^2-(2\kappa)^2} \quad (\kappa \neq 0) \tag{5b}$$

Thus, knowing the lift distribution $\ell(y)$, one knows the even numbered set $C_{2\kappa}$, and thereby knows the complete pressure function $\bar{p}$. Alternatively, it is sufficient to know the odd numbered set $C_{2\lambda+1}$.

The second physical condition for $\bar{p}$ is that the downwash $w(y)$ which is induced on the wing by the pressure distribution p should be physically meaningful. Ordinarily, one wants this induced downwash to equal a given downwash $w(y)$. The latter is described by an infinite set of numbers $\bar{w}_s$:

$$\bar{w}_s = (4s+3) \int_0^1 \{ \int_0^y w(\eta)d\eta \} P_{2s+1}(y)dy \tag{6}$$

[For example, for a planar wing at incidence α, one has $w(y) = \tan\alpha$ and

$$\bar{w}_0 = \tan\alpha \; ; \; \bar{w}_s = 0 \quad \text{for } s \neq 0 \tag{6a}]$$

One then obtains an infinite system of linear equations for determining the amplitudes $C_{2\kappa}$:

$$\bar{w}_s = 2 \sum_{\kappa=0}^{\infty} b_\kappa^s \, C_{2\kappa} \tag{7}$$

This completes the Kinner formulation of the problem.

From here, Kinner[4] proceeded as follows: He truncated the system Eq. (7) and calculated approximate values for the first N (say) amplitudes $C_{2\kappa}$. The corresponding $C_{2\lambda+1}$ (all of them!) then followed from Eq. (5a). This procedure we will refer to as (K).

Unfortunately, due to some oversight, Kinner wrote the matrix coefficients $b_\varkappa^s$ in an unnecessarily cumbersome form (see Eq. (60) of ref. 4). This seems to have been the reason why the Kinner work has not been followed through earlier. Actually, a well known transformation reduces the Kinner formula to the following form

$$2b_\varkappa^s = \frac{1}{2\varkappa+2s+1} + \frac{1}{2\varkappa+2s+2} - \frac{2}{2\varkappa-2s-1} \qquad (7a)$$

The simplicity of this form invites our further analysis.

An obvious alternative form of (K) is to use the reversed relation Eq. (5b) to eliminate the $C_{2\varkappa}$ and to write the linear system, Eq. (7), in the form

$$\bar{w}_s = \pi \sum_{\lambda=o}^{\infty} c_\lambda^s C_{2\lambda+1} \qquad (8)$$

This reversed procedure was used by Turowski[5]; we denote it by (T). It has the advantage that its results converge faster, as N is increased, than those of (K); the reason for this will become obvious below. It has the disadvantage that the coefficients c_λ^s remain more cumbersome to calculate than the $b_\varkappa^s$, even though the Turowski procedure has been improved by van Spiegel[6,7]. It is thus a matter of weighing advantage against disadvantage whether one chooses (K) or (T). Mathematically, the two courses are dual and equivalent; in particular, both fulfill the Kutta condition identically (notwithstanding assurances to the contrary in refs. 5, 6 and 7).

2b. The Wing Tip Problem

The key to a complete solution of the circular wing problem lies in the wing tip problem. As the wing tip is approached, $y^2 \to 1$. A look at Eq. (3) shows that any finite set of approximate amplitudes C_n, obtained from either (K) or (T), will be of little value in determining $\bar{p}$ near or at the tip. Indeed, none of the earlier work[4-7] makes an attempt to   discuss the tip problem.

An intriguing aspect of the tip problem lies in the distinct difference between the two types of chordwise pressure distributions involved, the smooth 2-D type (Birnbaum-Glauert) which must prevail over the central part of the wing span, and the kinked slender wing type which must appear at any curved wing tip[8]. We wish to find out, among else, how the transition occurs.

We note again that for any l.e. point (r=1, $0<\emptyset<\pi$) the two sums in Eq. (3) must be equal. From this, and going to the limit $\emptyset \to 0$, we find the important tip condition

$$\sum_{\varkappa=0}^{\infty} C_{2\varkappa} = 0 \tag{9}$$

Combination of Eqs. (5a) and (9) leads to

$$\pi C_{2\lambda+1} = \frac{4}{(2\lambda+1)} \sum_{\varkappa=1}^{\infty} \frac{(2\varkappa)^2 C_{2\varkappa}}{(2\lambda+1)^2 - (2\varkappa)^2} \tag{10}$$

Inspecting Eqs. (5b) and (10), one comes to expect that the $C_{2\varkappa}$ will converge (roughly) like $\varkappa^{-2}$ and the $C_{2\lambda+1}$ like λ^{-3}. We try the following:

We introduce elementary pressure functions $\bar{p}_r$ by defining

$$c_o^r = \zeta(r) \; ; \quad c_{2\varkappa}^r = -\frac{1}{\varkappa^r} \quad (\varkappa \neq 0) \tag{11}$$

Each individual set $c_{2\varkappa}^r$ fulfills Eq. (9). To each belongs a set $c_{2\lambda+1}^r$ by means of Eq. (10), and $\bar{p}_r$ is defined by Eq. (3). Using these elements, we expect to build up the desired solution $\bar{p}$ by means of linear superposition:

$$\bar{p} = \sum_{r=2}^{\infty} a_r \bar{p}_r \tag{12}$$

We are going to show that the coefficients a_r here are not independent but have to fulfill certain conditions if the downwash w(y) is to be physically meaningful.

First, however, we list a few important properties of the $\bar{p}_r$. One has to show, obviously, that r=2 is the correct number for the first element. This is readily done. Namely, one finds along the leading edge

$$\bar{p}_{2,1.e.} = \frac{8\emptyset}{\sin\emptyset} (\pi-\emptyset) \tag{13}$$

This function remains regular in the limit $\emptyset \to +0$, and its tip value is

$$\bar{p}_{2,T} \equiv \lim_{\emptyset \to 0} \bar{p}_{2,1.e.} = 8\pi \tag{14}$$

For no other value r does $\bar{p}_{r,T}$ have a finite value:

$$\bar{p}_{r,T} = \begin{cases} \infty & \text{if } r<2 \\ 8\pi & \text{if } r=2 \\ 0 & \text{if } r>2 \end{cases} \qquad (14a)$$

(In Eq. (14a), r need not be integer.) But, from physical arguments, $\bar{p}_T$ has to be a finite value. It follows that r=2 is the
correct number for the first element. Also, it follows that no
other element contributes to $\bar{p}_T$; hence

$$\bar{p}_T = 8\pi a_2 \qquad (14b)$$

Furthermore, it can be shown: as the field point approaches
the wing tip along any half ellipse ξ = const., the limit value of
$\bar{p}$ is

$$\lim \bar{p} = \begin{cases} |\xi| \, \bar{p}_T & \text{if } \xi \le 0 \\ 0 & \text{if } \xi \ge 0 \end{cases} \qquad (15)$$

This law represents the limit form of the kinked slender wing distribution.

The nature of the tip singularity in $\bar{p}$ is illustrated in Fig. 2.
The pressure distribution over the planar circular wing with $\tan\alpha=1$
is shown. At the tip point, $\bar{p}$ jumps from its l.e. value $\bar{p}_T \approx 3.1862$
to its t.e. value $\bar{p} = 0$.

We note finally that

$$c_{2\lambda+1}^2 = \frac{8}{\pi(2\lambda+1)^3} \qquad (16)$$

As expected from Eq. (10), the $c_{2\lambda+1}$ converge more rapidly than the
$c_{2\kappa}$. This explains why the (T) procedure converges more rapidly
than the (K) procedure (see section 2a).

Corresponding to Eq. (12) we build up the span loading function
$c_\ell(y)$, Eq. (4), as follows:

$$c_\ell(y) = \sum_{r=2}^{\infty} a_r c_\ell^r(y) \qquad (17)$$

The elementary span loading functions $c_\ell^r(y)$ are the span loading
functions of the elements $\bar{p}_r$. In the neighborhood of the wing tips
$y^2=1$ these functions can be written

$$c_\ell^2(y) = 4\pi \left[1 + \frac{1}{8}(1-y^2)^{\frac{1}{2}}\log \frac{4}{1-y^2} \right] + 0(1-y^2)^{\frac{1}{2}}$$

$$c_\ell^3(y) = \pi(1-y^2)^{\frac{1}{2}}\log \frac{4}{1-y^2} + 0(1-y^2)^{\frac{1}{2}}$$

$$c_\ell^r(y) = 0(1-y^2)^{\frac{1}{2}} \quad (r \geq 4) \tag{18}$$

From Eq. (18) one reads, for example, that an elliptic span loading (that is, $C_\ell(y)$ a constant) requires that

$$a_3 = -a_2/2 \quad \text{(elliptic span loading)} \tag{19}$$

Having briefly described certain properties of the elementary pressure distributions $\bar{p}_r$, we now take a look at the downwash distributions $w_r(y)$ which are induced by the p_r. According to Eq. (6), these downwash distributions are described by certain sets $\bar{w}_s^r$ of numbers; the latter are related to the $C_{2\kappa}^r$ by the operator

$$\bar{w}_s^r = 2 \sum_{\kappa=0}^{\infty} b_\kappa^s c_{2\kappa}^r \tag{20}$$

compare Eq. (7). The summation in Eq. (20) is readily performed due to the simple form of the b_κ^s, Eq. (7a). One finds

$$\bar{w}_s^2 = -\frac{1}{2s^3} \left[(\log s + \gamma) - \frac{5}{2} \right] + 0(s^{-4})$$

$$\bar{w}_s^3 = -\frac{1}{2s^3} \left[4(\log s + \gamma) + \zeta(2) \right] + 0(s^{-4})$$

$$\bar{w}_s^r = -\frac{1}{2s^3} \left[\zeta(r-1) + 4\zeta(r-2) \right] + 0(s^{-4}) \quad (r \geq 4) \tag{21}$$

Here, as already in Eq. (11), ζ denotes the Riemann function.

From Eq. (21) a very significant conclusion can be drawn. Namely, reversing Eq. (6), one has

$$w_r(y) = \sum_{s=0}^{\infty} (s+1)(2s+1)\bar{w}_s^r \bar{P}_{2s}(y) \tag{22}$$

The $\bar{P}_{2s}$ are orthogonal polynomials. They may be interpreted as derivatives of the Legendre polynomials P_{2s+1}, or as Gegenbauer or ultraspherical polynomials. The point here is that they are normalized in the same manner as the P_{2s}:

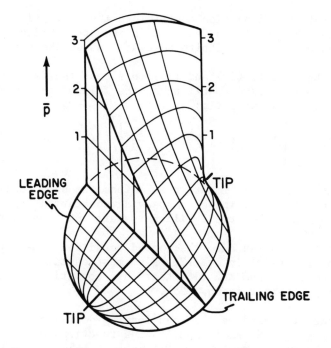

Fig. 2   Pressure Distribution over Planar Circular Wing

$$\bar{P}_{2s}(1) = \bar{P}_{2s}(-1) = 1 \tag{22a}$$

From Eqs. (21), (22) and (22a) one reads that all functions $w_r(y)$ diverge to minus infinity at the tip:

$$\lim_{y^2 \to 1} w_r(y) = -\infty \quad \text{(all r)} \tag{22b}$$

Thus none of the elementary pressure functions $\bar{p}_r$ is by itself a physically meaningful solution.

In order for the pressure function $\bar{p}$ itself to be physically meaningful, the downwash

$$w(y) = \sum_{r=2}^{\infty} a_r w_r(y) \tag{23}$$

must remain finite.   This implies two separate conditions.   First, the terms of order $O(s^{-3}\log s)$ in Eq. (21) must cancel each other; thus

$$a_3 = -a_2/4 \tag{24}$$

One sees that this condition contradicts the assumption of elliptic span loading, Eq. (19); in fact, it is seen that elliptic span loading induces a <u>positive</u> infinite downwash inside the wing tip.

Using Eqs. (17), (18) and (24), one sees that any physically acceptable span loading function has the following form in the neighborhood of the wing tip:

$$c_\ell(y) = \left[1 + \frac{1}{16}(1-y^2)^{\frac{1}{2}}\log\frac{4}{1-y^2} \right] c_\ell(1) + 0(1-y^2)^{\frac{1}{2}} \qquad (25)$$

This result is of major importance. Since the span loading cannot be elliptic, it cannot induce a spanwise constant downwash, and in consequence the vortex trail will not start out as a flat sheet.

We have of course also the second condition , namely, the terms of order $0(s^{-3})$ in Eq. (21) also must cancel each other. However, one sees from Eq. (18) that this condition affects only terms of higher order in Eq. (25).

Our result regarding the span loading distribution is illustrated in Fig. 3. Several span loading functions are shown, all normalized to the correct end value $C_\ell(1) = 1.5931..$ of the planar circular wing with $\tan\alpha = 1$. The elliptic distribution appears as a straight line; it induces a <u>positive</u> infinite downwash at the tip. The first elementary distribution $C_\ell^2(y)$ overshoots the correct curve at the tip; it induces a <u>negative</u> infinite downwash. The curve "w(1) finite" was calculated using Eq. (25), and adding a third element (a_4) so that the second condition also could be fulfilled. This curve goes asymptotically into the "correct" curve.

One may add further elements to fulfill further conditions (note that so far we have not made any assumption how w(y) is distributed over the span). If one adds a fourth element (a_5) and makes the condition that the overall lift coefficient C_L should have the correct value, Fig. 3, one obtains a curve which, on the scale of Fig. 3, cannot be distinguished from the "correct" curve.

With this illustration we conclude our brief description of the span loading solution for the circular wing. We had to omit many details; for example, we did not describe how the "correct" curve in Fig. 3 was obtained, nor how C_L and $C_\ell(1)$ were calculated. As aforesaid, for these and other details we have to refer the reader to ref. 3.

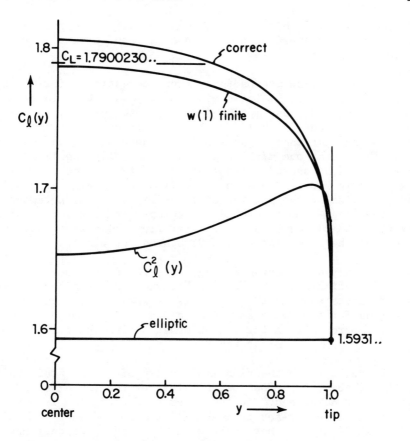

Fig. 3   Span Loading of Planar Circular Wing

3.   WAKE FORMATION

3a.   Preliminary Discussion

We start our preliminary discussion of wake formation with a
look at Fig. 4, a collection of (a few) curves concerning the downwash at and behind a wing. All curves refer to the downwash in the
plane of the wing (z=0) and at the center (y=0) of the wing span.

Two of the curves refer to 2-D flow (AR=∞). For 2-D flow the
exact downwash is well known. Its curve, marked "correct", has a
kink at the t.e., and it continues with a vertical tangent. From
this it follows that the correct 2-D wake has an infinite curvature
where it starts out from the t.e.

The "lifting line" curve for AR=∞ was calculated using the
standard approximative assumption of a single lifting vortex at the

1/4 chord line.   This curve, which crosses the correct $w(x)$ of the
wing at its 3/4 chord point, approaches the "correct" curve asymptotically behind the wing.

One obtains an unexpectedly accurate estimate for the total
lift coefficient of elliptic wings of any aspect ratio if one uses
the "lifting line" assumption (and assumes, furthermore, elliptic
span loading; see Helmbold[9] or Jordan[8] ; the key for the success of
this approach is that it uses only the downwash in the center plane,
$y=0$). Therefore, one might expect that the "lifting line" $w(x)$
curves would be useful approximations to the correct downwash for
any elliptic wing. One such curve is shown in Fig. 4, namely, the
"lifting line" curve for the circular wing. It is noteworthy that
this curve has almost reached its asymptotic value, w_{∞}, already at
$x=2$ (both $AR=\infty$ curves are still far from their asymptotic value,
$w_{\infty}=0$, at the same distance, $x=2$, behind the wing).

The only other curve in Fig. 4 is the dashed curve; it is an
artist's estimate of the "correct" curve for the circular wing.
This estimate assumes that there is an analogy between the two pairs
of curves for the two cases, $AR=\infty$ and circular wing. If this assumption is correct, then the vortex sheet behind the circular wing also
has infinite curvature at the t.e.

We now turn to the classical theory of vortex trail roll up.
In this theory, the flowwise curvature of the vortex sheet at the
t.e. is not discussed. In the paper by Kaden[1] , it is assumed that
the vortex sheet arises from an elliptic span loading, that it is
originally flat but that somehow the rolling up process has started.
The progress of this process is analyzed. In the paper by
Westwater[2] , a flat vortex sheet, extending to infinity fore and aft
(and its vortex distribution corresponding again to an elliptic span
loading) is assumed to move downward as a flat sheet, due to its
self-induced downwash w_{∞}, up to the instant $t=0$ at which instant
it is released. At $t=0$, the rolling up process is supposed to start,
somehow, and its progress is calculated step by step.

There is a point in this concept which is not quite obvious.
It is clear, of course, that the vortex sheet would not remain flat
in real life. While ideally it is self-perpetuating, it is unstable
with respect to external disturbances. Also, viscous effects of the
upwash outside the edges of the sheet would slowly start the rolling
up even in the absence of distinct disturbances. However, Westwater
does not mention any external disturbance, and he does not consider
viscous effects. Why, then, does his sheet start rolling up at $t=0$?

The explanation is that the Westwater procedure does in fact
introduce an artificial disturbance. Namely, in it the continuous
sheet is replaced by 20 individual vortices. Marking one of them
as k, one removes k, calculates the induced speed $v_{i,k}$ at the locus

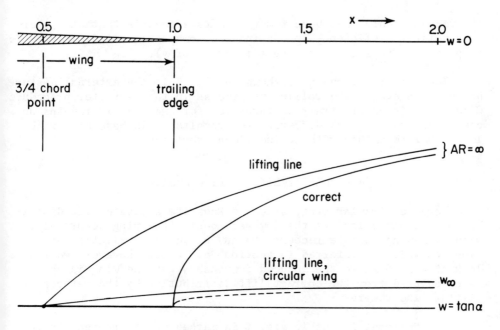

Fig. 4   Downwash At and Behind Wing Center

of k due to the remaining 19 vortices, replaces k and lets k move
with the speed $v_{i,k}$ during the next time step.   This procedure would
appear to be natural and innocuous enough; however, it is not innocuous.   Figure 5 explains why.   Shown is the downwash due to an
"elliptic" planar vortex trail which is not complete but has a gap
of width ε at some point y.   Let this gap represent the position of

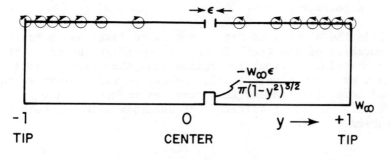

Fig. 5   Downwash of "Elliptic" Planar Vortex Trail

the vortex k which has been removed. Its effect is that at y there is a definite deficiency in the induced downwash; this deficiency is proportional to ϵ (and is also a function of y).

The deficiency which is shown in Fig. 5 is the external disturbance which starts the rolling up process in the Westwater procedure. Clearly it is an arbitrary disturbance; for example, it remains an open question how much different the results would have been with 10 or 40 rather than with 20 individual vortices.

3b. A Result of Lifting Line Analysis

For the circular wing, since we know its accurate lift distribution, we can calculate the downwash behind the wing accurately, without having to take recourse to any of the approximating assumptions to which we referred in section 3a. First, however, we take a look at Fig. 6 where the induced downwash α_i at the wing is shown as calculated from the correct lift distribution by the methods of lifting line theory.

The horizontal line in Fig. 6 is marked C_L and represents the equivalent elliptic span loading of the correct total lift. The curve $C_\ell(y)$ describes the actual distribution of this lift over the wing span. The incidence which is induced at the wing by the vortex trail due to $C_\ell(y)$ is denoted by α_i; this curve goes to minus infinity at the wing tip. This fact is not to be interpreted, of course, that there would be an infinite upwash at the wing itself. Rather, lifting line theory in reality calculates the downwash w_∞ in the Trefftz plane; the hypothetical assumption is then made that $\alpha_i = w_\infty/2$. However, Fig. 6 does show the destabilizing effect of lifting line analysis (as one calculates α_i from a given span loading; resp., the stabilizing effect as one reverses the process): to the infinite tangent at a finite end value $C_\ell(1)$ corresponds an infinite end value $\alpha_i(1)$.

The factor 2 to w_∞ in Fig. 6 simplifies the comparison; if we would have inserted the elliptic distribution instead of $C_\ell(y)$, we would have found $2w_\infty = C_L$.

While the α_i curve of Fig. 6 cannot be taken as an accurate representation of the trail-induced downwash at the wing, it nevertheless contains a strong indication that the correct trail will not start out as a planar trail. Below, we will return to the original interpretation of this curve as representing w_∞, and will compare this result of lifting line analysis with the correct downwash behind the wing.

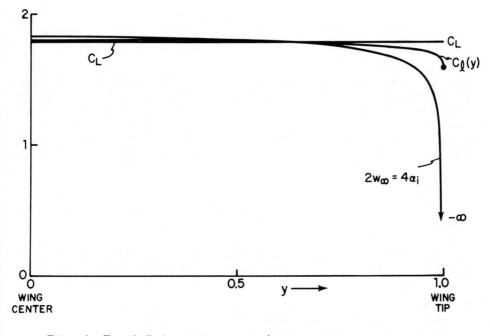

Fig. 6   Trail Induced Downwash (Lifting Line Analysis)

3c. The Downwash Behind the Circular Wing

Not only the downwash behind the circular wing but the complete flow field around this wing can be calculated accurately, in principle at least (and the wake formation process, the rolling up of the vortex trail, can thus be followed in accurate detail). Namely, the complete flow field is known[4] individually for each one of the elementary solutions n in Eq. (3). For any given downwash w(y) on the wing, since we can determine (the pressure function $\bar{p}$ by determining) the sequence of amplitudes C_n, the remaining task of calculating the complete flow field is merely one of performing summations. In practice, this task may be somewhat cumbersome; convergence may be slow, and singularities at the wing edges require special considerations. In principle, however, the answers are available.

In the present paper, we limit our task to that of calculating the downwash behind the planar circular wing in its plane (z=0), and only up to the station x=2 (see Fig. 4). Also, we do not extend our calculation beyond $|y| \leq 0.9$.

According to Ref. 4, Eq. (22), we obtain the downwash at a point (x,y,0) by evaluating the integral

$$w(x,y)_{z=0} = -\int_{-\infty}^{x} \left(\frac{\partial \phi_a}{\partial z}\right)_{\substack{y=y \\ z=0}} du \qquad (26)$$

Here ϕ_a is the acceleration potential. If the point $(x,y,0)$ lies behind the wing and inside the wing span (and this is the case in which we are here interested), then performance of the integration involves dealing with singularities where the path of the integration crosses a wing edge. We can avoid much of this difficulty by starting the integration at the trailing edge. The downwash at the t.e. itself is given, and

$$w(x,y)_{z=0} = w_{t.e.}(y) - \int_{(1-y^2)^{\frac{1}{2}}}^{x} \ldots du \qquad (26a)$$

Alternatively, since, corresponding to Eq. (3), the potential ϕ_a is an infinite sum of elementary potentials $\phi_{a,n}$, we can also wri

$$w(x,y)_{z=0} = w_{t.e.}(y) - \sum_{n=0}^{\infty} C_n w_n(x,y) \qquad (26b)$$

The C_n are again the amplitudes of the elementary solutions; the $w_n(x,y)$ are elementary integrals which are defined by correlating Eq. (26b) with Eq. (26a).

Using the formulas developed in ref. 4, one obtains the following formal result

$$w_n(x,y) = \pm \int_{o}^{\eta_o} \frac{\cos n(\phi+\pi/2)}{ur^n} \frac{d\eta}{\eta^2} \qquad (27)$$

where

$$\eta^2 = u^2 + y^2 - 1 \; ; \; \eta_o^2 = x^2 + y^2 - 1$$

$$r^2 = u^2 + y^2 \; ; \; \cos(\phi+\pi/2) = u/r \qquad (27a)$$

The + sign applies when $n = 0$ or $3 \pmod 4$, the - sign when $n = 1$ or $2 \pmod 4$.

The formal result Eq. (27) cannot be used without precaution: one readily sees that the individual integral w_n does not exist. This is because, as mentioned earlier, the individual elementary solutions are not physically meaningful. We did achieve a meaningful sum, the pressure function $\bar{p}$, Eq. (3), by introducing the Kutta condition. We now make use of this fact, of the fact that our

complete solution does fulfill the Kutta condition, to show that
the sum of the w_n, Eq. (26b), does exist even though the individual
w_n do not exist.

Writing $f_n(\eta)$ for part of the integrand of Eq. (27), we split
the integral into two parts

$$w_n(x,y) = \int_o^{\eta_o} f_n(\eta)\frac{d\eta}{\eta^2} = \bar{w}_n(x,y) + \lim_{\epsilon \to o} \left\{ f_n(0) \int_\epsilon^{\eta_o} \frac{d\eta}{\eta^2} \right\} \qquad (28)$$

From this follows

$$\bar{w}_n(x,y) = \int_o^{\eta_o} \left[f_n(\eta) - f_n(0) \right] \frac{d\eta}{\eta^2} \qquad (28a)$$

The integrals $\bar{w}_n(x,y)$ exist individually and can be evaluated (some
care is required, obviously, near the lower limit).

The { } bracket within the second part on the right of Eq. (28),
inserted into the sum in Eq. (26b), yields

$$\sum_{n=o}^\infty \frac{\pm C_n}{(1-y^2)^{\frac{1}{2}}} \cos n(\phi + \pi/2)_{t.e.} \cdot \int_\epsilon^{\eta_o} \frac{d\eta}{\eta^2}$$

The sum itself is zero because of the Kutta condition.   Hence the
contribution to Eq. (26b) of the complete { } bracket is zero for
all positive ϵ; it follows that it is also zero in the limit $\epsilon \to 0$.
Interpreted physically, this result states the following: the downwash becomes infinite, for each elementary solution, as the t.e. is
approached from behind the wing; however, it remains finite for the
complete solution.   This is because in the complete solution the
pressure on the wing becomes zero at the t.e. (i.e., the complete
solution fulfills the Kutta condition).

Concerning the task of evaluating $w(x,y)$ behind the wing
numerically, our result means that we can use Eq. (26b), with w_n
replaced by $\bar{w}_n$.

Results of the numerical evaluation for the planar wing are
shown in Fig. 7 for a number of spanwise stations y.   Note that in
all cases the $w(x,y)$ curves are not smooth continuations of the
downwash $w = \tan\alpha$ on the wing: all have vertical tangents at the
t.e.   This result, which is in agreement with what one would have
expected from Fig. 4, comes about because each elementary downwash
$\bar{w}_n$ is of order $d^{\frac{1}{2}}$ in the distance $d = x - x_{t.e.}$ behind the wing when
d is small (this is easily verified from Eq. (28a)).

3d.   Interpretation

In section 3a we mentioned two tentative assumptions concerning
the downwash behind a lifting wing of finite span: one, the concept
of a downwash curve which has a vertical tangent at the t.e., Fig. 4;
two, the concept of an initially planar vortex trail[1, 2] .   Figure 7
shows, at least for the planar circular wing, that the actual downwash distribution presents a truly three-dimensional picture: its
curves do have vertical tangents at the t.e., but there is also a
pronounced spanwise variation.   The deviation from the horizontal
line $w = \tan\alpha$ increases rapidly as the wing tip is approached.

It remains to compare the results of the exact theory, Fig. 7,
with the approximate prediction of lifting line analysis, Fig. 6.
This comparison is given in Fig. 8.   Compared are the exact results
for x=2 (circles) with the curve w_{∞} from Fig. 6; this curve, of
course, refers to x=∞.

Near the wing center, the x=2 circles lie slightly below the
w_{∞} curve.   The difference here corresponds closely to the difference between x=2 and x=∞ in the lifting line result for the circular wing, Fig. 4.   We thus see from Fig. 8 that in the wing span
range close to the wing center the lifting line result is acceptable
already at a fairly small distance behind the wing.   This, indeed,
is in accordance with experience concerning downwash distributions
at the usual place of the tailplane of an aircraft.

The new and unexpected result of the comparison is that the
lifting line result ceases to be reliable as the wing tip is approached.   The lifting line curve turns upward in Fig. 8, as we
discussed in connection with Fig. 6, to indicate an infinite upwash at the inside of the wing tip.   The result of exact theory
shows again an upward turn, but here the upward turn is considerably
more pronounced.

The few numerical results which are presented in Figs. 7 and 8
are not sufficient to describe the complete flow field.   Interesting
details remain to be explored (for example, the exact nature of the
upwash singularities inside and outside the wing tip).   We have not
yet accumulated enough numerical results to be able to follow the
rolling up process in detail.   On the other hand, it is clear already
from Figs. 7 and 8 that rolling up starts in a much more positive
and forceful manner than classical theories[1,2] have assumed.

This conclusion, even though it has been illustrated here by
numerical results which have been obtained specifically for the circular wing, does apply to a much more general class of wing planforms.   Namely, the lift distribution in the immediate neighborhood
of any curved wing tip must conform to Eq. (25) (with the radius of
curvature providing the length scale).   Furthermore, a logarithmic

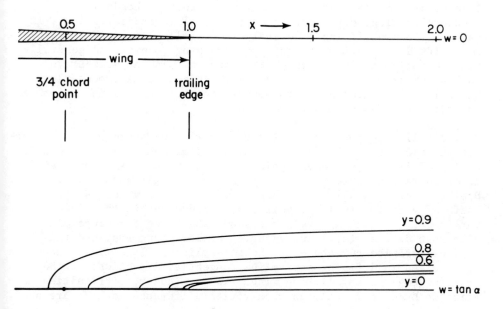

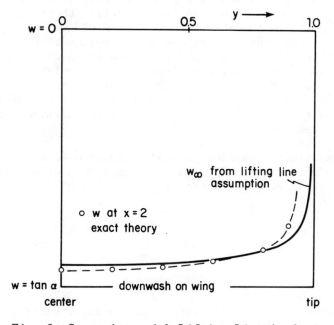

Fig. 7   Downwash Behind Circular Wing

Fig. 8   Comparison with Lifting Line Analysis

term of similar form appears also in the work of Landahl[10] on
rectangular wings with control surfaces. This indicates that, if
one were to use lifting line analysis to calculate w_∞ from the correct lift distribution for any ordinary wing, one should generally
find an infinite upwash inside the tip; in turn, an accelerated
rolling up process could be expected. Indeed, this expectation is
in qualitative agreement with measurements of the rolling up process
on ordinary wings by McCormick et al[11].

Finally, we note in passing that our results are of some interest in relation to a discussion which has arisen recently. Consider
a wing of finite span having also a finite thickness. Upper and
lower surface of this wing form an enclosed angle at the t.e. point.
Does the vortex trail leave this t.e. tangentially to its upper
surface? - or to its lower surface? - or perhaps to the bisec-
tor?[12,13]. Now our analysis has shown that also for wings of finite span do the $w(x,y)$ curves have vertical tangents, and this
implies infinite curvature of the streamlines. It follows that care
is required in defining streamline tangential directions at the
trailing edge of any lifting wing. In particular, difficulties may
arise in numerical analyses when finite difference methods are used.

4. CONCLUSION

Our discovery of a solution of the exactly formulated problem
of a lifting wing of finite span (specifically: the circular wing)
has made it possible to determine the flow field around this wing
without making the usual approximative assumptions. The downwash
field behind the planar circular wing has been calculated. This
field is found to be highly non-planar; in particular, and unexpectedly, it contains a range of large negative downwash (i.e., upwash) next to each wing tip. In consequence, the rolling up process
of the vortex trail starts in a much more forceful manner than was
assumed in classical theory.

This qualitative conclusion can be expected to be valid for
other wing planforms as well, and is in agreement with experimental
results for ordinary wing planforms.

ACKNOWLEDGEMENT

The research reported in this paper was sponsored under Contract
F44620-69-C-0096 by the Aeromechanics Division, Air Force Office of
Scientific Research, Air Force Systems Command.

REFERENCES

1. H. Kaden, Aufwicklung einer unstabilen Unstetigkeitsfläche.
   Ing. Archiv 2 (1931) pp. 140-168.

2. F. L. Westwater, Rolling Up of the Surface of Discontinuity
   Behind an Aerofoil of Finite Span, British R&M 1692 (1935).

3. P. F. Jordan, The Parabolic Wing Tip in Subsonic Flow. To be
   presented at the AIAA 9th Aerospace Sci. Mtg. (New York, Jan.
   1971).

4. W. Kinner, Die kreisförmige Tragfläche auf potentialtheoreti-
   scher Grundlage. Ing. Archiv 8 (1937) pp. 47-80.

5. W. Turowski, Über die kreisförmige Tragfläche mit vorgegebenem
   Abwind. Diplom paper, T. H. Karlsruhe (1957).

6. E. van Spiegel, Theory of the Circular Wing in Steady Incompressible Flow. NLL-TN F. 189 (1957).

7. E. van Spiegel, Boundary Value Problems in Lifting Surface
   Theory. NLL-TR W 1 (1959).

8. P. F. Jordan, Remarks on Applied Lifting Surface Theory.
   Jahrbuch 1967 der WGLR, pp. 192-210.

9. H. B. Helmbold, Der unverwundene Ellipsenflügel als tragende
   Fläche. Jahrbuch 1942 der Dtsch. Luftfahrtforschg. pp. I
   111-113.

10. M. Landahl, Pressure-Loading Functions for Oscillating Wings
    with Control Surfaces. AIAA J. 6 (1968) pp. 345-348.

11. B. W. McCormick, J. L. Tangler & H. E. Sherrieb, Structure of
    Trailing Vortices. AIAA J. Aircraft 5 (1968) pp. 260-267.

12. K. W. Mangler & J. H. B. Smith, Behavior of the Vortex Sheet at
    the Trailing Edge of a Lifting Wing. RAE TR 69049 (1969).

13. Th. E. Labrujere, W. Loeve & J. W. Slooff, An Approximate Method
    for the Calculation of the Pressure Distribution on Wing-Body
    Combinations at Subcritical Speeds. Paper 11 in "Aerodynamic
    Interference", AGARD Conf. Proc. No. 71 (pre-print) (1970).

AN EXPERIMENTAL INVESTIGATION OF TRAILING VORTICES BEHIND A WING WITH A VORTEX DISSIPATOR

Victor R. Corsiglia, Robert A. Jacobsen, and
Norman Chigier*

Ames Research Center, NASA, Moffett Field, Calif. 94035

ABSTRACT

An experimental study was carried out on a rectangular wing in
the NASA-Ames 7- by 10-Foot Wind Tunnel. Flow visualization studies
were made using a tuft grid and smoke. Preliminary studies showed
that the introduction of a bluff body into the trailing vortex downstream of the wing resulted in modification of the vortex. Further
studies showed that a small vertical panel, termed a vortex dissipater, mounted on the wing upper surface near the wing tip also
caused modification of the vortex. Both the smoke and tuft grid
visualization studies indicated that the dissipater caused a significant reduction in the maximum tangential velocities in the trailing
vortex. Additional studies using a hot wire anemometer showed significant reductions in the magnitude of the tangential velocities,
increases in the cross-sectional dimensions of the core of the dissipated vortex and changes in the turbulence structure. Limited
flight tests with a dissipater fitted to a Convair 990 wing tip and
using a Lear jet aircraft as a probe indicated that the rolling
acceleration and the degree of roll control required was less in the
modified vortex than in the unmodified vortex.

INTRODUCTION

Trailing vortices from wing tips have been observed to persist
for distances of 10 miles and more behind large aircraft.[1]** These
vortices present a hazard to following aircraft, especially when the

---

*National Research Council Research Associate.
**Numbers refer to references.

following aircraft are smaller in size.[2] The exact nature of trailing vortices has not been clearly established. Most of the evidence
is based on photographs of the vortices made visible by the introduction of smoke from wing or tower mounted smoke generators or
from engine exhaust. Circumferential velocities of the same order
of magnitude as the flight speed have been observed, and there is
clear evidence of axial velocity components largely in a direction
toward the aircraft. The assumption has been made that the maximum
circumferential velocity decreases with distance downstream, and
McCormick[3] concluded, on the basis of model and flight tests, that
at large distances downstream of the aircraft this decrease is
approximately in inverse proportion to the square root of the distance. He also concluded that the core size of the vortex increases
with distance. This conclusion of McCormick is in conflict with the
more recent observations of pilots[4] flying in the wake of the
Lockheed C-5A and the Boeing 747 that, under conditions of smooth
stable air with no flap deflection, no attenuation in wake intensity
could be detected in the range between 3 and 10 miles behind the
generating aircraft. Further, photographs of smoke trails in the
wake of a Boeing 747[5] show little change in the diameter and concentration of smoke for time intervals between 20 seconds and 80 seconds after the passage of the aircraft.

The aim of the investigations reported in this paper was to
seek a means of dissipating the trailing vortices in order to allow
closer approach of following aircraft. Pilots in chase planes had
observed[4] that, when flaps are deflected in the leading aircraft,
the vortices appeared to be weaker and the highly intense vortex
core was not detectable. From this observation it was concluded
that flaps, spoilers or other modifications of the wing contour can
cause a reduction in wake velocities. An exploratory wind-tunnel
investigation at NASA-Ames Research Center showed that the introduction of objects into the vortex caused changes in the core structure.

A further indication that it is possible to cause changes in
the structure of vortices is the phenomenon of vortex breakdown.
Lambourne and Bryer,[6] in studying vortices over delta wings, showed
that the concentrated core can be dissipated abruptly. The axial
flow in the core of the leading edge vortex was decelerated from a
value several times the free-stream velocity to rest while the whole
core expanded as if it had suddenly encountered a solid obstacle.
The phenomenon is shown clearly by a photograph of a pair of such
vortex breakdowns. Maltby[7] has shown, further, that the position of
vortex breakdown can be moved downstream of the aircraft by rolling
and yawing a slender wing aircraft. It is not clear at this stage
whether there is a connection between the phenomenon of vortex
breakdown and that of dissipation of vortices at large distances
downstream of an aircraft.

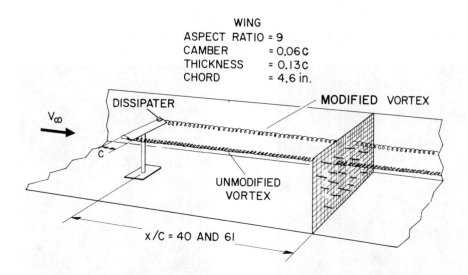

Figure 1.- Experimental setup in the Ames 7- by 10-Foot Wind Tunnel.

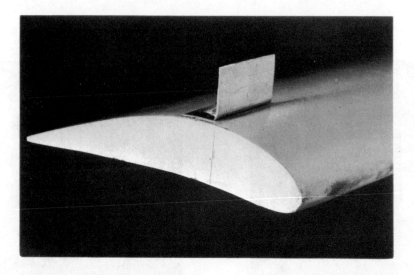

HEIGHT = 12% OF CHORD
SPAN   = 4% OF WING SEMISPAN
POSITION: 0.25 CHORD FROM LEADING EDGE; WING TIP

Figure 2.- Vortex dissipater mounted on wing.

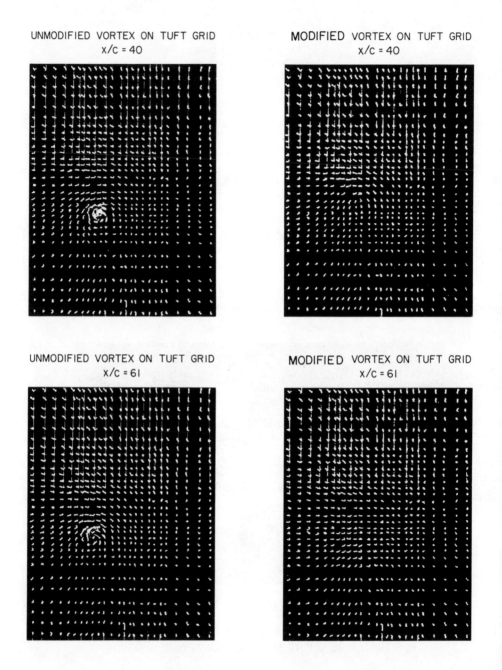

Figure 3.- Unmodified and modified vortex
shown by tuft grid.

On the basis of the observations it appeared possible to cause modification of the vortices downstream of the aircraft without any resulting large effect on aircraft performance. An experimental program was initiated in order to demonstrate that it was, in fact, possible to cause this modification, to find the optimum means of dissipating and to obtain some quantitative measurement of the changes.

This paper reports on a wind-tunnel and flight investigation of the effects of a vortex dissipater plate fitted to the wing tip of an aircraft. The wind-tunnel investigation consisted of flow visualization studies using smoke and a tuft grid as well as measurements with a hot wire anemometer. Flight tests were conducted with a dissipater plate fitted to the NASA-Ames Convair 990 jet transport with a Lear jet model 23 used as a probe aircraft.

WIND-TUNNEL TESTS

Wind-tunnel tests were carried out in the NASA-Ames 7- by 10-Foot Wind Tunnel. A rectangular planform wing of aspect ratio 9, chord (c) 4.6 in., camber 0.06c and thickness 0.13c, was installed in the wind tunnel as shown in figure 1. Wind-tunnel speed was generally maintained at 100 feet per second (30.5 meters per second) but some tests were carried out at higher speeds.

Flow Visualization Studies

A tuft grid was located downstream of the wing at distances of 40 and 61 chord lengths from the wing trailing edge. Smoke visualization (with the tuft grid removed) was effected by introducing smoke through a probe in the region of the vortex core. Initially, small bluff objects were placed in the vortex core at a number of positions downstream of the trailing edge. These were found to produce significant modifications in the vortex as visualized on the tuft grid and in smoke trails. Placing the objects on the wing surface was also found to be effective. By progressively reducing the size of the object and moving its position on the wing surface it was found that a simple plate mounted on the wing surface as shown in figures 1 and 2 was as effective as the larger objects. A dissipater-plate of height = 12% wing chord and span = 4% wing semi-span was fitted at one wing tip, on the top surface, at a distance of 1/4 wing chord from the leading edge (see figure 2). It will be noted that the dissipater plate is placed across the path of the vortex which is generated near the leading edge at the wing tip.

The effect of the dissipater is presented in figure 3 for two axial stations $x/c = 40$ and $x/c = 61$ (x refers to longitudinal

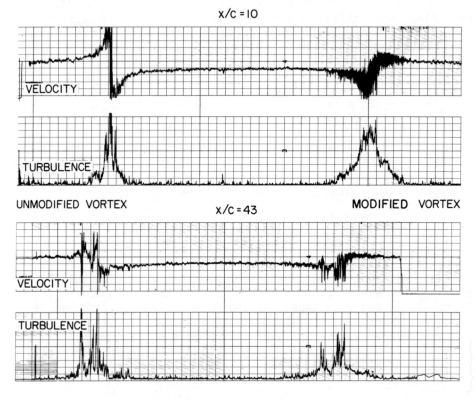

Figure 4.- Hot wire anemometer spanwise traverse
through vortex centers.

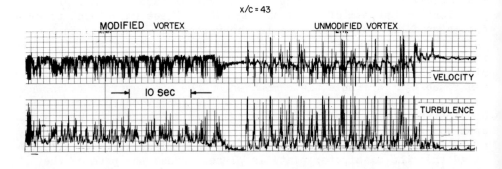

Figure 5.- Velocity and turbulence variation with time: Hot wire
held fixed at vortex center; x/c = 43.

distance downstream of the wing trailing edge). This figure compares the characteristics of the unmodified vortex with those of
the modified vortex. With the dissipater installed the high
intensity swirl associated with the vortex core has disappeared
from the tuft grid at each of the axial stations. However, the
outer regions of the vortex are still detectable in the modified
vortex. Visualization with smoke showed that the unmodified vortex
core was relatively clear of smoke, indicating that the smoke particles were centrifuged out of the vortex core. In the "modified"
vortex the central clear region was not observed. Further, the
cross-sectional area over which smoke was observed and photographed
was greater in the modified vortex than in the unmodified vortex.
Since both the tuft grid and smoke trails had demonstrated a significant effect of the dissipaters, it was decided to make more
quantitative measurements by using a hot wire anemometer.

Hot Wire Anemometer Measurements

Flow in a trailing vortex is three-dimensional so that determination of the velocity field requires measurement of axial, circumferential, and radial velocity components. The radius of the
vortex core was approximately 0.25 in. (0.635 cm) for the 4.6 in.
(11.68 cm) chord wing used in this study. In order to provide minimum disturbance to the vortex, a constant-temperature hot wire
anemometer with wires of diameter 0.00015 in. (0.00381 mm) and
length 0.050 in. (1.27 mm) was used. By varying the orientation of
the wire and measuring the DC and RMS time mean average voltages at
each measuring point for each wire orientation, a set of equations
was obtained from which the velocity components were calculated.

For the particular configuration used in the wind tunnel it was
found that a probe set at 45° to the main stream flow direction was
most sensitive to circumferential velocity changes. The hot wire
probe was fitted to a traversing mechanism and the output from the
anemometer and RMS meters were connected to a pen chart recorder so
that continuous sweeps could be made at various axial stations.
Figure 4 shows typical spanwise traverses made at axial stations
x/c = 10 and x/c = 43 downstream of the wing. (The dissipater had
been fitted to the right-hand wing tip only.) The DC output from
the anemometer represents changes in velocity and the RMS output
represents changes in turbulence level (turbulence scale is magni-
fied). Examination of the anemometer trace from left to right shows
the main stream wind-tunnel velocity followed by the outer "potential" region of the vortex (upwash) through the vortex center with a
high circumferential velocity gradient and change of sign of circumferential velocity, followed by the potential region of the vortex
(downwash). The modified vortex also shows the changes from downwash to upwash but the nature of the modified vortex is clearly

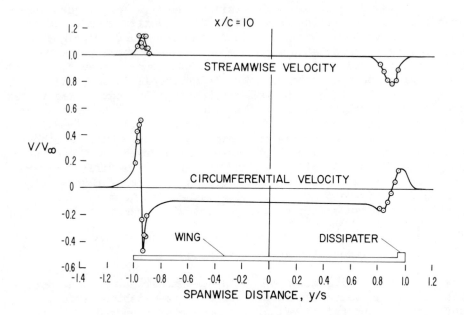

Figure 6.- Spanwise variation of circumferential and streamwise
velocity components.

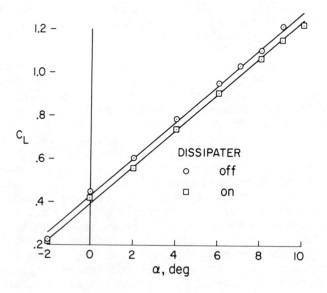

Figure 7.- Effect of dissipater on wing lift.

different from that of the unmodified vortex. Turbulence levels
are seen to reach peaks in the regions of circumferential velocity
maxima. The differences between modified and unmodified vortex
remain at $x/c = 43$.

In order to obtain further evidence of differences in character of the two vortices the probe was set at each vortex center and
continuous recording of voltage variation with time was made as
shown in figure 5. Because of the high turbulence levels within
the core and the movement of the vortex center the instantaneous
positions of circumferential velocity maxima are seen to move with
respect to the probe. Examination of the traces (figure 5) shows
the modified vortex to have lower circumferential velocity maxima,
and lower maximum turbulence levels than those measured in the
unmodified vortex.

For calculation of velocity magnitudes, traverses were made
with hot wires at three different orientations to the main stream
floor, one with the hot wire normal to the flow and the other two
with the wires inclined at 45° to the flow with different orientations. Results of these calculations are shown in figure 6 for a
spanwise traverse through the vortex centers at $x/c = 10$. Due to
the small dimensions of the vortex core and the fact that measurements were made with each wire consecutively, it was not possible to
synchronize measurements made in the vortex core. However, the magnitudes of the velocities shown in figure 6 are considered to be
sufficiently accurate to indicate the differences between the modified and unmodified vortices. Figure 6 shows the reduction in magnitude of the circumferential velocity maxima and increase in radius
of core of the modified vortex when compared to the unmodified
vortex. The streamwise velocity component of the modified vortex
has the form of the velocity defect in the turbulent wake downstream
of a bluff body.

Force Measurements

Changes in vortex structure and modifications to the flow patterns over the wing surface may cause changes in the aerodynamic
characteristics of the wing. The contribution to total lift of surfaces in the vicinity of the wing tips is small and thus modifications such as installing a dissipater plate were not expected to
result in significant alterations to lift. Measurements of lift
made for the wing with the dissipater fitted to one wing are shown
in figure 7 and compared with measurements made with the dissipater
removed. It is seen that with the dissipater installed an increase
in angle of attack of 1/2° is required to maintain the same lift
coefficient as obtained without the dissipater. Drag coefficients
are expected to be increased due to the bluff body effect of the

Figure 8.- NASA-Ames Convair 990 with vortex dissipater
installed on left wing tip.

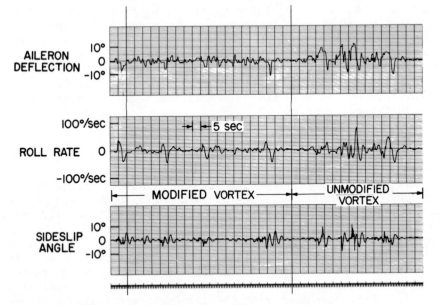

Figure 9.- Aileron deflection, roll rate, and sideslip angle
variation with time in Lear Jet Probe aircraft, 3-mile
separation from Convair 990.

dissipater plate. However, the changes in drag forces were too small to be detected on the wind-tunnel balance.

FLIGHT TESTS

The effectiveness of the dissipater on the near flow field in the wind tunnel needed to be demonstrated in the far flow field of concern in flight operations. Since the region of interest is up to about 10 miles behind a large aircraft, it was decided to carry out flight tests using the NASA-Ames Convair 990 jet transport of gross weight approximately 170,000 lb (77,100 kg). Preliminary wind-tunnel tests of a model of this aircraft with a dissipater plate installed were conducted and hot wire measurements at $x/c = 60$ showed that some modification of the vortex on the aircraft model was achieved. However, a larger dissipater was required to achieve the effectiveness found in the rectangular wing tests discussed previously.

A dissipater similar to that shown in figure 2 except for a 50% greater height was then installed on one wing tip of the aircraft as shown in figure 8. For structural reasons the dissipater was fitted at a chordwise position of 20% chord at the tip. On the basis of the wind-tunnel tests, the dissipater was positioned in a plane perpendicular to the main flow direction. The dissipater was mounted on the left wing tip of the Convair 990 while the right wing was unaltered. High-speed taxi tests indicated that little change in trim was required to overcome the moments introduced by installation of the dissipater.

The Lear jet-probe aircraft was equipped with linear accelerometers and angular rate gyros about three axes. Tests were carried out at an altitude of 12,500 ft (3,810 m) and at an air speed of 185 knots indicated. The Lear jet was flown nearly axially along the vortex. The pilot's visual observation of engine smoke entrained in the vortex was the only indication of aircraft position relative to the vortex. Measurements were made of the aircraft's dynamic response in both the left- and right-hand trailing vortices. Qualitative assessments based on pilot's opinion as to change of response of the aircraft to the vortex were made in the range from 5 to 1.5 miles (8.05 to 2.42 km) behind the Convair 990. Measurements of dynamic response, including aileron position, of the Lear jet were made at a range of 3 miles (4.83 km) while the pilot attempted to remain in the vortex for a period of approximately 1-1/2 minutes.

Motion pictures of tufts on the wing surface and natural condensation trails showed the formation of the vortex over the wing surface of the unmodified wing. This "condensation" vortex was not

visible on the wing with the dissipater, and tufts showed reverse
flow over the wing surface, indicating the presence of a turbulent
bluff body wake.

Visual observation of the engine smoke, which was entrained in
the vortices, indicated that no differences existed between the vortices when viewed from several positions of the probe aircraft.
Several short encounters with each vortex were made by the probe
aircraft and the pilot reported that he could tell no difference
between the modified and the unmodified vortices from the dynamic
response of the aircraft.

In contrast to the pilot's opinion, the recorded data did indicate differences between the modified and the unmodified vortices.
An example of the data recorded during flight is shown in figure 9.
In reducing the dynamic data, the rolling acceleration due to the
vortex was calculated for each encounter. Taking the maximum rolling acceleration on each side as being a measure of the strength of
the vortex, a reduction of about 50% (from 4.4 to 2.4 rad/sec^2) was
obtained in the modified vortex. In addition, the RMS average of
aileron position was calculated and showed a reduction of about 60%
(from 4.8° to 2.0°). A similar flight test was conducted with a
smaller dissipater (height 12% of wing chord, span 4% of wing semi-
span). These results did not show the marked reduction in roll
acceleration and aileron control cited above. However, there was
no attempt to remain in the vortex for a prolonged period as was
done with the larger dissipater.

The flight tests were terminated after only one prolonged
encounter had been made. Since most of the data presented was from
that encounter, it is evident that more tests are required to
resolve the disagreement between pilot observations and recorded
data, and to try correlating probe aircraft dynamic response with
dissipater size and range from the generating aircraft.

DISCUSSION

On the basis of classical aerodynamics it is generally predicted that the total vortex strength is equal to the circulation
around the median section of the wing; and also that the distribution of vortex strength on the wing is directly related to the spanwise distribution of lift. It might be deduced, therefore, that any
attempt to alter the vortex velocities will affect the lift and circulation of the wing. For a given total vortex strength, however,
there can be a variety of circumferential velocity distributions,
and there is, as yet, no means of predicting the radius of the vortex core or the velocity distributions in the vicinity of the core.
The results reported in this paper show that it is possible to

effect a change in the distribution of circumferential velocity without affecting the total vortex strength to any appreciable extent (as judged by the lift measurements).

The persistence of vortices reported over the long distances behind aircraft has not been explained, but it is probable that the centrifugal force field stabilizes the flow and acts against forces of diffusion. The evidence of reduction in vortex velocities under unstable weather conditions and as a result of deflecting flaps or installing a dissipater shows that the vortex flow field is subject to the conditions under which it is generated and the environment in which it exists. Further, since the dimensions of the vortex core are relatively small, the judicious introduction of a distur- bance within the vortex core can be expected to produce local effects which will alter the distribution of the circumferential velocity and, under certain conditions, may destabilize the flow and cause rapid dissipation of the vortex flow field.

On the basis of the tests carried out to date, it is not clear whether the dissipater acts as a destabilizer of the flow field or whether its main effect is to generate a bluff body turbulent wake which causes entrainment of air with low angular momentum into regions of higher angular momentum. Detection of the effects of the dissipater on following aircraft in the flight tests suggests that the disturbing effect of the dissipater in the near field persists in the far flow field.

CONCLUSIONS

Fitting of a dissipater plate to the tip of a wing in a wind tunnel has been shown to result in a reduction in maximum circum- ferential velocities and an increase in vortex core radius.

Installation of a dissipater plate on the wing of a Convair 990 aircraft has shown that the far flow field can be altered in struc- ture. However, its effects on the probe aircraft's response with respect to handling and ride qualities could not be detected by the pilots.

Further research is required to determine the optimum technique for dissipation of trailing vortices. Further measurements in wind tunnels and in flight are required in order to show quantitatively the magnitudes of changes in velocity and vortex core size, caused by a dissipater, in both the near and far flow fields.

REFERENCES

1. Andrews, W. H.: Flight Evaluation of the Wing Vortex Wake
   Generated by Large Jet Transports. Symposium on Aircraft
   Wake Turbulence, Seattle, September, 1970.
2. McGowen, W. A.: Trailing Vortex Hazard. Society of Automotive
   Engineers Business Aircraft Meeting, Wichita, Kansas, April,
   1968, Paper G80220.
3. McCormick, B. W., Tangler, J. L., and Sherrieb, H. E.: Struc-
   ture of Trailing Vortices. Journal of Aircraft, Vol. 5,
   May–June, 1968, pp. 260-267.
4. Drinkwater, F.: Flight Operations, NASA-Ames Research Center,
   Moffett Field, Calif., Flight Test Reports.
5. Crow, S. C., and Murman, E. M.: Trailing-Vortex Experiments at
   Moses Lake, Boeing Scientific Research Laboratories Technical
   Communication 009, February, 1970.
6. Lambourne, N. C., and Bryer, D. W.: The Bursting of Leading
   Edge Vortices. Brit. Aeronautical Research Council R and M
   No. 3282, 1962.
7. Bisgood, P. L., Maltby, R. L., and Dee, F. W.: Some Work on
   the Behaviour of Vortex Wakes at the Royal Aircraft Estab-
   lishment. Symposium on Aircraft Wake Turbulence, Seattle,
   September, 1970.

VORTEX WAKE DEVELOPMENT AND AIRCRAFT DYNAMICS

J. E. Hackett and J. G. Theisen

Lockheed-Georgia Company

ABSTRACT

The written paper describes and augments material delivered semiformally in Session III of the Symposium, as commentary to a movie. The computed aircraft response results were reported at the end of Session VI of the Symposium.

Calculations of the sinuous instability of a trailing pair are extended, using a nonlinear vortex modeling technique, and the effects of initial parameters are examined, particularly the initially assumed wavelength. Realistic-looking tresses obtained were demonstrated using computer graphics.

More complex modes of vortex decay may be demonstrated by flow visualization in water, using the hydrogen bubble technique. Vortex bursting is shown to be possible before and/or after the "wavy" mode. These small-scale measurements are shown to be consistent with flight measurements.

In another series of experiments, smoke injected into the core of the trailing vortex behind a 13-foot-span C-130 wind-tunnel model, mounted in a 23-foot by 16-foot working section, showed remarkable coherence until the adverse pressure gradient in the wind-tunnel diffuser caused vortex bursting. The effects, on vortex burst position in the diffuser, of auxiliary blowing into or near the vortex core are discussed qualitatively.

The results of a computer simulation are discussed concerning the dynamic response and loads induced on an aircraft which enters the trailing vortex of a lead aircraft of large size. In the study,

McCormick's or Owen's semi-empirical theories were judged as
preferable for vortex decay estimates with distance behind the
aircraft. However, the simulation results were only available
using vortex core magnitudes from classical theory. The calculated normal load factors and maximum angular motions (with and
without control) were close to those reported for recent FAA and
NASA tests with a CV-990 or DC-8 flying into the wake of a DC-9
or a jumbo jet.

1. INTRODUCTION

The problem of a small aircraft entering the vortex wake of a
much larger one has received renewed emphasis with the advent of the
over-500,000-pound class of commercial and military aircraft. The
problem falls into two parts: first, regarding the nature of the
vortex trail and its decay, and second, concerning the aerodynamic
and dynamic consequences as they affect a penetrating aircraft.
Some theoretical and experimental aspects of vortex wake decay,
particularly the later stages, will be discussed in Sections 2, 3,
and 5. Section 4 will be devoted to the motions and loads calculated for a typical aircraft after it enters a relatively young
vortex. The results of these studies and of some recent flight
tests underline both the seriousness of the problem and the need to
understand the various possible decay processes. Section 5 includes
a description of some preliminary and qualitative wind-tunnel experiments designed to investigate means of making the trailing vortex
less dangerous.

2. VORTEX DISTORTION AND DECAY

A wide variety of vortex decay modes may be seen in full-scale
vortex trails. Some decay in a fairly continuous manner, remaining
fairly straight. Others display a burst phenomenon which sometimes
leaves a vortex ring wrapped around one vortex in a vertical transverse plane. Still others adopt a sinuous pattern which may develop
into vortex rings in a horizontal plane, each containing part of the
original left- and right-hand vortices. Even in still air, it is
currently only possible to predict which mode will be selected under
fairly extreme conditions; for example the occurrence of vortex
bursting behind a slender delta at high incidence or more continuous
dissipation for low lift at low Reynolds number.

The intermediate cases include modes which are characterized
by preferred wavelengths along the vortex axis. Some of these are
essentially single-vortex phenomena (References 6 and 7) involving
long-wave axisymmetric core distortions. Others rely on the mutual
interaction of the trailing pair (Reference 1) and exhibit sinuous
form. However, quantitative results are frequently lacking.

Some analytical research being done at Lockheed-Georgia involves the equations of motion for fluid particles moving in a steady vortex field which is unaccelerated, except for the centripetal components. It is assumed that the energy sources and potential interactions are present which are necessary to create unstable, antisymmetric motions. Equations having the form given in References 6 and 8 and in the Appendix of Reference 7 may then be used to solve for a characteristic vortex (wavy core) frequency. This method, being based substantially upon the vorticity equation and energy concepts, yields a single natural frequency, which is dependent upon the assumed boundary conditions. This contrasts with that by Crow[1], which is a two-vortex, stability treatment from the outset. However, both methods, when applied to the B-47 case quoted by Crow, produce a most critical wavelength of about 8 times the vortex spacing. In the case of the vorticity transport method, however, there is some indication that, in tests with poorly-controlled boundary conditions (such as in wind tunnel tests where there are adverse pressure gradients in the diffuser), the energy expression can be modified as in Reference 9 to allow reasonably accurate predictions of such wavelengths from scaled model tests.

Calculated Motion of an Initially Sinuous Vortex Pair

Photographs of condensation trails are given in References 1 and 2 for a B-47 at normal cruise speed and altitude and for a Comet at 150 knots at 10,000 feet, respectively. Both trails have initially sinuous deformations which include a range of wavelengths, the B-47 case varying from about 6 to 9 and the Comet from about $4\frac{1}{2}$ to 7 times the rolled-up vortex span.

In his corresponding analysis, Crow[1] shows that both symmetric (Figure 1) and antisymmetric natural modes are possible, at wavelengths which are a function of an integration cutoff radius which leads to a wavelength of 8.5 times the converged vortex spacing for the B-47 case just mentioned. It seemed desirable to extend these treatments beyond the linear range but, because of the associated analytical difficulty, a flow-modeling, finite-element, time-stepping technique was used.

"Point" vortex elements were used, with strengths in ft^3/sec, which concentrate a potential line vortex at a point. Ten were placed on each half-wave, oriented appropriately. Velocities were summed at each of the 10 points, the effects of at least three whole waves of both vortices being included by using imaging methods. Simple stepping was then carried out with vortex vector reorientation and allowance for stretching after each step. Figure 1 shows the initial geometry.

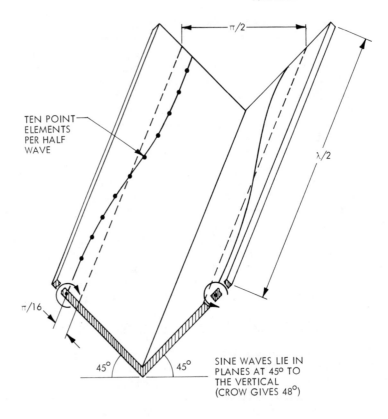

FIGURE 1   INITIAL GEOMETRY
CIRCULATION STRENGTH = 0.2 FT^2/SEC

Some representative results are shown in Figure 2 for three values of initial wavelength. The "ripples" in some examples indicate a need for more elements per half wave in some cases. Despite the simple nature of the calculation method, the shapes are similar to observed contrails. However, there is no program logic in the computation to accommodate the "pinching-off" process, by which complete vortex rings are formed in an old contrail. This lack reduces the flattening tendency of the rings, evident in Figure 3. In a real flow, downward mutual convection at necking-in point of the vortices is instantaneously replaced by upward mutual convection when they break and change partners.

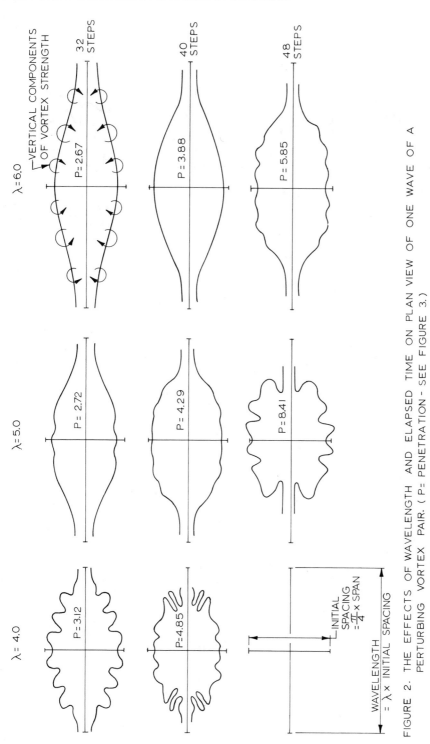

FIGURE 2. THE EFFECTS OF WAVELENGTH AND ELAPSED TIME ON PLAN VIEW OF ONE WAVE OF A PERTURBING VORTEX PAIR. (P= PENETRATION - SEE FIGURE 3.)

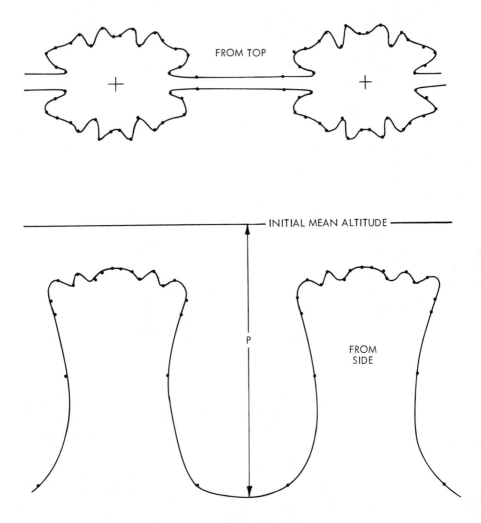

FIGURE 3. VORTEX TRESSING FOR SHORTER WAVELENGTH INITIAL PERTURBATIONS
$\lambda = 5.0$ 48 STEPS

The more rapid progression of shorter waves toward the ring state, evident in Figure 2, can also be found in contrails. This may be attributed to the closer spacing of the vertical components of circulation, sketched on the upper example for $\lambda = 6$.

Although they are seldom discussed for potential flow, both radial and axial velocities occur whenever a change of vortex curvature exists in the presence of a mean flow. An illustration of this, discussed in Reference 4 (Figure 15) and Reference 7, is the flow in the vertical center plane of a horseshoe vortex. Here, a segment of fluid is "captured" as a pair of streamlines from the mainstream spiral into the center of the bound vortex. There is spanwise axial flow away from each side of the center plane. Some regions of a tressed trail can experience a similar effect, leading to axial flows away from the low points. The associated radial inflow plays a part in the "necking" process. The axial flow of low-energy core air towards the more widely spaced parts of the ring provides a potential flow mechanism which might lead to vortex bursting at the crests of the tresses.

3. SMALL-SCALE FLOW VISUALIZATION IN WATER

Experimental Methods

A series of experiments was carried out in a towing tank of 5-foot-square cross-section, using first a unit aspect ratio delta wing of 4-inch span and then later an aspect-ratio-4 unswept rectangular wing of 8-inch span. The hydrogen bubble electrolytic technique was used to visualize the trailing vortex system, with the model edges as electrodes. With sufficiently small bubbles, they did not rise to the surface for a considerable time. Typical forward speeds were about 1 ft/sec.

Though excellent appreciation of the flow is possible by looking at the bubbles directly, adequate black-and-white still photography proved difficult. Figure 4 shows a sample result, taken near the starting position for the delta where more intense trails formed. Motion pictures in color gave good results, however. Many of the measurements quoted later were obtained by analysis of movie records. The use of the electrolytic dye technique is now being considered, which eliminates the possibility of the hydrogen bubbles modifying the flow, particularly under incipient vortex burst conditions. Nevertheless, the bubbles do have the desirable property of not mixing continuously with the fluid. Because of this, and their ability to collect in a vortex core, it has been possible to identify the core again after vortex bursting has occurred. Under certain conditions, early bursting behind the delta wing was followed by an apparent reorganization of the core (at least as depicted by the bubbles) and subsequent looping. Second bursting events were often responsible for the ultimate decay of the

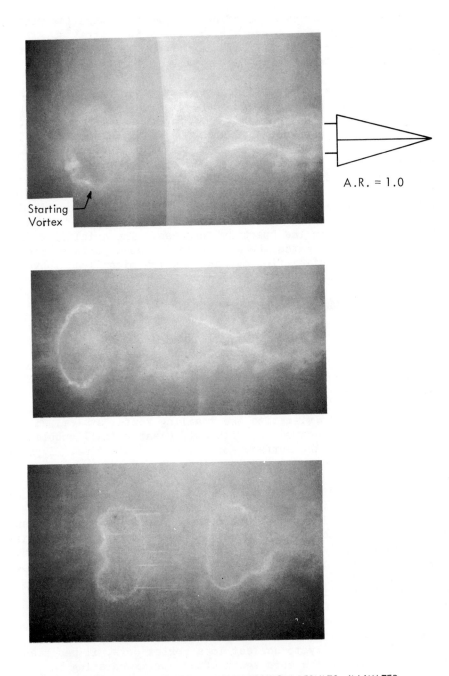

FIGURE 4. SOME TYPICAL VISUALIZATION RESULTS, IN WATER,
USING THE HYDROGEN-BUBBLE TECHNIQUE

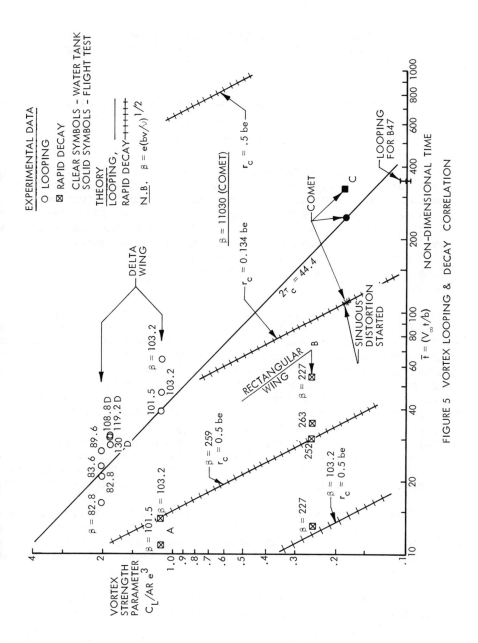

FIGURE 5   VORTEX LOOPING & DECAY   CORRELATION

loop, provided the (stable) ring-state did not become established.

Looping

In Reference 2, a dimensionless time constant, τ_c , is discusse
which can be used to describe major events such as looping. Its
definition is:

$$\tau_c = \Gamma t/b_e^2 \qquad\qquad (1)$$

where b_e is the rolled-up vortex span, Γ is the circulation strength
and t is real time in seconds. An alternative form is:

$$\tau_c = \frac{1}{2}(C_L/AR\,e^3)(Vt/b) \qquad\qquad (2)$$

where C_L, AR and e are respectively lift coefficient, aspect ratio,
and the ratio of rolled-up vortex span to wing span. V is forward
speed and b is wing span. It is convenient to plot experimental
results against axes corresponding to the bracketed parameters, sinc
this allows decay correlations to be attempted on the same plot. On
a double logarithmic scale, lines of constant non-dimensional time,
τ_c , have unit negative slope. We are indebted to C. J. Dixon, who
also carried out the water-tank experiments, for suggesting these
parameters.

The solid circular point in the lower right part of Figure 5
corresponds to the first looping event, seen in Reference 2, for the
Comet. The B-47 data from Reference 1 have also been added, but the
flight parameters are not accurately known. A line of constant dime
sionless time, $2\tau_c = 44.4$, corresponding to the Comet result, passes
among points corresponding to the slender delta results measured in
the water tank. Those marked with a 'D' were taken further from
the starting vortex and should compare best. Others may have looped
early as the result of chain reaction propagating from the starting
vortex. For similar reasons, no looping results will be quoted for
the 8-inch rectangular wing; there was only 22 feet of carriage trav
and it was never possible to be further than 16 wingspans away from
starting or a stopping vortex, which is too few looping wavelengths.

It has already been remarked (Section 2) that the time to loopi
seems to be a function of wavelength, λ , so that any comparison oug
to be made on the basis of the same wavelength loop. This is possib
in general, but no attempt will be made here because of insufficient
data.

Figure 5 shows that the Comet vortex trail and that from the
slender delta share the same non-dimensional time to looping, despit
the widely differing geometries and lift coefficients, differing
vortex generation mechanisms (with consequent doubts about choice of
time scale zero) and two orders of magnitude difference in Reynolds
number. Wider-ranging and more definitive tank testing is needed
before attempts should be made to draw meaningful conclusions.

Viscous Decay

Of the several theoretical and semi-empirical methods which have been proposed over the past few years, that by Owen (Reference 5) appears the most appropriate in the present context, since the range of Reynolds numbers to be covered is large. He derives the following result for the maximum circumferential velocity, U_1, in a free vortex as a function of time:

$$U_1 = 0.065(\Gamma/\nu)^{3/4}(\nu/t)^{1/2} \tag{3}$$

where ν is kinematic viscosity and the other symbols are as before. We may rearrange this to determine a dimensionless time scale which includes the parameters of Equation (2):

$$\bar{t} = Vt/b = 0.001498(C_L/AR_e^3)^{3/2}(bV/\nu)^{1/2}{}_e{}^3(V/U_1)^2 \tag{4}$$

Equation (4) may be further developed to replace the circumferential velocity U_1 by core size r_c corresponding to the beginning of irrotational flow, by using

$$U_1 = \Gamma/2\pi r_c \tag{5}$$

After some algebra, we obtain

$$\bar{t} = Vt/b = 0.001498(4\pi^2)(C_L/AR_e^3)^{-1/2}(bV/\nu)^{1/2}{}_e{}^{-1}\left(\frac{r_c}{b/2}\right)^2 \tag{6}$$

where the last parameter is now the ratio of core radius to semi-span; unit value corresponds to the vortex cores converging at the center plane.

Experimental points characterizing three different viscous decay histories may be found in Figure 5. The group marked 'A' correspond to well-defined vortex burst behind the slender delta in the water tank; looping occurs next at about 40 time units, followed some time later by further decay which propagates around the vortex loops, as suggested in Section 2.

The vortex strength for group 'B,' which corresponds to the rectangular wing, is too low either for well-defined bursting or for looping. Instead, viscous dissipation takes place continuously and, according to Equation (6), the vortex cores probably intermingle

and the vortices tend to destroy themselves as the core of one
becomes strained by the potential field of the other. Careful measurements will be needed to confirm this, however.

At the Reynolds numbers and lift coefficient corresponding to
the Comet results, on the other hand, looping occurs long before
steady dissipation would cause core overlap. It seems probable
that the rapid decay shown in Figure 5 is due directly to looping,
possibly by means of the mechanism suggested in Section 2. Possibly, if the rings were formed in such a way as to avoid bursting,
they would last a very long time - though Reference 5, which per-
tains only to a straight vortex cannot predict this.

If the ideas expressed previously are correct, Reynolds simulation must be attained. Otherwise, the decay events may be completely uncharacteristic. Indeed, urepresentative small models,
tested at higher lift coefficients, may provide a better simulation
of shear stress in the vortex cores at full scale.

4. AIRCRAFT DYNAMICS IN VORTEX WAKES

The motions of a following aircraft obliquely entering the
vortex trail of a large leading aircraft have been calculated using
equations of motion with six degrees of freedom. This computer
program is an outgrowth of previous simulation studies (References
10 and 11) with proved capability to duplicate flight measurements.
Special features include the use of nonlinear aerodynamics and
variable rotary inertias for large-amplitude motions in a fixed-
axis system. Provision is made for pilot control inputs.

The vortex contrail of the lead airplane is assumed to be in a
stable state, except for normal dissipation with distance behind
the aircraft. This simplifies the three-dimensional velocity field
in the wake. The maximum velocities from the classical theory tend
to be smaller, and the radii larger, than those predicted by either
McCormick, et al. (Reference 12) or Owen (Reference 5). The latter
two theories tend to agree in velocity but because of empirical
assumptions, they differ in predicted radii. From a dynamics viewpoint, larger core velocities or radii will produce greater energy
for excitation. Semi-empirical arguments given by McCormick and
Owen are quite convincing that the vortices decay inversely pro-
portionally to the square root of normalized distance behind the
generating aircraft. However, because of simplifying assumptions
made during programming (computer storage limitations) it was
necessary that the vortex velocity be distributed over a larger
percentage of span, more consistent with the radii and velocities
from potential theory. Such wake inputs should yield conservative,
dynamic responses, so the program is to be modified for other vortex fields using a larger computer. The necessary input parameters
for each simulation are summarized in Table I.

TABLE I

Parameter	Lead Aircraft		Following Aircraft
	Case A	Case B	
Weight	760,000 lbs.	190,000 lbs.	175,000 lbs.
Span	220 ft.	110 ft.	____
Height	*38,000 ft.	38,000 ft.	38,000 ft.
Mach/C_L	0.72/.7*	0.72/.7	0.72/.4
Time ahead	5.5 secs.	5.5 secs.	____
Max. vortex velocity	46 ft./sec.	31 ft./sec.	} at encounter
Vortex radius, R_c	30 ft.	20 ft.	
Initially			
Descent angle	5.4°	3.6°	} flight path
Track angle towards outside	4.6°	3.0°	

*Arbitrarily held constant for each case.

For the first case, with the jumbo jet for the lead aircraft, the maximum core velocity is conservatively predicted to be about 46 fps, at a 30-foot radius. The time histories of the simulation results are shown in Figures 6A and 6B. In case A it is assumed that the following aircraft penetrates the right-hand vortex contrail at less than 1 mile behind the lead aircraft, entering the core from above at top dead center. The jet transport approached obliquely with a flight path at approximately 5 degrees or 3 degrees (cases A or B, respectively) in both descent angle and lateral angle, heading to the right. The pitch rate on entry was varied in the simulation until the deepest penetration was achieved. An initial angle of attack of about 3 degrees nose-up and a pitch rate of about 4 degrees per second seemed to result in very severe responses. Such rates appear to be achievable, based on turbulence/upset incidents reported in Reference 11.

The flight simulation for this case with a 760,000-pound lead airplane is graphically depicted in Figure 7 (solid line, uncontrolled airplane). The trailing aircraft enters the solid rotation core initially from the top side at a slight nose-up incidence. At first, the wings are imbedded nearly symmetrically in the vortex rotational field, and the aircraft starts to roll anticlockwise in the first second. Almost simultaneously, the vertical fin becomes subjected to a lateral velocity field to the left, yawing the airplane to the right, to a maximum of about 12° at 1.3 seconds. The normal acceleration passes through zero almost simultaneously. The aircraft moves to the right so that the left wing is mostly in the high-velocity upwash of the vortex right half plane. This initiates a rather severe, clockwise roll rate which persists through most of the remainder of the upset incident. Upwash on the horizontal tail also causes nose-down pitching, so the aircraft descends, sideslips to the right, gains speed, and finally leaves the vortex at about 2 seconds in a severely rolled condition. Subsequently, the natural stability tends to reduce the right yaw, and a tendency for roll recovery at 4 seconds is indicated. However, because of poor roll damping and the lack of restoring control action by the pilot, the roll angle continues to increase past 60°. This is very severe, since the following airplane is a typical jet transport of 175,000 pounds flying at 38,000 feet and at Mach 0.72. This roll angle and the corresponding maximum roll rate of 41 degrees/second are in substantial agreement with preliminary NASA flight-test data (Reference 13) on a DC-9 transport flying through the wake of a C-5A. However, no uncontrolled upsets were reported in those tests.

For case B, the lead airplane was reduced to one-half size, holding $C_L = 0.7$ (Figure 8). According to classical theory, this decreases the hard-core radius to 20 feet, and the maximum tangential velocity reduces to 31 fps. The upset maneuver is very similar in nature, except for some reduction in maximum roll rates and a stronger tendency for recovery which result from the lower induced

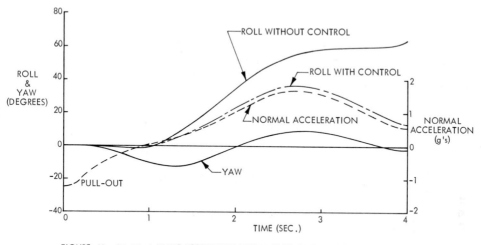

FIGURE 6A. (CASE A) WAKE UPSET SIMULATION FULL-SIZE JUMBO JET IN LEAD

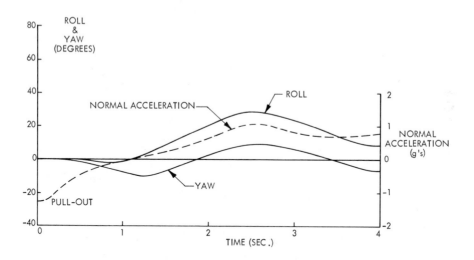

FIGURE 6B. (CASE B) WAKE UPSET SIMULATION HALF-SIZE JUMBO JET IN LEAD

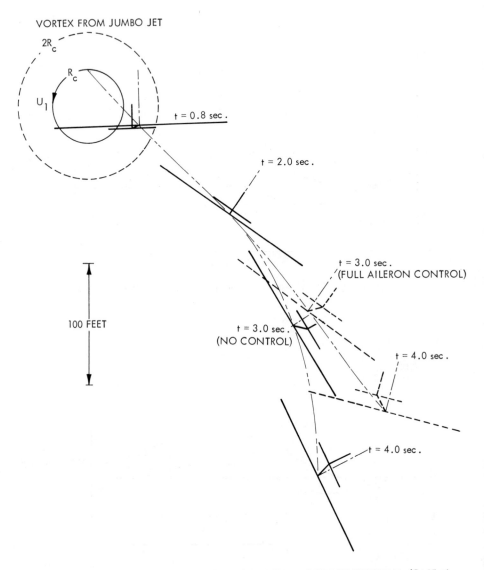

FIGURE 7.   WAKE UPSET SIMULATION - END VIEW   LOOKING UPSTREAM (CASE A)

VORTEX FROM HALF-SIZE LEAD AIRCRAFT

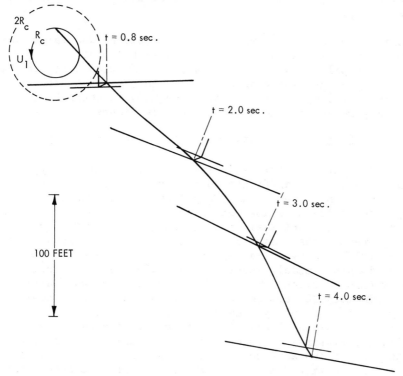

FIGURE 8. WAKE UPSET SIMULATION - END VIEW LOOKING UPSTREAM (CASE B)

velocities. As shown in Figure 6B, the normal load factors are
somewhat lower for this latter case. Also, the maximum roll excur-
sion (about 30°) appears to be slightly less than that from test
data shown in Reference 13 for a DC-9 penetrating the wake of a
CV-990. However, the simulation conditions were not intended to
match these tests. Another run (not shown) was made with the core
radius increased arbitrarily to 80 feet. Under these conditions,
load factors in excess of design limit and roll angles of over 60°
were experienced, demonstrating the importance of the physical size
of the vortex.

Simulation case A (jumbo jet lead airplane) was later repeated
with maximum roll control inputs initiated at about 2.3 seconds,
assuming the pilot reacts to clockwise motion with a reasonable lag
time. This reduced the peak roll angle to 37° and continued to
attenuate roll response to 15° at 4 seconds, as shown by the broken
line in Figure 6A and the dotted traces in Figure 7.

5. QUALITATIVE TESTS ON A LARGE WIND TUNNEL MODEL

The life of a vortex pair generated by (particularly) a large
aircraft can be both difficult to predict and, under certain undefined circumstances, embarrassingly long. Consequently, it is
appropriate to try to endow trailing vortices with a self-destruct
property or to prevent their ever becoming too tightly organized.
Some exploratory tests have been carried out by the Lockheed-Georgia
Experimental Fluid Dynamics Department, using a 13-foot-span STOL
C-130 wind-tunnel model, mounted in a 16.25- by 23.25-foot wind
tunnel working section. Smoke-flow visualization was carried out
at approximately 40 feet/second forward speed, using a heated-tube
smoke generator of the type developed at the University of Washington. Extra lighting illuminated the vortex core as it passed down
the wind tunnel first diffuser, which precipitated early instabilities because of the adverse pressure gradient. It has already been
indicated in Section 2 that it might be possible to make analytical
allowance for this effect.

Although entirely qualitative, the results obtained were
encouragingly similar to events observed in flight. In particular,
single-vortex bursting was observed in the diffuser, including the
intermittent formation of vortex rings around each vortex in a
transverse vertical plane. The vortices from the wing tips had
remarkably coherent cores, which persisted until bursting. Because
of the relative motion, however, looping events could not be identified, and it is not known whether they occurred. However, periodic convergence of the vortex pair could not be seen.

Some limited configuration changes showed that, with flaps in
the takeoff position, the secondary flap vortex was fairly diffuse
and appeared to interfere only slightly with tip vortex decay.

A small compressed-air jet introduced near the wing tip had a notice-able effect for some combinations of blowing angle and position.

6. CONCLUSIONS

Limited numerical and flow-visualization experiments have demonstrated and, to a small extent, explained a number of the many possible modes for the decay of the trailing vortex pair behind an aircraft. The looping phenomenon has been studied in a flow-modeling computer study that predicted more rapid decay at shorter wavelengths, which was borne out by reexamination of flight photographs. Several modes of decay, including steady dissipation, looping, and bursting, were observed in a water tank using the hydrogen-bubble technique, sometimes with multiple events within one time history. Comparison with full-scale events showed looping at the same value of an appro-priately defined value of non-dimensional time, despite vast differ-ences in model size, lift coefficient, and Reynolds number. However, wide-ranging and more definitive tests are needed before meaningful conclusions may be drawn. The importance of scale, however, has been made evident. At too-low Reynolds numbers, decay events are possible which are totally uncharacteristic of the corresponding full-scale conditions. Tests in a large wind tunnel, on the other hand, at least produced realistic-looking burst events, although some tentative theoretical means for correcting for diffuser effect are not yet sufficiently developed.

A number of six-degree-of-freedom dynamic response calculations have been made for a limited class of vortex encounters, in which the following aircraft descended so as to intersect an anticlockwise vortex obliquely, from above, at a small angle. Trials with geo-metrically similar large and half-size lead aircraft confirmed the importance of the physical size of the vortex, as well as the maximum velocity present. Another significant point which may be made from these simulations concerns probable pilot reactions. Since the initial roll motion was counterclockwise in both cases, there will be a tendency to counteract with aileron control in an opposite sense, thus aggravating the violent clockwise roll induced by the vortex during the subsequent motion. However, there are other actions pos-sible by the pilot, particularly rudder and elevator motion, which could be favorable. The roll rates in the first case exceed the roll-control capability of most commercial jet transports, so even correct actions would have limited recovery capability. In a trial simula-tion run, maximum aileron input reduced the maximum roll angle from 60° to 37°.

Emphasis should be made on the necessity for including non-linear/stall aerodynamics in such violent maneuver simulations. Sequential penetration of the various surfaces is essential, as well as profile drag effects (cf. Reference 10). Many other classes of encounter need to be investigated, even with straight vortices.

Others, in which the following aircraft experiences significant velocity changes at aircraft natural flight or at structural frequencies, may be particularly dangerous. Vortex looping wavelengths are of an appropriate magnitude for excitation in this regard.

7. ACKNOWLEDGMENTS

The authors wish to express sincere thanks to C. J. Dixon for carrying out the water tank flow visualization and data analysis, and to C. V. Pierce and G. F. Hart for carrying out the computations for Sections 2 and 4.

REFERENCES

1. Crow, S. C., "Stability Theory for a Pair of Trailing Vortices," B.S.R.L. Document D1-82-0918.

2. Bisgood, P. L., Maltby, R. L., and Dee, F. W., "Some Work on the Behavior of Vortex Wakes at the Royal Aircraft Establishment," Paper presented in Session IV of the 1970 Aircraft Wake-Turbulence Symposium.

3. Parks, P. C., "A New Look at Vortex Dynamics," Paper presented in Session V of the 1970 Aircraft Wake Turbulence Symposium.

4. Hackett, J. E., and Evans, M. R., "Vortex Wakes behind High Lift Wings," AIAA Paper 69-740, 1969.

5. Owen, P. R., "The Decay of a Turbulent Trailing Vortex," The Aeronautical Quarterly, February 1970, pp. 69-78.

6. Scruggs, R. M., and Marris, A. W., "Axial Periodicity in Vortices," Lockheed-Georgia Research Memorandum ER-9098, 1968.

7. Theisen, J. G., "Vortex Periodicity in Wakes," AIAA Paper No. 67-34, 1967.

8. Scorer, R. S., "Local Instability in Curved Flow," Journal of Fluid Mechanics, Vol. 25, Part 3, pp. 557-576.

9. Piercy, N. A. V., Aerodynamics, English Universities Press, Ltd., London, England, 2nd Ed., 1947.

10. Theisen, J. G., "V/STOL Stability and Control in Turbulence," AMS/AIAA Paper at National Aerospace Meteorology Conference, New Orleans, Louisiana, May 1968.

11. Theisen, J. G., and Haas, J., "Turbulence Upset and Other
     Studies on Jet Transport Aircraft," Journal of Aircraft, July-
     Aug. 1969.

12. McCormick, B. W., Tangler, J. L., and Sherrieb, H. E., "Struc-
     ture of Trailing Vortices," Journal of Aircraft, May-June
     1968.

13. Boswinkle, R. W., "Vortex Wake Turbulence," Paper presented at
     the Flight Safety Foundation 23rd International Air Safety
     Seminar, Washington, D. C., October 1970.

MEASUREMENTS OF BOEING 747, LOCKHEED C5A AND OTHER AIRCRAFT VORTEX

WAKE CHARACTERISTICS BY TOWER FLY-BY TECHNIQUE

Leo J. Garodz

Federal Aviation Administration

National Aviation Facilities Experimental Center

ABSTRACT

Flight tests have been conducted by the Federal Aviation
Administration (FAA) at both the Environmental Science Services
Administration/Atomic Energy Commission (ESSA/AEC) facility, Idaho
Falls, Idaho, and the National Aviation Facilities Experimental
Center (NAFEC), Atlantic City, New Jersey, during the period
18 February 1970 through 3 August 1970 to gather quantitative data
on aircraft vortex wake characteristics using the tower fly-by technique. Aircraft tested included the Boeing 747, 707-300 and 727-100,
the Douglas DC-8-63F, DC-8-33 and DC-9-10, the Lockheed C5A, the
Convair 880, and the Learjet 24. A 200-foot and a 100-foot tower
was used at the ESSA and NAFEC test sites, respectively. Vortex
flow visualization for vortex characteristics and movement was provided by colored smoke grenades mounted on the towers and by
injecting CORVUS-type smoke oil into the outboard jet engine exhausts
of certain aircraft with wing mounted engines.

Vortex flow velocities were obtained using hot-film/hot-wire
sensors.

Measured tangential velocities were approximately double those
velocities predicted by certain theory (which assumes an elliptical
lift distribution).

Distinct vortex characteristics peculiar to certain model aircraft and configurations per aircraft model were noted: (1) T - tail
aircraft with engines mounted on the fuselage, i.e., the B-727 and
DC-9 were observed to produce much higher tangential velocities, on
the order of 175 - 200 feet/second, than aircraft with engines mounted

on the wing, e.g., the B-707 and B-747 which were on the order of
130 feet/second; (2) For "clean" configuration and small flap
deflections, the vortex systems of all the aircraft were observed
to be of a tubular form, relatively small in diameter, very clearly
structured and very persistent. For greater flap deflections,
vortices of tubular form were much less evident. With the exception
of the B-727 and DC-9, when maximum flap deflections were employed
this characteristic was not observed, the vortex flow field appearing much larger in diameter, i.e., field of influence; (3) From both
recorded data and visual flow observations of the tubular-type
vortex system when highest vortex tangential velocities were
recorded, the core diameters as outlined by the tower smoke appeared
to be small: from approximately 5 - 6 feet for the larger B-747 and
C5A aircraft to about 1 - 3 feet for the CV-880, B-727, and DC-9
type aircraft.

Particular attention was given vortex axial flow phenomena by
observing and photographically recording entrainment of the colored
smoke. When tubular-type vortex systems passed the tower, smoke
was clearly entrained and spontaneously moved within the vortex
system along its axis in one or both directions, i.e., up and/or
down flight path. A vortex has a low-pressure area within its
cylindrical wall analogous to a tornado or waterspout. Once the
wall is penetrated, as by the instrumented tower upon vortex passage,
there is an immediate injection of relatively higher pressure
ambient air which carries the colored smoke in one or both axial
directions in the attempted pressure equalization process.

It is possible that the tower technique may cause premature
vortex instability onset when the vortex tube passes through the
tower with subsequent valid data acquisition on that particular
vortex made questionable.

This is an interim report on the data processed and analyzed
to date.

INTRODUCTION

In January 1970, a directive was issued by the FAA restricting
the airspace behind the Boeing 747 and Lockheed C5A aircraft within
60 degrees either side and 2000 feet below their flight paths to a
distance of 10 nautical miles. These procedures were based on the
potential, but unknown in extent at the time, vortex hazard
associated with operating other aircraft behind these large aircraft.
These procedures were interim and were to be modified, as appropriate, when additional data were to be made available from full-
scale aircraft flight tests underway and planned. In view of the
increased airspace requirements generated by this restriction and
because of the interest expressed by concerned aviation officials,
the pace of the Wake Turbulence Program to measure aircraft vortex

wake turbulence characteristics and their effects on penetrating aircraft was significantly accelerated by FAA priority action to three aviation agencies in February 1970.

The overall flight test program was divided into three geographical areas and phases, all of which were conducted concurrently. The first involved jointly the Western Region of the FAA, NASA, and the Air Force. The NASA Lockheed 990 and Learjet and the FAA Cessna 210D were specially instrumented for in-flight vortex penetrations. In addition, an instrumented FAA DC-9 was also used for vortex penetrations. Concurrent arrangements were made with the U.S. Air Force for use of the Edwards, California, flight test range and for test flights of a Lockheed C5A aircraft as vortex generator. The results of this phase are covered in Reference 1.

The second phase was initiated by the Boeing Company and was conducted in the Seattle, Washington, area. In these tests, a Boeing 737 and North American F86 fighter aircraft penetrated the trailing vortices of a B-747 and B-707-300 aircraft. In addition, the instrumented CV-990 proceeded from Edwards to Seattle and it also was used in penetrating the vortices behind the B-747 and B-707 aircraft. The results of the second phase are covered in Reference 2.

The third phase of the Wake Turbulence Program was conducted by NAFEC.

The objective of the third phase of this Wake Turbulence Program was for NAFEC to conduct flight tests in both the Southeastern Idaho and Atlantic City areas to gather quantitative data on the vortex wake turbulence characteristics of large, medium, and small jet transport aircraft by low-altitude tower fly-by techniques. These aircraft were the Boeing 747, 707-320C, and 727-100, the Douglas DC-8-63F, DC-8-33, and DC-9-10, the Lockheed C5A, the Learjet 24 and the Convair 880.

The third phase in turn was divided into three tasks:

1. Measurements of the vortex wake characteristics of the jumbo jet Boeing 747, Lockheed C5A and other aircraft. These flight tests and data processing were performed at the Air Resources Field Research Office (ARFRO) of ESSA, located at the National Reactor Test Station (NRTS) of the AEC, Idaho Falls, Idaho.

2. Investigation of jet aircraft vortex systems descending into ground effect and those generated within ground effect. These flight tests were performed at NAFEC, Atlantic City, New Jersey. However, the data was processed by ESSA and Idaho Nuclear Corporation Computer Complex at Idaho Falls.

3. Investigation of the relatively long time-history

TABLE I.   PERTINENT PHYSICAL AND PERFORMANCE CHARACTERISTICS OF TEST AIRCRAFT

	BOEING 747	BOEING 707–320C	BOEING 727–100	DOUGLAS DC–8–63F	DOUGLAS DC–8–33	DOUGLAS DC–9–10	LOCKHEED C–5A	LEARJET MOD 24	CONVAIR 880
WING SPAN (FEET)	195.67	145.75	108.00	148.42	142.42	89.30	222.71	35.58	120.00
WING AREA (FEET^2)	5500	2892	1650	2927	2884	934	6200	232	2000
ASPECT RATIO	6.95	7.36	7.67	7.52	7.03	7.40	7.20	5.02	7.00
MEAN AERODYNAMIC CHORD (in)	327.8	272.3	180.0	272.8	275.9	141.5	370.5	84.5	2268
TAKEOFF WEIGHT (max) (lbs)	710,000	336,000	160,000	355,000	315,000	90,700	728,000	13,000	193,000
TAKEOFF SPEED (kts) FLAP SETTING	170 10°	170 14°	138 25°	163 23°	162 25°	144 20°	140 10°	132 20°	165 22°
LANDING WEIGHT (max) (lbs)	564,000	247,000	142,500	275,000	207,000	81,700	635,850	11,880	155,000
LANDING SPEED (kts) FLAP SETTING	142 40°	137 50°	133 40°	150 50°	133 50°	134 50°	131 41°	127 40°	140 55°

vortex characteristics of the Convair 880 jet aircraft in terminal area-type operations. These flight tests and data processing were performed at NAFEC. The data acquired during the third phase are listed in References 3 - 5.

This paper covers this third phase and provides some of the highlights of the flight test program and test results based on the data processed and analyzed up to the time of this writing.

TEST AIRPLANES

Nine different model jet transport-type aircraft were used in this phase of this test program. These aircraft were the Boeing 747, 707-320C and 727-100, the Douglas DC-8-63F, DC-8-33, and DC-9-10, the Lockheed C5A, the Learjet 24, and the Convair 880. For these aircraft, geometric, gross weight and performance data, as applicable to vortex theoretical calculations, are listed in Table 1.

For task one (at Idaho Falls), flight tests included all of the above aircraft except the CV-880. For task two, the test aircraft were the B-727-100, B-707-300, the DC-9-10, and the CV-880. For task three, only the CV-880 was used.

With the exception of the 2.75-degree landing approach glide path operations, it was planned to gather vortex data in level flight for three different aircraft configurations: (1) takeoff (normally partial flaps extended, landing gear down); (2) holding or clean (normally no flaps, gear up); and (3) landing (normally full flaps, landing gear down).

It was the intent to have the test aircraft arrive at the test site at maximum permissible gross take-off weight, minus fuel consumed enroute from points of departure to the test site, so as to insure generating the highest circulation and vortex tangential velocities.

TEST PROCEDURES

The tower fly-by technique shown in Figure 1 was used for this third phase. Although Figure 1 depicts the operation for the CV-880 vortex measurements conducted at NAFEC (Task 3 cited above), this scheme is generally applicable to all three flight test projects conducted by NAFEC, the one at Idaho Falls and two at Atlantic City.

The technique consists of flying the test aircraft perpendicular to the ambient surface wind, as measured at the top of the tower, at an appropriate distance vertically and upwind laterally from the tower under low-ambient wind conditions. Proper positioning permits the aircraft vortex system to drift laterally into the instrumented tower. By manipulating vertical and lateral positioning of the test

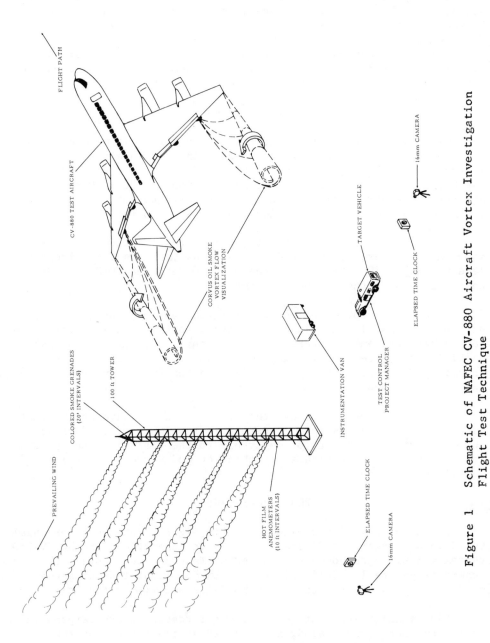

Figure 1   Schematic of NAFEC CV-880 Aircraft Vortex Investigation
Flight Test Technique

aircraft, the "age" of the vortex system can be varied. Miscalculating the wind or mispositioning the airplane can easily result in unsuitable vortex movements, i.e., the system could float over the tower or settle prematurely into ground effect.

It was important to calculate the vortex settling rate before each test run. However, because classical theory on vortex settling rates is incomplete, particularly for vortices generated by various aircraft configurations very close to the ground, and because of atmospheric stability and buoyancy effects on vertical motions of vortices, a combination of theory, observation of previous fly-bys, and professional judgement was used to predict vortex-tower intercept and consequent track and altitude selection.

Task two flight testing differed from the other two tasks in that the majority of the flight paths were on a 2.75-degree glide slope to a runway in aircraft approach or landing configurations. The test aircraft were flown abreast of the tower at 20, 40, 60, 80, and 100-foot altitudes above the ground. Very few vortices missed the tower. However, this operation was "boxed in" in that neither the tower nor, obviously, the runway could be moved thereby necessitating testing only on days when the ambient wind direction was along a line extending perpendicularly from the runway heading to the tower. Glide slope information was provided to the pilots by either VASI (Visual Approach System Indicator) or by TALAR[1] Landing Approach System which is a portable microwave ILS.

Marking of B-747, B-707-320C, and CV-880 aircraft vortices improved the predictability of vortex-tower intercept and aided in timely tower smoke activation. This marking was accomplished by injecting CORVUS-type smoke oil into the exhaust systems of the outboard engines of the three aircraft with a resultant dense, white smoke as noted in Figure 2. The smoke became entrained in and adequately marked the rolled up vortex system, particularly its vertical and lateral movement. The C5A engine exhaust smoke alone aided in visualizing vortex flow from that aircraft. For the remaining aircraft, vortices were not visible until tower smoke was entrained.

The majority of the flight tests were conducted during the early morning hours immediately following sunrise. This period was found to be most conducive to providing the type of atmospheric conditions required for this type of flight testing; namely, low, steady, ambient winds of approximately 3 to 5 knots and a stable atmosphere with little, if any, thermal activity and which conditions are generally accepted as being most conducive to providing hazardous vortex conditions in terminal area type operations.

[1]Product of Singer-General Precision, Inc., Pleasantville, New York.

Figure 2   Boeing 747 Vortex Marking By Injection of CORVUS-Type
            Smoke Oil Into Exhaust Systems of Outboard Jet Engines

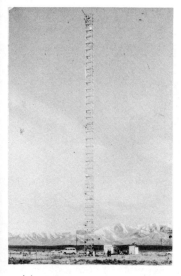

(a) TOWER – FULL VIEW

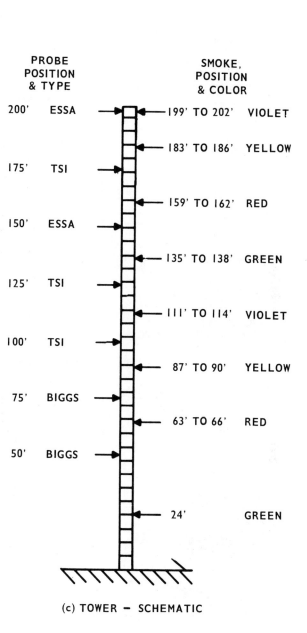

PROBE POSITION & TYPE		SMOKE, POSITION & COLOR	
200'	ESSA	199' TO 202'	VIOLET
		183' TO 186'	YELLOW
175'	TSI		
		159' TO 162'	RED
150'	ESSA		
		135' TO 138'	GREEN
125'	TSI		
		111' TO 114'	VIOLET
100'	TSI		
		87' TO 90'	YELLOW
75'	BIGGS		
		63' TO 66'	RED
50'	BIGGS		
		24'	GREEN

(c) TOWER – SCHEMATIC

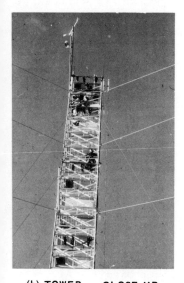

(b) TOWER – CLOSE UP

Figure 3   Instrumented ESSA Tower Used for Aircraft Vortex Wake Turbulence Measurements

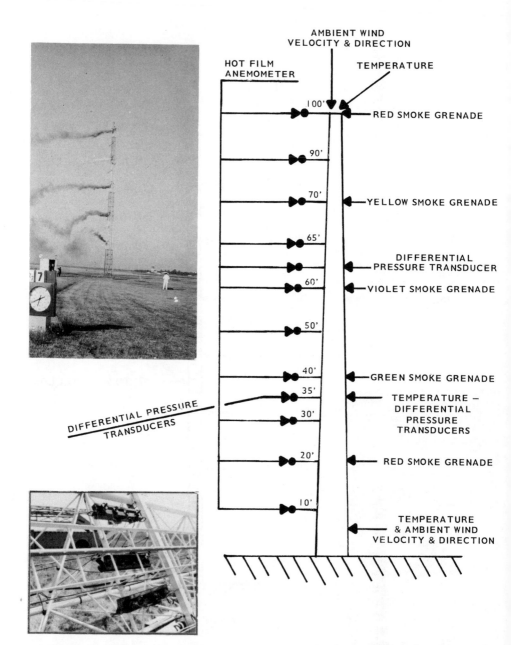

Figure 4   Schematic of Instrumented NAFEC Tower Used for Aircraft
Vortex Wake Turbulence Measurements

The flight tests required that the aircraft be in stabilized flight when passing abreast of the tower. All flights past the tower were controlled by the "Range" Safety Officer who was also the Project Manager. His position was in the vicinity of the tower at a ground-air radio command control vehicle. This vehicle, which had a rotating beacon installed on top, and/or strobe lights were used to facilitate final aircraft alignment to the test track.

At NAFEC, phototheodolites were used to derive accurate space position information on the test aircraft when they were abreast of the tower.

At Idaho Falls, tape measure was used to determine aircraft lateral position from the tower and the pilot's radar altimeter reading was recorded and used for height above the ground determination when abreast of the tower.

TEST TOWERS

ESSA, Idaho Falls, Idaho: The tower used for these tests was the ESSA Grid III Research Tower. The tower is constructed of aluminum, is 200 feet high with its base 4915 feet above sea level, Figure 3. The tower is a rectangular upright scaffold with a 4- by 6-foot cross-section especially designed for meteorological research and presents extremely low blockage to air currents from any direction. Accordingly, this causes very little interference with airflow measurement, as can be seen in Figure 3. Because of the relatively small cross-section to height geometry, guy wires are required at various level to insure structural rigidity. The test site is very level, extending approximately 10 miles in all directions from the tower used for these tests, thereby minimizing possible distortion of the trailing vortex flow by strong vertical motions of air due to adverse terrain features.

NAFEC, Atlantic City, New Jersey: The tower used for these tests was 100 feet high with its base 76 feet above sea level, Figures 4 and 5. It is an equilateral triangle tower tapering from 5 feet on a side at its base to about 2 feet on a side at the top. This tower is constructed of steel, of cantilever design and requires no guy wires. It was erected 125 feet from the edge (200 feet from the center) of Runway 4/22 which is 150 feet wide. The adjacent area was relatively flat composed mostly of taxiways and mowed grass.

INSTRUMENTATION

Aircraft: The test aircraft were not specially instrumented for the tower fly-by testing nor was there any need for them to be. Pilot knee-pad data on time abreast of the tower, aircraft configuration, weight, airspeed, radar altitude above the ground, pressure

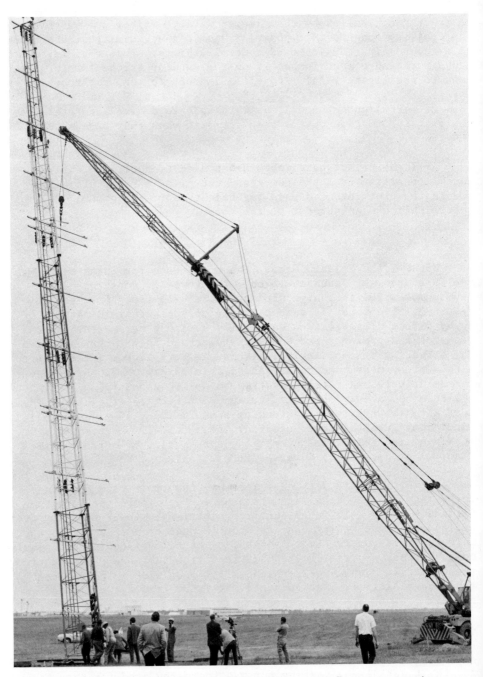

Figure 5   Photographs of NAFEC Tower and Instrumentation

altitude, track, lateral distance from tower, and engine performance
was found to be sufficient for subsequent data correlation and
analysis. When available, phototheodolite data were used in lieu
of pilot recorded data for aircraft altitude and position
information abreast of the tower.

Towers: Both the 200-foot and 100-foot towers were instrumented by ESSA and NAFEC personnel with hot-wire and hot-film
anemometers to measure aircraft vortex airflows. Colored smoke
dispensers were also installed. The smoke emitted from the dispensers entrained in the passing vortex system producing a visual
indication of vortex movement and structure as shown in Figure 6.
Airflow sensor and smoke grenade installation levels are shown in
Figure 3 for the ESSA test site and Figure 4 for the NAFEC tower.

The airflow sensors on the ESSA tower were both hot-film
and hot-wire anemometers and were capable of recording three-
dimensional vortex data, i.e., tangential, radial, and axial
velocities. The NAFEC tower sensors only had a capability of
recording scalar airflow components perpendicular to the axis of
the hot-film element, i.e., in this case, when properly aligned,
they would measure vortex tangential velocities. For both towers,
more sensors spaced at closer intervals vertically on the tower were
required but were not readily available. Instrumentation for
measuring ambient atmospheric conditions existing at the time of
flight testing was also installed along the vertical span of the
tower.

A typical instrumentation arrangement for the tower fly-
by technique is as follows, in which the NAFEC tower is covered.
Fast response hot-film sensors for measuring the vortex-induced
airflows were mounted on 10-foot horizontal sliding booms. These
booms were bolted to the tower at the 100-foot height level and at
successive 10-foot lower levels and could be telescoped parallel to
Runway 4/22. The sensors were TSI[2] quartz-coated, hot platinum
film elements operated at a constant temperature of 400°F. These
sensors were mounted at the outer end of the telescoping booms in a
clamp which could be rotated through a 180° horizontal sector. The
hot-film sensors were a single cylindrical rod about 1/2 inch long.
The longitudinal axes of the rods were aligned parallel to the
planned aircraft flight path and horizontal so as to measure vortex
tangential velocities.

Atmospheric measurements were made with slow response
instrumentation to provide a first indication of the state of the
atmosphere and aircraft positioning input. Temperatures were
measured on the tower at 10, 30, and 100 feet above the height of
the runway. The temperature was sensed by Honeywell nickel-A-type

[2]A product of Thermo-Systems, Inc., St. Paul, Minnesota.

resistance bulbs within Climet aspirated shields. Wind direction
and speed were measured by a Bendix-Friez aerovane located at the
top of the tower.

Photography: This is part of the data acquisition system and
is considered a mandatory requirement for this aircraft vortex flow
measurement technique. Photography combined with vortex flow
visualization provided a means of determining if and when either or
both wing vortex systems intercept the tower, tower intercept level
for correlation with sensor data, and vortex characteristics
analysis. Bell and Howell 16mm motion picture cameras with 400-foot
magazines and synchronous motors were used. Fifteen millimeter
focal length lenses were selected. Twenty-four frames per second
rate was used. Kodachrome II Film was selected to provide the
maximum color saturation of the four colors of smoke.

Time: Central time was recorded automatically on the analog
magnetic tape data and phototheodolite data. For data correlation
with the photographic coverage, an event marker switch was activated
when the test aircraft was abreast of the tower. This was recorded
on the magnetic tape and concurrently ignited a flash bulb in the
field of view of the movie camera for recording on film.

DATA PROCESSING

Automatic digital computer data processing was mandatory due
to the sheer volume of recorded data. All of the tower data were
recorded on magnetic tape in analog form and subsequently digitized
for automatic data processing, through special software programs
written for this phase, on either the IBM 360-75 or 7090 Computers.
The data processed at Idaho Falls was digitized at a rate of one
data point per 104 milliseconds wherein the extreme values on the
vortices were obtained from analog records. The NAFEC processed
vortex data were digitized at 1000 points per second. Vortex core
diameters were obtained from expanded analog data for the Idaho
Falls processed data and from expanded digital data for the NAFEC
processed data. Various data filering and smoothing techniques
were employed at both data processing complexes.

RESULTS

A total of 386 aircraft tower fly-bys were flown in support of
this phase as follows:

            Task 1 (Idaho Falls) - 104
            Task 2 (NAFEC)   - 153
            Task 3 (NAFEC)   - 129

Of this total, about 80 percent of the passes provided tower data
of sufficient significance to warrant further processing and
analyses.

Figure 6   Photograph of CV-880 Tubular Vortex Structure

Figure 7   Photograph of CV-880 Tubular Vortex Structure

TABLE II - PRELIMINARY RESULTS/CONCLUSIONS

<u>Aircraft Models Tested:</u> Boeing B-747, B-707, B-727; Douglas DC-8-63, DC-8-33, DC-9;
Lockheed C5A; Learjet-24; Convair-880
(NOTE: No data acquired on Learjet)

I. AIRCRAFT VORTEX CHARACTERISTICS

    A. <u>Specific:</u>

Aircraft Model	Configuration	VORTEX Characteristics

All except B-727 and DC-9	"Dirty", Full-Flap, Gear down (Landing)	Non-tubular Flow
		"Cores": Not noticeable as such
		Relatively large field of influence
		   (1/3 to 1/2 Wing Span)
		$V\theta$: 60 to 100 FPS
		Breaks up rapidly from "ordered" flow
		   to turbulent type flow:
		   (up to 1 1/2 minutes age)

All Aircraft	Partial-Flap, Gear up/down (T.O. or holding)	Semi-Tubular
		"Cores": Random (Diameter: .04 to
		   .06 Wing Span)
		Medium Field of Influence
		   (1/3 Wing Span)
		$V\theta$: Approx. 140 FPS
		More persistent than dirty configura-
		   tion (up to 1 1/2 - 2 minutes age)

	"Clean", No Flaps (Holding, Enroute)	Tubular
		"Cores": Consistently noticeable
		   (Diameter: .02 to .04 Wing Span)
		Small to Medium Field of influence
		   (1/6 to 1/3 Wing Span)
		$V\theta$: Approx. 140 FPS
		Very persistent: (up to 2 minutes of age)
		Artillery Shell Whine - C5A & B-747

B-727, DC-9	"Dirty", Full Flap	Tubular
		"Cores": Consistently Noticeable
		   (1 - 2 Feet Diameter)
		Medium Field of Influence
		   (1/3 Wing Span)
		$V\theta$: Approx. 190 - 200 FPS
		Very persistent: (up to 2 minutes of age)
		Artillery Shell Whine - B-727

B.   General

    1.   Vortices generated by aircraft in T.O. or landing configuration, that is with some flap deflection, tended to become unstable and degenerate to discontinuous, segmented type vortices or a random type motion in about 90 seconds.

    2.   A pulsating axial flow appears to signal the onset of vortex core disintegration.

    3.   When recorded, no significant vortex velocities were noticed for vortex ages >2 minutes, regardless of whether generated in or out of ground effect.

II.   TEST TECHNIQUES

A.   Test towers should be about 250 - 500 feet high except for vortex investigations in ground effect.

B.   More airflow sensors (3-dimensional) and closer vertical spacing (2 feet or less) on towers are required.

C.   Only relatively short time-history vortex data, <2 minutes, could efficiently be gathered with the tower fly-by technique with the existing towers.

D.   A minimum of two towers is required for efficient vortex decay rate determination.

E.   The tower may cause premature vortex instability onset when it "ruptures" the cylindrical wall upon vortex passage making subsequent time-history analysis of that particular vortex questionable since it no longer represents an undisturbed vortex.

F.   Vortex marking on test aircraft is mandatory to minimize tower misses and flight test time.

G.   Tower smoke (colored) is mandatory for flow visualization and data correlation.

H.   Insufficient data (other than wind velocity) was gathered on the atmosphere to attempt any correlation between atmospheric conditions existing during the tests and the decay of the vortices. Richardson Number and power spectral density analysis of atmospheric turbulence indices were considered but had to be abandoned.

Data output includes data tables and graphical depictions as follows: plots of vortex cross-sectional size and shape in the form of isopleths as a function of time, tangential speed versus distance from the vortex center, plots of vortex flow velocities as a function of time (three-dimensional data for the Idaho Falls' flight test program and tangential velocity for the NAFEC tests), radius of maximum velocity, peak recorded tangential velocity and age, rate of atmospheric transport and supplemental atmospheric characteristics.

Only limited analysis has been made to date because of the large volume of data accumulated and the fairly recent completion (3 August 1970) of the flight testing in support of this phase.

The significant findings to date for all of the phase three tests are summarized in Table II. However, it is felt pertinent to expound on some of the obvious deviations on vortex characteristics noted therein from some of the generally well known, and used, theories on the aircraft vortex flow phenomena.

For the "clean" configuration aircraft and small flap deflections, the vortex systems of all the aircraft were observed to be of tubular form, relatively small in diameter, very clearly structured and very persistent. A typical example of this type of vortex system is shown in Figure 7 for the CV-880 in holding configuration (no flaps). For larger flap deflections, vortices of tubular form were much less evident. When maximum flap deflections were employed this characteristic was not observed, the vortex appearing much larger in diameter and more closely resembling random turbulence.

Such random turbulence would dissipate rapidly compared with the well structured vortex system. As a general observation, one would say that the "dirtier" the aircraft configuration is for a given weight and airspeed, the less dangerous its wake turbulence should be for identical vortex ages downstream of the aircraft. Analytical consideration tends to support this contention.

To further explain, vortices appearing downstream and parallel to the aircraft flight path are a direct consequence of lift. For a constant configuration and altitude, vortex strength varies as aircraft gross weight divided by airspeed. These vortices represent the aggregate of all the vorticity shed along the wing trailing edge. This trailing edge vorticity wraps up into two large vortices at a distance downstream dependent on aircraft span, aspect ratio, gross weight, altitude, and airspeed. The local direction of rotation of vorticity first appearing at the trailing edge depends on whether the local lift is increasing or decreasing toward the tip. Irregularities in lift distribution, such as are caused by large flap deflections, produce alternately left and right

hand rotation, whereas a smoothly varying lift distribution associated with small or zero flap settings produces a constant
direction of rotation along the wing. The former results in a
less well-defined vortex structure, with more rapid decay, even
though a dirty configuration generates more wake energy initially.

These observations were very clearly substantiated by our
photographic coverage and to a lesser degree, because of an inadequate number of vortex flow sensors and lack of a second tower for
determination of vortex dissipation rates on the same vortex, by our
time-history data recorded from the hot-film sensors. Higher
tangential velocities and smaller fields of influence were found for
the clean configuration aircraft as compared to lower velocities and
larger fields of influence for the landing configuration.

The tubular-type cores were, in general, found to be of such
small diameter as to result in many misses of the hot-film sensors,
even with the 10-foot vertical spacing interval, i.e., the vortex
cores passed between the sensors.

The notable exceptions to the above findings were the vortex
characteristics of aircraft with T-tails and fuselage-mounted engines,
i.e., the B-727 and DC-9 which produced tubular type vortex systems
and high-tangential velocities even in the landing configruation.
Peak vortex tangential velocities on the order of 200 feet per
second were recorded for the B-727. The above vortex characteristics
as noted visually and in the data output were not random. Indeed,
the same results were obtained in the data acquired from two
different test sites, towers, altitudes, and instrumentation systems.
The same aircraft, DC-9 and B-727, were used at both Idaho Falls
and NAFEC which, as noted before, are 5000 and 76 feet above sea
level, respectively.

This can be attributed to one of three factors or a combination thereof: (1) The T-tail, (2) lack of high-engine exhaust
velocity injection into the vortex flow field, or (3) lack of
engine pods on the underside of the wing, thereby providing a "clean"
wing with a relatively free and uninterrupted outboard flow of the
vortex sheet. The latter is suspected for two reasons: (1) Several
tower fly-by passes were flown at NAFEC with the CV-880 aircraft
with both jet engines on one side at zero thrust. No noticeable
differences were found in either the observed or recorded data when
compared with the data acquired on the CV-880 passes wherein all
four engines were at normal power for that configuration, airspeed,
and weight; (2) The B-727 and DC-9 type aircraft engine exhaust
smoke (and contrails when the aircraft is at high altitude) has been
found not to become entrained in the trailing vortex pair as it does
for aircraft with engines mounted on the wing. That is, the engine
exhaust trails are above the field of influence of the wing vortex
system and does not merge with the vortex sheet.

Certain other characteristics of aircraft trailing vortices were noted which are worthy of note: (1) Double or triple cylinders were observed to form as the tower smoke became intrained in the vortex as it passed through the tower. Whether this is a basic physical characteristic of the tubular-type vortex system or caused by different density smoke particles, or due to the disturbance caused by the tower is not clearly known at this time. Some of the vortex tangential velocity time history plots appear to show a possibility of multiple cylindrical flow but further analyses are required; (2) The tubular-type vortex systems, particularly for the C5A and B727, produced a whine like that heard from an artillery shell going overhead. Based on monitoring this whine, one could actually walk with such a vortex system as it left the aircraft and drifted toward the instrumented towers; and (3) Vortex axial flow-- this is an area which certainly requires further investigation. A first observation is that in a continuous unidirectional flow it either does not exist or is fairly neglible in magnitude for rela- tively long-age vortex systems. What appears to be an axial flow in the basic vortex structure, as produced visually by tower-mounted smoke, is believed to be caused by the tower. The basic tubular- type vortex structure has a low-pressure area within its cylindrical wall analagous to a tornado or waterspout. Once this wall is pene- trated, as by the instrumented tower upon vortex passage, there is an immediate injection of relatively higher pressure ambient air which carries the colored smoke in one or both axial directions in the attempted pressure equalization process. However, and this has been noted frequently for tubular-type vortex systems, at times a pulsating axial-flow phenomena is observed which appears to signal vortex flow instability onset and subsequent rapid vortex disinte- gration from ordered to random type flow. One could be tempted to say that the instrumented tower always causes this instability onset. However, many tubular-type vortex systems passed through the tower and, as outlined by the smoke, appeared to decay through diffusion only as they drifted away. It can be that the instability onset is not only a function of a disturbance, as caused by the tower, but also of vortex age. It is also worthy of note, that this instability onset is followed by a "necking-down" of certain longi- tudinal segments of the vortex cylinder and concurrent bursting of other sections. This same phenomena has been noted in photographs of vortices, outlined by smoke oil, at test flight and vortex passage heights above the top of the tower.

No conclusions have yet been reached on the vortex in/out of ground effect flight test investigation (Task two). Although the majority of the recorded data are good data, the data may not be sufficient. The results thus far indicate no significant difference in peak recorded tangential velocities at the tower for vortex systems generated in ground effect when compared with those generate out of but descending into ground effect for identical vortex ages. It is advised, however, that the majority of these data are of relatively short age in that, as cited earlier, we were "boxed in" on our lateral distance between the tower and aircraft. Accordingly, the primary means of changing the vortex age at the tower was by wind velocity changes. The results of this analysis will be reported at a later date.

REFERENCES

1. Andrews, William H., Robinson, Glenn H., and Krier, Gary E., Flight Research Center, and Drinkwater III, Fred J., Ames Research Center. Flight Test Evaluation of the Wing Vortex Wake Generated by Large Jet Transport Aircraft. NASA FWP-18, April 1970.

2. Condit, Philip M., Results of the Boeing Wake Turbulence Test Program. Boeing Document Number D6-30851, April 1970.

3. Garodz, Leo J., Measurements of the Vortex Wake Characteristics of the Boeing 747, Lockheed C5A, and Other Aircraft. FAA Data Report Project 177-621-03X (Special Task), April 1970.

4. Garodz, Leo J., Investigation of Aircraft Vortex Systems Descending Into and Generated In-Ground Effect, FAA Data Report Project 177-621-03X (Special Task No. 2), to be published.

5. Garodz, Leo J., Investigation of the Relatively Long Time-History Vortex Characteristics of the Convair 880 Airplane in Terminal Area Type Operations. FAA Data Report Project 177-621-03X (Special Task No. 3), to be published.

FLIGHT EVALUATION OF THE WING VORTEX WAKE GENERATED BY LARGE JET TRANSPORTS

William H. Andrews

NASA - Flight Research Center

Edwards, California 93523

ABSTRACT

A flight program has been conducted to update the current level of knowledge relative to the behavior of wing wake vortices generated by existing large transport aircraft and future jumbo jet transports. The tests were conducted to evaluate the wake location, persistence, apparent intensity, and influence out of ground effect and primarily under terminal area configuration and operational conditions.

The generating aircraft used were considered to be a cross-sectional representation of the transports in service and covered a gross weight range from 165,000 to 610,000 pounds. This group of test aircraft included a Douglas DC-9, Convair 990, and Lockheed C-5A.

The wake behavior was evaluated by measuring aircraft response and controllability of a series of probe airplanes flying in the generating aircraft wake at separation ranges from 1 to 15 nautical miles. The probe airplanes included a Lockheed F-104, Cessna 210 and 310, Lear Jet 24, DC-9, and the Convair CV-990. These airplanes were instrumented to record standard handling qualities parameters; however, the primary analysis was based on the probe airplane roll response and control inputs to counteract the induced rolling moment produced by the wake of the generating airplane. The generator-to-probe airplane separation ranges were resolved from simultaneous radar tracking of the aircraft with FPS-16 radars.

A written version of this paper was not available. For further details on this work see "Flight Test Evaluation of the Wing Vortex Wake Generated by Jet Transport Aircraft", G. H. Robinson, W. H. Andrews, and R. R. Larson, to be published as a NASA Technical Note.

AN ASSESSMENT OF DOMINANT MECHANISMS IN VORTEX-WAKE DECAY

P. B. MacCready, Jr.*

Meteorology Research, Inc.

ABSTRACT

Predicting the transport and decay of the organized vortex
system for various aircraft types, flight modes, and meteorological
conditions requires clear identification of the various significant
factors and mechanisms involved in the transport/decay process.
Outside of the ground effect, the dominant factors include the circulation and core characteristics of the vortices, the turbulence
in the wake and in the environment, and the thermal stability of
the environment. The environmental factors are sometimes dominant;
the vortex wakes from an airplane operating in two different
meteorological regimes can differ by an order of magnitude in decay
and by an order of magnitude in descent distance. Thus any theory
of vortex-wake decay in the real atmosphere must consider the
meteorological factors, and field observations on decay must be
interpreted with consideration of these factors.

There are two distinct mechanisms advanced for vortex-wake
decay-- the slowing down of the vortices by mixing action of eddy
viscosity, and the interaction of the vortices with each other,
which disorganizes the organized vortex flow field. The slowing
down by eddy viscosity is conceptually simple, but is difficult to
treat quantitatively because of uncertainty as to how the eddy viscosity coefficient depends on the aircraft characteristics or on
the atmospheric turbulence. The vortex interaction mechanism
commonly operates by the development of perturbations in the vortex

*   Present affiliation: Technical Consultant, 1065 Armada Drive,
    Pasadena, California

lines on a scale an order of magnitude greater than the vortex
separation distance. Existing theory seems useful in predicting
the wavelength and shape of the breakdown, but not the time of
breakdown. Factors involved in initiating breakdown may include
atmospheric turbulence, wake turbulence, periodic lift variations,
Benard cell-type structure arising from buoyancy, and core characteristics (especially size and axial flow).

Atmospheric stability causes relative buoyancy of the wake,
with subsequent strong effects on the vertical transport of the
vortices and on their horizontal separation. The effects also
depend intimately upon the mixing between the wake and the environment. Thus the environmental turbulence and the wake turbulence
both play a role in the mechanism.

Operational rules for terminal operations of aircraft in
strong turbulence should be rather easy to establish, even with
the present limited stage of understanding. However, greatly
improved understanding is required for predicting the worst (slow-
est decay) cases. Present reasoning suggests these will occur
during the climbout of heavy aircraft, flying slowly in "clean"
configuration, in marine air conditions where zero turbulence and
near-neutral stability may often coexist.

1. INTRODUCTION

This paper represents one approach to trying to view the
transport/decay of vortex wakes, from an overall perspective which
has evolved from field observations, experiments, and theoretical
work beginning 12 years ago. Most of the work involved the effects
of wakes on meteorological experiments, rather than being concerned
with the hazards which the organized vortex wakes presented to
subsequent aircraft. The research approach was usually to try to
illuminate dominant physical mechanisms, so that apparently conflicting observations from various experiments could be assembled
into a coherent picture.

From this particular series of inputs, we have been forced
to conclude that the atmospheric turbulence and atmospheric
stability often have much to do with determining the descent,
spread, interaction, and decay of the vortices. Such factors can
sometimes be dominant and, unfortunately, are almost impossible to
duplicate in model experiments. As will be seen, a conclusion which
then arises from this background is that the atmospheric environment must be considered in establishing the mechanisms of vortex
interaction and decay, field experiments should include measure-

ments of environmental factors and, in spite of the costs and difficulties of doing quantitative experiments in the atmosphere, the atmospheric laboratory should have a strong emphasis in any total attack on the wake problem.

2. RESULTS OF THE BACKGROUND INVESTIGATIONS

Involvement with wake decay began at MRI in 1958 during a contract with AFCRL for evaluation of contrail suppression techniques.[1] The program emphasis was on the short period evaporation of contrails, but one's scientific interest could not help but be stimulated by the related qualitative observations of the interaction of the vortex pair. Figure 1 shows the breakup sequence of a B-47 contrail taken on this project. The wavelength of the interaction instability tended to be semi-regular, and an order of magnitude larger than the core-to-core spacing. Figure 2 shows more detailed views of a B-47 contrail breakup, emphasizing variations in core stretching and a microstructure of spiraling. In contrast with the short time intervals for these particular B-47 vortices to break up, Fig. 3 is presented to depict the longest-lasting contrail we observed. The single sinuous vortex had very sharp characteristics five minutes after its formation, and some portions were even rather distinct after seven minutes. This B-52 trail segment was associated with turning flight. It appeared that in straight flight the vortices interacted and destroyed themselves in ordinary fashion more quickly. In a turn, however, the two vortices acted independently, with one getting short wave (<1 vortex pair spacing) sinusoidal perturbations which quickly became exaggerated and destroyed that vortex while the other vortex remained. This surviving vortex eventually interacted with itself at rather long wavelength (Fig. 3), with large waves and axial variations.

From these and other observations on the contrail program, we got clues that (a) the time for vortex interaction varied day to day, as though the air mass characteristics were of importance, (b) the breakup would occur in a fairly distinct wavelength range, and (c) the persistence and characteristics of the vortex pairs varied greatly between generating aircraft. Fighter-type aircraft had engines in or close to the fuselage, and their trailing vortices were poorly marked by contrails. We presumed, without quantitative proof, that if the vortices were well marked by contrails, then whenever the visible trails dissipated the

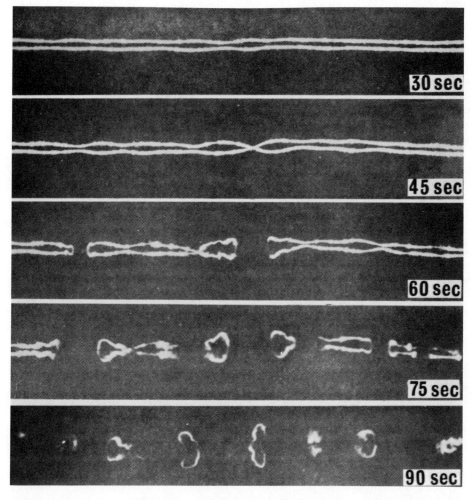

Fig. 1.   B-47 Contrail Breakup

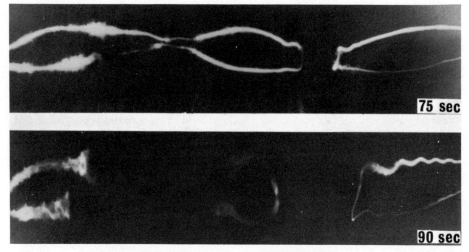

Fig. 2. B-47 Contrail Breakup, Detail

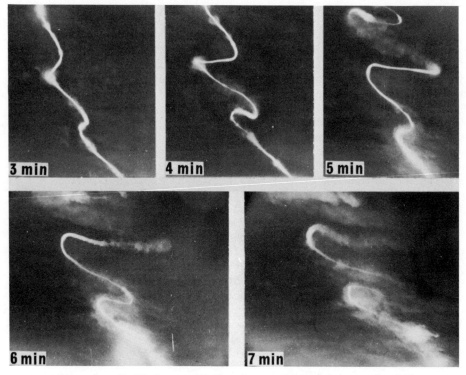

Fig. 3. B-52 Contrail Breakup. Elevation angle 22°

vortices were well broken down and would not bother other aircraft
except as conventional turbulence. Sharply delineated cores
indicated still strong vortices. The visible contrail diameter
remained smooth until the pair interaction became advanced.

    The above considerations were derived from observing "ships
of opportunity." One informal experiment was later made for us.
A B-47 circled back through its own wake to measure the wake's
vertical descent after two minutes. The three experiments yielded
observations showing one wake descended, one stayed the same, and
one rose. Wave upcurrents in this mountainous region were
evidently overwhelming the local contrail descent.

    More controlled experiments became possible during work on
diffusion for the Army.[2,3] In experiments to verify numerical
models for the diffusion of material released from aircraft, it
was necessary to see how far the airplane pushed the material
down before its initial descent stopped and atmospheric transport
and diffusion took over. Tests with small and medium size aircraft
(Cessna 180, Twin Bonanza, DC-3, and several fighters) gave
typical descent distances of 25-100 feet. Quantitative tests
with the Cessna 180 showed the wake initially would descend at
the theoretical velocity, but then quickly broaden and slow up.
On the other hand, releases over the ocean sometimes (but not
always) moved material down as much as 600 feet to the water.
On later smoke release tests over land in Alaska with a Cessna 180
on one day, the descent was about 30 feet, stopping well short of
the ground, while on another day it was 200 feet, reaching the
ground. In summary, the atmosphere apparently could dominate the
descent picture-- and presumably the vortex-pair longevity picture.
The most logical explanations then seemed to be that (a) turbulence
in the atmosphere, added to that in the wake from the airplane
propulsion and drag, would slow the wake's vertical motion and
help spread it, and (b) descent into a stable atmosphere would
slow the descent speed, this being somewhat analogous to the
slowing and spreading of a vortex pair when approaching the
ground. Over land there is usually turbulence or stability, so
the vortex life and descent distance is slowed. In the low levels
over the aerodynamically "smooth" ocean, commonly the atmospheric
stability is almost neutral and the turbulence is negligible,
permitting long vortex lives and large descents. The same ocean
conditions also are associated with gentle convective cells which
would help to move the vortices both down and up; just how much
of the observed descent came from the wake motion through the
air and just how much arose from convective flows was not determined
in the crude experimentation. The main conclusion seemed evident
and logical: (a) atmospheric turbulence can slow the vortex wake
descent, and (b) atmospheric stability can also slow it. As an

interesting sidelight,[4] in a calm nighttime atmosphere, the wake
of a Twin Bonanza was observed as a turbulent mass one-half hour
after the aircraft passage.

In addition to the early field trials, later Army support
permitted MRI to treat the subject of wakes a little more funda-
mentally. As a consultant to MRI, Dr. Steven Crow put the concept
of perturbing parallel line vortices into theoretical terms,
initiating the work which he subsequently finished at Boeing[5]
predicting the wavelength of the long wave instability. We conducted field experiments during which cores and wakes were marked
in various ways, and the aircraft were systematically maneuvered.
Qualitative field verification was made of two concepts: (a) the
wake from an aircraft during a pullout would descend faster and
further than for normal flight, and (b) perturbations in roll or
pitch induced by the pilot at the period around that of the long
wavelength instability (say about five to seven wingspans) would
speed up the interaction and decay of the vortex pair.

In endeavoring to develop methods of marking wake vortices,
we hypothesized that buoyant balloons would be drawn into the
vortex core because of centripetal force. Flight experiments
showed this to be true in dramatic fashion. A Piper Apache was
flown so that its wake descended into a cluster of slowly rising
balloons. One after another, the balloons whipped into the core,
each maneuvering less than one complete circle, and then moved
axially along the core. Axial motion toward the plane was about
15 mph, 11 percent of the aircraft speed, for the 20 seconds the
sequence was photographed. Once in the core, the balloons rarely
wobbled, implying each moved almost to the center of a smooth
core. This marking technique visually delineates the core
positions without requiring something to be released from the
airplane or to be released from a tower directly into the core.
We have experimented with buoyant, permanent, plastic bubles,
generated automatically[6] -- these might prove useful in wake
studies or even terminal operations. It is worth noting that
bubbles serve somewhat as corner reflectors for light and thus
can sometimes be made visible at extreme distances.

In 1967, MRI considered aircraft wakes in weather modification programs. The wake can carry seeding material down with it,
or just serve to transport air downward. Thus an airplane or
helicopter can move dry air down into a fog to promote fog dissipation. The many experiments conducted with aircraft of C-123
size and smaller usually included tracers to make the wake
visible. The main results from the standpoint of wake phenomena
were to note once again that wakes over the ocean commonly
descended further than those over land, and that, in many instances,

over fog tops the descent was very shallow.

The large magnitude of the direct effect of stability in inhibiting wake descent is not generally appreciated. To give an example, if a mass of air of any size starts a descent at 2 m/sec into a 6°C temperature inversion, it will go down only 10 meters. This example ignores turbulent mixing which inhibits the descent and the vortex "powering" of the descent which increases it, but serves to provide a feel for the magnitudes to be expected.

In 1970 at MRI, Dr. Ivar Tombach began a study,[7] sponsored by AFOSR, directed at improving our understanding of the actions of a vortex wake descending through a stable environment. The study considered how turbulent mixing brings environmental air into the top and center of the wake, decreasing its buoyancy and eroding the vortex strength. The study emphasized the importance of a particular factor in determining the descent and spread of vortex pair. The factor depends on the initial wake character- istics, the environmental stability, and the mixing between the environment and wake. Its value determines whether the downward velocity decays more rapidly than the vortex circulation, or vice versa. In the former case, the wake will slow its descent and spread, in a manner somewhat analogous to the action of a vortex pair descending in the presence of the ground. In the latter case, the vortices will try to move closer together. The theory represents a simplified model ignoring many secondary factors, but still serves to emphasize the differences in vortex decay and transport arising from meteorological factors.

3. SUMMARY OF DECAY MECHANISMS

From the perspective of the aforementioned investigations, together with the standard vortex wake generation and character- istics concepts from aerodynamic theory, a picture emerges of wake decay which is simple in concept but suggests a complexity in the interactions of mechanisms. The following brief discussion is directed at conventional fixed wing aircraft with aspect ratios exceeding 3. Helicopters and delta wings are thus outside the present scope; they are similar in principle, but complicate the picture.

Table I is a listing of the factors involved in determining and describing the initial vortex-wake system, and the additional factors which determine the subsequent evolution of the system. From the table it can be seen that many factors may be significant.

Table I

Characteristics/Factors on Which Wake Decay and
Transport Depend

1.   Initial Characteristics of Developed Vortex System

A.   Overall characteristics from lift/energy theory:

Circulation $\Gamma = \dfrac{W}{\pi b \rho V}$ $W \equiv$ weight
                                    $\rho =$ air density
                                    $b =$ span
                                    $V =$ velocity

Vortex spacing $b' = \dfrac{\pi}{4} b$

Wake sinking speed $w_s = \dfrac{\Gamma}{2\pi b'} = \dfrac{2\Gamma}{\pi^2 b} = \dfrac{8W}{\pi^3 b^2 \rho V}$

Dimensions of enclosed cylinder:
        height = 1.33 b
        width = 1.62 b

B.   Core diameters $\sim 0.2 b'$

Actual cores may be $< 0.2\ b'$, have axial flow,
and have complex configurations.

C.   Turbulence situation:
        Eddy viscosity in environment (small scale
            "eroding" turbulence)
        Eddy viscosity in wake cylinder (from
            environment, and from vehicle drag
            and propulsion)
        Eddy viscosity in cores (from vehicle drag
            and propulsion, as modified by core
            dynamics)

D.   Buoyancy factors:
        Temperature (mean temperature of wake
            relative to environment)
        Temperature distribution in wake outside
            cores
        Core temperatures (relative to mean wake)

E.   Non-constant vortex strength due to lift
     variations:
        From maneuvering
        From atmospheric turbulence

2.   Environmental Characteristics Determining Subsequent
     Events

     A.   Environmental density (temperature) for
          buoyancy forces

     B.   Environmental turbulence:
               Small scale (eroding)
               Larger scale (modifying relative vortes
                    positions)

     C.   Mean upcurrents/downcurrents

Within the first ten span lengths behind the airplane, the
vortex sheet system wraps up into what we consider the initial
condition. The wake size, circulation, vortex spacing, and descent
speed are determined by the generating airplane. The core
dimensions and characteristics (size, axial flow) are also a
function of the airplane, depending on gross parameters but also
on details of wing and nacelle drag and the position and type of
thrust devices. Anything altering the thickness, turbulence, and
longitudinal velocities in the vortex sheet out of which the core
is formed can be expected to alter its characteristics. The core
is generally deemed simple and non-turbulent, but it may sometimes
be a more complex structure of tight spiraling vortices. Assuming
the classical picture of wake flows, calculations have indicated
that the diameter of a "solidly rotating" core should be about
one-fifth of the core spacing if the wake energy is to agree with
the energy going into induced drag. Actual cores are sometimes
much smaller than this, implying that the classical picture is
only a crude first approximation to the real situation. Knowledge
of core details is definitely quite limited. The turbulence outside the cores but still within the wake is the result of the
combined effects of environmental turbulence and that generated by
the parasite drag and the thrust systems of the aircraft. The
subsequent evolution of the wake will involve additional exterior,
environmental factors-- the thermal stability, the turbulence
which can both erode each vortex and foster interaction of the
vortices, and mean vertical flows which displace the vortex system.

Vortex decay will take place from eddy viscosity in the wake,
in analogous fashion to the viscous decay of a vortex at low
Reynolds number. The details of how this erosion serves to erode
the core requires more knowledge of core characteristics than we
have. The Rayleight stability theory[8] for rotating flows requires
that the core, with rotational flow, should have no velocities
higher than those extrapolated from the exterior potential vortex

field. As the exterior vortex erodes, erosion of the outer shells
of the core would take place, with a smaller stable core remaining.

Vortex interaction typically takes place as shown in Figs. 1
and 2. Crow[5] has developed the theory giving the shape and wavelength details of the initial perturbation growth. Once the
perturbation is finite, its further development follows logically
from vortex theory. The cause of the initial perturbation which
makes the vortices locally non-parallel must be some turbulence
in the wake or environment, or maneuvering of the aircraft.
Turbulence should be able to do the job in two distinct ways: by
modifying established vortices, or altering the lift generation as
the wing moves through the air.

The elements of the effects of environmental stability and
turbulence on the descent, separation, and strength of the vortices
have already been cited in Tombach's theory.[7] How long wavelength
interaction of vortices is modified by stability has not yet
received attention, either in observation or theory.

4. COMMENTS ON FORECASTING

Present estimates of the decay of vortex wakes are derived
primarily from summarizing the results of observational programs.
When the roles of environmental factors are understood, and when
these factors are monitored in future experiments, the estimates
can be improved. Even at the present stage, it is useful to
consider three aspects of wake duration: (1) what environmental
turbulence value assures fast decay, (2) when decay will be least,
and (3) how vortex intensity relates to vertical displacement.

By either vortex erosion or vortex interaction, strong
turbulence will certainly speed the dissipation of wakes.
It is reasonable to hypothesize that the wake of any jet transport
would decay in one minute to where it only has turbulence comparable to the environment if the environmental turbulence intensity
exceeds a certain level, say $\varepsilon^{1/3} = 5$ $cm^{2/3}sec^{-1}$.* This particular
choice of a turbulence level may be incorrect, but there will
certainly be some turbulence intensity value above which vortex
wakes are inconsequential after one minute. This number can be

---

* See P. B. MacCready, Jr., "Operational Application of a Universal
   Turbulence Measuring System," AMS/AIAA Paper No. 66-364, 1966,
   for a description of the rating concept. $\varepsilon^{1/3} = 5$ $cm^{2/3}sec^{-1}$
   would be called "moderate" by many pilots in typical aircraft in
   cruising conditions.

quickly established experimentally, and then with simple airborne
or tower instrumentation one can define conditions when wakes pose
little practical hazard.

As to long duration wakes, one can hypothesize the conditions
most likely for their existence. We cannot, at the moment, provide quantitative estimates of maximum durations since complete
atmospheric data are not available for interpretation of existing
wake decay observations. For a given airplane type, duration may
be longest when the plane is heavily loaded, is flying slowly,
when it is clean (not producing extra drag from landing gear or
sharply deflected flaps), and when the atmosphere is non-turbulent
and has neutral stability. The critical aircraft configuration
occurs primarily right after takeoff, and secondarily in holding
patterns. The meteorological factors are critical often in the
lowest kilometer over the ocean or in regions very close to the
coast. The conditions should also be found routinely at certain
altitudes in the late afternoon over land when ground heating is
over so heat flux upward ceases, and after the turbulence caused
earlier by convection has died out.

One can suggest some correlation between vortex strength and
descent speed or vertical position. If the wake is descending at
the initial rate, it is probably not decaying rapidly since it is
not interacting strongly with the environment. If the descent has
slowed, the wake is probably experiencing turbulence or atmospheric
stability effects which also speed its erosion. Thus, ignoring
mean upcurrents which can upset the reasoning, an airplane should
not encounter a really strong wake far behind a generating aircraft unless the following aircraft is fairly near the altitude
separation defined by the separation time and the initial wake
sinking speed.

5.   SPEEDING UP VORTEX WAKE DISSIPATION

There are two obvious "brute force" methods of altering wakes
to minimize their effect on subsequent aircraft-- adding turbulence throughout the atmosphere, or sucking the wake away. The
growing understanding of the whole wake decay phenomenon suggests
more practical alternatives, wherein only small energies are
required for triggering wake instabilities which thus harness the
wake energy for its own destruction.

One instability is the long wavelength (5-8 wingspans) vortex
pair interaction. Since perturbations at this scale grow readily,
the goal is to induce them more strongly and more quickly than they
would develop naturally. The experiments cited earlier where strong

oscillations in pitch or roll were excited by the pilot at this wavelength demonstrated the principle, although crudely. However, in commercial aircraft, these maneuverings in the 1-4 second period range would not be acceptable. As a more appealing approach, Dr. Crow has suggested using coordinated inner flaps and outer aileron-flaps in a manner which keep the lift constant while vortex spacing is increased and decreased at the desired frequency. The long wavelength instability could also be induced at low altitude exterior to the aircraft by air jets, blocking structures, suction devices, or other flow-altering techniques operating with time or space scales to induce variations at this wavelength.

Another instability is inherent in the tiny, tight cores. Is there a way of causing this highly organized flow to degenerate into turbulence and diffuse the vorticity? The NASA experiments[9] show some modification of the core is feasible with drag devices at the tips, although there is some question of the significance of the alteration when viewed on a long time scale. Garodz[10] noted large differences between cores from different aircraft, which verifies that core characteristics are alterable. Variations of drag, turbulence, or exhaust entering the vortex sheet and hence vortex core system must be expected to cause core alterations. The thrust system details should be of special importance, since some jet engine positions put high turbulence and axial momentum directly into the vortex sheet region, while other engine positions are more remote from the vortex sheet.

Considering both the long wavelength instability and core modifications brings up the concept of doubling up on trigger mechanisms so that very tiny modifications cause large results. An obvious approach is to modify the core cyclically, at the long wavelength instability frequency. For instance, intermittently add surface roughness to a portion of the wing. The resulting axial flow variations in the vortex, as well as the added turbulence, may speed the vortex interaction while having a negligible effect on aircraft performance.

For the sake of completeness in discussing core stability, we should note here that the core temperature with respect to its exterior has some bearing on stability. A cold core may be expected to diffuse more rapidly than a neutral or warm one. Spraying a volatile liquid into the core will cool it. For decreases of several degrees, the material requirements are quite small.

6. RESEARCH CONCEPTS

There is probably no single cure for the problem of coping with the vortex wake hazard to aircraft operations. Rather, the problem will be handled by a combination of improved forecasting and monitoring of wake characteristics and evolution, improved pilot understanding of the dangers and of corrective actions, improved air traffic procedures, and the application of some artificial techniques to speed vortex wake decay. All of these will benefit from an improvement in the understanding of wake characteristics and wake decay. Thus research is a key to the practical goal of decreasing wake hazards. All facets of research deserve continuing attention-- theory, laboratory studies, and field experiments. However, the evident importance of atmospheric effects, which cannot readily be modeled in the laboratory, gives special emphasis to the field work. Economic reality dictates compromises in field programs, and requires that the programs be carefully designed so as to be as rewarding as possible.

Field studies should be true experiments rather than just collections of observations. "Experiments" means there exists some model for the phenomenon in question, specific questions are being asked about the model, and the field tests are designed to answer those questions. The outdoor laboratory must be instrumented adequately to afford a description of the factors which may be important in the wake's evolution-- particularly the stability and turbulence throughout the whole region traversed by the wake. The wake's characteristics should also be monitored throughout the period of interest. Monitoring the circulation, core characteristics, and three-dimensional wake aspects requires both marking, as by smoke or balloons, and quantitative measurements, as by an instrumented aircraft or tower. The experiments should be simple and systematic, using the philosophy of studying simple cases and varying only one parameter at a time. For instance, first test where there is neutral stability and negligible turbulence. Then test at several specific stability conditions. Next treat varying levels of turbulence. Vary power settings, keeping all else constant. Maneuver to induce long wavelength instability, etc. As understanding and observational capability grow, then one can tackle the more complex situations, such as low altitude trials where the ground effect and steep gradients or turbulence and stability make it unlikely that a simple theory can prove satisfactory.

The various foregoing discussions suggest there are three areas of research which are especially vital. One area is developing the understanding of environmental influences. Another is getting a clear picture of the characteristics of cores, and the causes of these characteristics. The third, which is closely tied to the first two, is ascertaining the factors which determine the initiation of the long wavelength instability. Achievements in these areas will be the basis of significant achievements in ameliorating the hazards of aircraft wakes.

REFERENCES

1. Smith, T. B., and K. M. Beesmer, "Contrail Studies for Jet Aircraft." Meteorology Research, Inc., Altadena, Calif., Final Report to AFCRL, Cont. AF 19(604)-1495, AD 110 278, 1959.

2. Smith, T. B., and P. B. MacCready, Jr., "Aircraft Wakes and Diffusion Enhancement." Meteorology Research, Inc., Altadena, Calif., Part B, Final Report to Dugway Proving Ground, Dugway, Utah, Cont. DA-42-007-CML-545, 1963.

3. Smith, T. B., and M. A. Wolf, "Vertical Diffusion from an Elevated Line Source over a Variety of Terrains." Part A, Final Report to Dugway Proving Ground, Dugway, Utah, Cont. DA-42-007-CML-545, AD 418 599, 1963.

4. MacCready, P. B., Jr., T. B. Smith, and M. A. Wolf, "Vertical Diffusion from a Low Altitude Line Source - Dallas Tower Studies." Meteorology Research, Inc., Final Report to U.S. Army Dugway Proving Ground, Dugway, Utah, Cont. DA-42-007-CML-432, AD 298 260 (Vol. I) and AD 298 261 (Vol. II), 1961.

5. Crow, S. C., "Stability Theory for a Pair of Trailing Vortices." Boeing Scientific Research Laboratories Document D1-82-0918, 1970.

6. MacCready, P. B., Jr., B. L. Niemann, and L. O. Myrup, "Cluster Diffusion in the Inertial Subrange." Atmospheric Research Group, Altadena, Calif., Report to U.S. Public Health Service, Grants AP 00359-01, 02, 03, 1967.

7. Tombach, I. H., "Transport of a Vortex Wake in a Stably Stratified Atmosphere." Paper presented at Symp. on Aircraft Wake Turbulence, Seattle, September 1-3, 1970.

8. Rayleigh, Lord, "On the Dynamics of Revolving Fluids." Proc. Roy. Soc., Ser. A, 93 (1916) 148-154, 1916.

9. Corsiglia, V. R., R. A. Jacobsen, and N. A. Chigier, "An Experimental Investigation of Wing Trailing Vortices with Dissipation." Paper presented at Symp. on Aircraft Wake Turbulence, Seattle, September 1-3, 1970.

10. Garodz, L., "Measurements of Boeing 747, Lockheed C5A, and
    Other Aircraft Vortex Wake Characteristics by Tower Fly-By
    Technique." Paper presented at Symp. on Aircraft Wake
    Turbulence, Seattle, September 1-3, 1970.

THEORETICAL AND EXPERIMENTAL STUDY OF THE STABILITY

OF A VORTEX PAIR

S. E. Widnall, D. Bliss, and A. Zalay

Massachusetts Institute of Technology

ABSTRACT

The linear stability of the trailing vortex pair from an air-craft is discussed. The method of matched asymptotic expansions is used to obtain a general solution for the flow field within and near a curved vortex filament with an arbitrary distribution of swirl and axial velocities. The velocity field induced in the neighborhood of the vortex core by distant portions of the vortex line is calculated for a sinusoidally perturbed vortex filament and for a vortex ring. General expressions for the self-induced motion are given for these two cases. It is shown that the details of the vorticity and axial velocity distributions affect the self-induced motion only through the kinetic energy of the swirl and the axial momentum flux. The presence of axial velocity in the core reduces both the angular velocity of the sinusoidal vortex filament and the speed of the ring. The vortex pair instability is then considered in terms of the more general model for self-induced motion of the sinusoidal vortex. The presence of axial velocity within the core slightly decreases the amplification rate of the instability. Experimental results for the distortion and breakup of a perturbed vortex pair are presented.

1. INTRODUCTION

It has been established by both theory and observation that the pair of trailing vortices behind the wing of an aircraft in cruise undergoes a coupled sinusoidal deformation that grows in time until the two vortex cores come into contact. The pattern then breaks up, often into crude rings, and dissipates rapidly.

Since the persistence of trailing vortices behind a large aircraft
can pose a safety hazard to other aircraft, an understanding of the
fluid mechanics of this instability is of considerable practical
interest.

The instability of a pair of trailing vortices to sinusoidal
deformations was studied previously by Crow[1]. The general features
of this instability are sketched in Fig. 1.

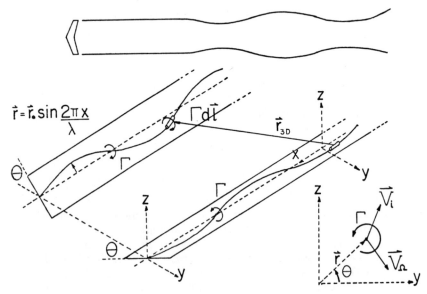

Figure 1. Vortex pair instability

The velocity $\vec{V}_i$ induced by one vortex on the other is independent
of the details of the vortex core structure and can be calculated
by applying the Biot-Savart law to a line vortex of circulation Γ.
In Crow's analysis, the self-induced motion of the sinusoidal vortex
due to the deformation, $\vec{V}_\Omega$, was calculated by the "cut-off line integral" technique. This involves placing the point where the velocity is evaluated on the vortex itself and then integrating the
Biot-Savart law over all of the vortex except the region near the
point of evaluation. The integral is cut off (stopped) at a distance, say ℓ_c, on either side of the point, to avoid divergence.
Now for some choice of ℓ_c, one gets the correct answer. The choice
of ℓ_c was based on the known solutions by Thomson for the rotating
sinusoid and the vortex ring, both cases having the vorticity constant in the core. This approach was quite successful for the study
of the instability of trailing vortices. However, the use of the
cut-off distance removes the fluid mechanics of the vortex core from
the problem, and the effect of different rotational and axial

velocity distributions is not directly apparent.

It is our purpose to present a general theory for the self-induced motion of a perturbed vortex of finite core size incorporating the effects of an arbitrary distribution of both swirl and axial velocity. Since the vortex pair instability depends on both $\vec{V}_i$ and $\vec{V}_\Omega$, such a solution for $\vec{V}_\Omega$ will give a more complete stability theory.

The structure of a tip vortex is more complicated than a simple line vortex with distributed vorticity in the core. It is a general feature of vortices created by a wing that they possess axial velocities different from the surrounding flow. In the case of a low aspect ratio delta wing, it can be shown from calculations based on conical flow that the rolling up process produces an excess of velocity in the downstream direction. For a high aspect ratio wing, the analysis is more difficult but experiments suggest that there is an axial velocity defect in the tip vortex flow field, due to the action of viscous forces in the tip region. Depending on the initial velocity defect of the vortex, several possibilities exist for subsequent development. If the initial defect is small, the axial velocity will gradually return to free stream. For a strong axial velocity defect, the acceleration along the axis due to viscous forces will cause a tightening of the vortex core due to the inflow required by mass conservation. (Tightening has been observed experimentally and measured vortex cores are often smaller then predicted on the basis of potential flow.) If the vortex has a high swirl so that the vortex is subcritical, stagnation of the axial flow will occur as described by Landahl and Widnall[2]. It is clear that a very complex interaction must take place in the final stages of the instability if vortex lines with initial axial velocities are to form into vortex rings. However, even in the linear phase of the instability the axial velocities influence the self-induced motion of the vortex and thus affect the stability and the persistence of the wake.

The general problem of the self-induced motion and detailed fluid mechanics of a curved rotational vortex flow is quite difficult. Fortunately, the limiting case in which the radius of curvature is everywhere large when compared with the size of the vortex core is of considerable physical interest. For this case, the self-induced motion of a vortex filament can be obtained directly by analyzing the effects of curvature on the velocity and vorticity fields within and near the core. This removes both the uncertainty of the cut-off method and the singularity which motivated its use, allowing the effects of the core structure details to be considered.

The deformation of an initially axisymmetric straight vortex core will produce a stretching (or contracting) of the vortex filaments within the core which is proportional to $\varepsilon = a/R$, the ratio of the core radius to local radius of curvature. The change in the velocity field within and near the perturbed vortex filament due to curvature has the dominant effect on the self-induced motion. The effect of velocity induced over the region of the vortex core by distant portions of the vortex line must also be included.

The solution will be obtained by aid of matched asymptotic expansions[3]. Two asymptotic solutions are sought: one approximate solution (the inner solution), valid in the limit $\varepsilon \to 0$, only within the rotational vortex core and its immediate surroundings and another approximate solution (the outer solution), valid in the limit $\varepsilon \to 0$, in all regions of the flow except the vortex core. In general, each of these solutions would contain undetermined parameters which must be found by matching.

Historically, the earliest work on this subject was reported by Lord Kelvin, who determined the motion of a vortex ring and a sinusoidally perturbed vortex line for special vorticity distributions in the core, see Thomson[4] and Lamb[5]. More recently, Tung and Ting[6] and Saffman[7] have studied the vortex ring. The relationship between these studies and the present analysis will be discussed.

2.   SELF-INDUCED MOTION OF A CURVED VORTEX LINE

2.1 Solution Within and Near the Vortex Core

A curved vortex filament, in addition to having very interesting patterns of flow created within and near the vortex core because of curvature, will also experience a net self-induced velocity which will move the vortex as a whole. To find a solution, valid in the immediate neighborhood of the rotational vortex core, it is advantageous to work in a coordinate system moving with the local velocity of the core. At the outset this velocity is unknown. The local solution to the conservation equations of mass and vorticity plus the effects of distant elements of the vortex filament determine the necessary self-induced motion.

The solution procedure is most straightforward if the coordinate system is chosen to account for the local curvature of the vortex line, as sketched in Fig. 2: the center of the vortex line is located at a radial coordinate, ρ, equal to the local radius of curvature R; the polar system, ρ and ϕ, lies in the local osculating plane of the vortex line; the z coordinate is parallel to the local binormal of the vortex line. In the analysis, a local r, θ, s curvilinear system, attached to the head of the R vector, is used with all lengths normalized by R. (Vector operations for this "local toroidal" system appear in Appendix A.)

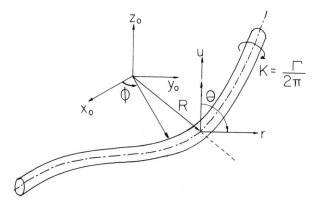

Figure 2.   Curvilinear coordinate system for a curved vortex filament.

        The initial swirl and axial profiles of the vortex core will
be assumed to have radial symmetry. A more difficult problem would
be the deformation of an asymmetric vortex flow with non-uniform
patches of vorticity swirling about -- as may exist in some aircraft
wake flows.   The present analysis should still be correct for the
large scale motion of such a flow in an average sense if the circulation, kinetic energy and axial momentum are modeled as a radially
symmetric vortex flow.

    Both the equation of conservation of mass

$$\nabla \cdot \vec{Q} = 0 \qquad\qquad (1)$$

and the definition of vorticity

$$\vec{\Omega} = \nabla \times \vec{Q} \qquad\qquad (2)$$

are kinematic conditions for the velocity and vorticity fields.
In the dynamic equation of vorticity one must consider that any
rotation of the coordinate system moving with the vortex will introduce an apparent vorticity.

The sinusoidally perturbed vortex filament, in the absence of any other influence, spins about its axis with angular velocity ω. However, in the vortex pair instability the influence of the other vortex is to suppress the rotation and induce only an accelerating divergence. Thus, in the aircraft wake stability problem the rotation need not be considered.

Although viscous forces play a role in determining the swirl and axial velocity profiles of the unperturbed vortex filament, a detailed analysis of the appropriate time scales for the instability shows that the perturbation problem may be treated as inviscid. Also, the slight variation of vortex properties along the core due to viscous forces may be ignored. (See Bliss[8].)

With these simplifications, the dynamic equation of vorticity becomes

$$\vec{Q} \cdot \nabla \vec{\Omega} = \vec{\Omega} \cdot \nabla \vec{Q} \tag{3}$$

merely relating changes in vorticity to stretching and convection.

To determine the effect of curvature on the flow within and near the vortex core, a solution will be sought in the form of an expansion in the small parameter $\varepsilon = a/R$.

$$\bar{Q} = \bar{Q}_o + \varepsilon \bar{Q}_1 + \varepsilon^2 \bar{Q}_2 + \cdots$$

$$\bar{\Omega} = \bar{\Omega}_o + \varepsilon \bar{\Omega}_1 + \varepsilon^2 \bar{\Omega}_2 + \cdots \tag{4}$$

where the components of the velocity and vorticity vectors are

$$\bar{Q} = u\bar{i}_r + v\bar{i}_\theta + w\bar{i}_s \tag{5}$$

$$\bar{\Omega} = \xi\bar{i}_r + \eta\bar{i}_\theta + \zeta\bar{i}_s \tag{6}$$

The velocity and vorticity are non-dimensionalized by κ/a and κ/a^2 respectively, $\kappa = \Gamma/2\pi$.

$\vec{Q}_o$ and $\vec{\Omega}_o$ are the velocity and vorticity of the straight vortex, $v_o(r)$ is the initial swirl velocity, $w_o(r)$ is the initial axial velocity; the initial radial components, u_o and ξ_o are zero. The initial components of vorticity are

$$\zeta_o = \frac{1}{r}\frac{\partial(rv_o)}{\partial r}$$

$$\eta_o = -\frac{\partial w_o}{\partial r} \tag{7}$$

$\vec{Q}_1$ and $\vec{\Omega}_1$ are the first order perturbations in velocity and vorticity due to curvature.

In the r, θ, s coordinates (Fig. 2) the appropriate length scale for changes along the vortex is s ∿ O[1], while the appropriate length scale for changes within the cross-section of the vortex core is r ∿ O[ε]. To identify the proper balance between these changes within and near the vortex core the variable r̄ = r/ε is introduced. The local geometry of a curved vortex tube is sketched in Fig. 3

a∿typical length in
vortex core

local toroidal
coordinate system

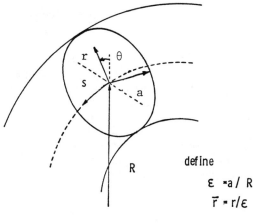

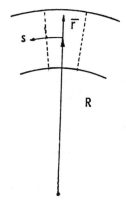

define

$$\varepsilon = a/R$$
$$\bar{r} = r/\varepsilon$$

Figure 3.   Local coordinate system for a curved vortex.

The governing equations for the flow in and near the vortex core are obtained by expressing (1), (2), and (3) in the local curvilinear coordinates, and transforming the variable r to εr̄.

The mass conservation equation then becomes

$$\frac{1}{\varepsilon}\left(\frac{\partial u}{\partial \bar{r}} + \frac{u}{\bar{r}} + \frac{1}{\bar{r}}\frac{\partial v}{\partial \theta}\right) + \frac{1}{1 + \varepsilon \bar{r}\sin\theta}\frac{\partial w}{\partial s} + \frac{u\sin\theta + v\cos\theta}{1 + \varepsilon \bar{r}\sin\theta} = 0 \quad (8)$$

The vorticity components are given by

$$\frac{\xi}{\varepsilon} = \frac{1}{\varepsilon \bar{r}}\frac{\partial w}{\partial \theta} + \frac{w\cos\theta}{1 + \varepsilon \bar{r}\sin\theta} - \frac{1}{1 + \varepsilon \bar{r}\sin\theta}\frac{\partial v}{\partial s}$$

$$\frac{\eta}{\varepsilon} = \frac{1}{1 + \varepsilon \bar{r}\sin\theta}\frac{\partial u}{\partial s} - \frac{1}{\varepsilon}\frac{\partial w}{\partial \bar{r}} - \frac{w\sin\theta}{1 + \varepsilon \bar{r}\sin\theta} \quad (9)$$

$$\frac{\zeta}{\varepsilon} = \frac{1}{\varepsilon}\left(\frac{\partial v}{\partial \bar{r}} + \frac{v}{\bar{r}} - \frac{1}{\bar{r}}\frac{\partial u}{\partial \theta}\right)$$

The components of the vorticity equation take the following form:

$$\bar{i}_r: \quad u\frac{\partial \xi}{\partial \bar{r}} + v\frac{1}{\bar{r}}\frac{\partial \xi}{\partial \theta} + \frac{\varepsilon w}{1 + \varepsilon \bar{r}\sin\theta}\frac{\partial \xi}{\partial s} = \xi\frac{\partial u}{\partial \bar{r}} + \eta\frac{1}{\bar{r}}\frac{\partial u}{\partial \theta} + \frac{\varepsilon \zeta}{1 + \varepsilon \bar{r}\sin\theta}\frac{\partial u}{\partial s}$$

$$\bar{i}_\theta: \quad u\frac{\partial \eta}{\partial \bar{r}} + v\frac{1}{\bar{r}}\frac{\partial \eta}{\partial \theta} + \frac{\varepsilon w}{1 + \varepsilon \bar{r}\sin\theta}\frac{\partial \eta}{\partial s} + \frac{\xi v}{\bar{r}} = \eta\frac{1}{\bar{r}}\frac{\partial v}{\partial \theta} + \frac{\varepsilon \zeta}{1 + \varepsilon \bar{r}\sin\theta}\frac{\partial v}{\partial s} + \frac{u\eta}{\bar{r}} + \xi\frac{\partial}{\partial}$$

$$\bar{i}_s: \quad u\frac{\partial \zeta}{\partial \bar{r}} + v\frac{1}{\bar{r}}\frac{\partial \zeta}{\partial \theta} + \frac{\varepsilon w}{1 + \varepsilon \bar{r}\sin\theta}\frac{\partial \zeta}{\partial s} + \frac{\xi\sin\theta + \eta\cos\theta}{1 + \varepsilon \bar{r}\sin\theta}w\varepsilon =$$

$$\xi\frac{\partial w}{\partial \bar{r}} + \eta\frac{1}{\bar{r}}\frac{\partial w}{\partial \theta} + \frac{\varepsilon \zeta}{1 + \varepsilon \bar{r}\sin\theta}\frac{\partial w}{\partial s} + \frac{u\sin\theta + v\cos\theta}{1 + \varepsilon \bar{r}\sin\theta}\zeta\varepsilon$$

$$(10)$$

The expansion (4) is substituted into these equations and terms with equal powers of ε are equated. The lowest order equations confirm (7). The next order terms give the governing equations for $\vec{Q}_1$ and $\vec{\Omega}_1$. The process is straightforward but cumbersome; the final results can be understood as the effects of the stretching and tilting of the vortex filaments due to curvature and the requirements of conservation of mass in the deformed vortex cross-section. (We refer to Bliss[8] for details.) The resulting equations for the perturbations in velocity and vorticity due to curvature are

$$\frac{\partial u_1}{\partial \bar{r}} + \frac{u_1}{\bar{r}} + \frac{1}{\bar{r}}\frac{\partial v_1}{\partial \theta} + v_0\cos\theta = 0 \tag{11}$$

$$\bar{i}_r: \quad v_0\frac{\partial \xi_1}{\partial \theta} = \eta_0\frac{\partial u_1}{\partial \theta}$$

$$\bar{i}_\theta: \quad u_1\frac{\partial \eta_0}{\partial \bar{r}} + v_0\frac{1}{\bar{r}}\frac{\partial \eta_1}{\partial \theta} + \frac{\xi_1 v_0}{\bar{r}} = \xi_1\frac{\partial v_0}{\partial \bar{r}} + \eta_0\frac{1}{\bar{r}}\frac{\partial v_1}{\partial \theta} + \frac{\eta_0 u_1}{\bar{r}} \tag{12}$$

$$\bar{i}_s: \quad u_1\frac{\partial \zeta_0}{\partial \bar{r}} + v_0\frac{1}{\bar{r}}\frac{\partial \zeta_1}{\partial \theta} + \eta_0 w_0\cos\theta = \xi_1\frac{\partial w_0}{\partial \bar{r}} + \eta_0\frac{1}{\bar{r}}\frac{\partial w_1}{\partial \theta} + v_0\zeta_0\cos\theta$$

$$\xi_1 = \frac{1}{\bar{r}}\frac{\partial w_1}{\partial \theta} + w_0\cos\theta$$

$$\eta_1 = -\frac{\partial w_1}{\partial \bar{r}} - w_0\sin\theta \tag{13}$$

$$\zeta_1 = \frac{\partial v_1}{\partial \bar{r}} + \frac{v_1}{\bar{r}} - \frac{1}{\bar{r}}\frac{\partial u_1}{\partial \theta}$$

The curvature of an initially straight vortex takes a slice of the cross-section with initially parallel sides and deforms it into a wedge as sketched in Fig. 3. Fluid particles are alternatively stretched and compressed as they swirl about the center of the vortex. In the mass conservation equation the term $v_0\cos\theta$ arises from

changes in thickness around the slice of the curved vortex line.
In the vorticity conservation equations and the definition of the
vorticity components, the terms involving products of $\vec{Q}_o$ and $\vec{\Omega}_o$
with $\sin\theta$ or $\cos\theta$ represent the stretching and compression of vorticity that results when the vortex cross-section is deformed.
This deformation also "tilts" vorticity initially in the s or θ
direction into other directions thus coupling the axial and swirl
motions.

The non-homogeneous solution to the perturbation equations is
found by assuming θ dependence of the form

$$u_1 = \hat{u}_1(\bar{r},s)\cos\theta \qquad \xi_1 = \hat{\xi}_1(\bar{r},s)\cos\theta$$

$$v_1 = \hat{v}_1(\bar{r},s)\sin\theta \qquad \eta_1 = \hat{\eta}_1(\bar{r},s)\sin\theta \qquad (14)$$

$$w_1 = \hat{w}_1(\bar{r},s)\sin\theta \qquad \zeta_1 = \hat{\zeta}_1(\bar{r},s)\sin\theta$$

To guarantee completeness of the solution a full Fourier series
should really be assumed in each case. However, the terms omitted
from the assumed solutions lead only to sets of homogeneous equations. Arguments can be made to show that these terms should be
either higher order or zero.

To this order then, the effect of axial velocity does not
appear in the equation for conservation of mass. Therefore, we
could have introduced a stream function ψ related to the velocity
field by

$$u = \frac{1}{1 + \varepsilon\bar{r}\sin\theta}\frac{1}{\bar{r}}\frac{\partial\psi}{\partial\theta} \quad\text{and}\quad v = \frac{-1}{1 + \varepsilon\bar{r}\sin\theta}\frac{\partial\psi}{\partial\bar{r}} \qquad (15)$$

(The scale for non-dimensionalizing ψ is κR.) The expansion for ψ
is

$$\psi = \psi_o(\bar{r}) + \varepsilon\psi_1(\bar{r},\theta) + \ldots \qquad (16)$$

The velocities are

$$v_o = -\frac{\partial\psi_o}{\partial\bar{r}}$$

$$v_1 = \bar{r}\sin\theta\frac{\partial\psi_o}{\partial\bar{r}} - \frac{\partial\psi_1}{\partial\bar{r}} \qquad (17)$$

$$u_1 = \frac{1}{\bar{r}}\frac{\partial\psi_1}{\partial\theta}$$

The appropriate θ dependence of ψ_1 is

$$\psi_1 = \hat{\psi}_1(\bar{r})\sin\theta \qquad (18)$$

where $\hat{\psi}_1 = \bar{r}\hat{u}_1(\bar{r})$.

The ordinary differential equation that governs $\hat{\psi}_1$ could have been obtained by introducing the stream function directly into the equation of vorticity conservation and into the definition of vorticity, eliminating vorticity between them. This approach was not used because of a desire to identify the coupling between the axial and swirl velocities and the effects of curvature on each vorticity and velocity component.

The equation governing $\hat{\psi}_1$ is obtained by a lengthy manipulation of (11) through (16). The result is

$$\frac{d^2\hat{\psi}_1}{d\bar{r}} + \frac{1}{\bar{r}}\frac{d\hat{\psi}_1}{d\bar{r}} - \left(\frac{d^2 v_o}{dr^2} + \frac{1}{\bar{r}\partial\bar{r}}\frac{\partial v_o}{}\right)\frac{\hat{\psi}_1}{v_o} = -2\bar{r}(\zeta_o - \frac{w_o\eta_o}{v_o}) - v_o \quad (19)$$

The equation determines perturbation vorticity ζ_1 due to three non-homogeneous source terms involving the initial velocity and vorticity distribution: the first is due to stretching of ζ_o, the second is due to tilting of η_o, the last is due to changes in swirl velocity required by mass conservation. By inspection, one solution to the homogeneous equation is found to be $\hat{\psi}_1 = v_o$. Once one homogeneous solution of a second order ordinary differential equation is known the complete solution can be found, see Hildebrand[9]. After some integration of the standard formula for the solution to (19) the result is

$$\hat{\psi}_1 = C_1 v_o + C_2 v_o \int^{\bar{r}} \frac{1}{\bar{r}v_o^2}\, d\bar{r} - \frac{1}{2}\bar{r}^2 v_o - v_o \int^{\bar{r}} \frac{1}{\bar{r}v_o^2}\left(\int^{\bar{r}} \bar{r}v_o^2\, d\bar{r}\right) d\bar{r}$$
$$- v_o \int^{\bar{r}} \frac{1}{\bar{r}v_o^2}\left(\int^{\bar{r}} r^2 \frac{d(w_o^2)}{d\bar{r}}\, d\bar{r}\right) d\bar{r} \quad (20)$$

For a vortex core with $v_o \propto \bar{r}$ at $\bar{r} = 0$, the condition that the velocity in the core be everywhere finite leads to the requirement $C_2 = 0$. All the integrals remaining in the solution are well behaved at $\bar{r} = 0$. The solution can be rewritten in terms of definite integrals.

$$\hat{\psi}_1 = C_1 v_o - \frac{1}{2}\bar{r}^2 v_o - v_o \int_0^{\bar{r}} \frac{1}{\bar{r}v_o^2}\left(\int_0^{\bar{r}} \bar{r}v_o d\bar{r}\right) d\bar{r} - v_o \int_0^{\bar{r}} \frac{1}{\bar{r}v_o^2}\left(\int_0^{\bar{r}} r^2 \frac{d(w_o)}{d\bar{r}} dr\right) d\bar{r}$$

$$(21)$$

(The arbitrary constant C_1 appearing above is not the same as the C_1 in (20).)

Now that $\hat{\psi}_1$ is known, the perturbations in velocity and vorticity, $\vec{Q}_1$ and $\vec{\Omega}_1$ can be determined. The general solution for the flow within and near a perturbed vortex core has now been obtained. If one wished to work out the streamline pattern and velocities for a particular initial vorticity distribution, the integrals in (21) would have to be evaluated. This may be difficult to do analytically for a general vorticity distribution. The details of such solutions for several distribtuions of vorticity and axial velocity appear in Appendix B.

However, the details of the vorticity distribution affect the self-induced motion only through the velocity field "seen" by the vortex at distances larger than the core radius, i.e. as $\bar{r} \to \infty$. The general solution gives the velocity field necessary to produce a steady flow within and near the perturbed vortex core with vorticity being convected around closed streamlines. When we examine this flow at large distances we shall find that a "uniform stream" is necessary to keep the flow in place. A portion of this apparent "free stream" is provided by distant elements of the vortex line. The rest of the "free stream" must be due to the actual self-induced motion of the vortex core. This is the physical meaning of the various mathematical operations we must carry out.

The various integrals appearing in the solution (21) for $\hat{\psi}_1$ are analyzed in Appendix B. For the conditions $v_o \sim 1/\bar{r}$ (i.e. $\zeta_o \sim 0$), $w_o \sim 0$ their limiting values as $\bar{r} \to \infty$ are as follows:

$$\lim_{\bar{r}\to\infty} \int_0^{\bar{r}} \bar{r}v_o d\bar{r} \sim A + \ln\bar{r}$$

$$\lim_{\bar{r}\to\infty} \int_0^{\bar{r}} \bar{r}^2 \frac{d(w_o^2)}{d\bar{r}} d\bar{r} = -2\int_0^{\infty} \bar{r}w_o^2 d\bar{r} = -C$$

$$\int_0^{\bar{r}} \frac{1}{\bar{r}v_o^2}\left(\int_0^{\bar{r}} \bar{r}v_o^2 d\bar{r}\right)d\bar{r} \sim B\frac{\bar{r}^2}{2}\ln\bar{r} + \frac{\bar{r}^2}{2}\left(A - \frac{1}{2}\right)$$

$$\int_0^{\bar{r}} \frac{1}{\bar{r}v_o^2}\left(\int_0^{\bar{r}} \bar{r}^2\frac{d(w_o^2)}{d\bar{r}}d\bar{r}\right)d\bar{r} \sim D - \frac{\bar{r}^2}{2}C$$

(22)

The swirl and radial velocities can be obtained by applying (15) to (21). In the limit $\bar{r} \to \infty$, these velocities (rewritten in the variable $r = \varepsilon\bar{r}$) become

$$u = -\varepsilon\frac{\cos\theta}{2}\left[\ln\frac{r}{\varepsilon} + \left(A - C + \frac{1}{2}\right)\right]$$

$$v = \frac{\varepsilon}{r} + \varepsilon\frac{\sin\theta}{2}\left[\ln\frac{r}{\varepsilon} + \left(A - C - \frac{1}{2}\right)\right]$$

(23)

The parameters A and C evidently contain all the effects of the details of both vorticity and axial velocity distribution. The interpretation of C is quite simple; C is proportional to twice the kinetic energy or the momentum flux in the axial flow. The interpretation of A is more subtle; the first integral in (22) expresses the behavior of the total kinetic energy of the swirl motion for the fluid within $\bar{r}$. However, the kinetic energy of a single point vortex is singular as $\log \bar{r}$ as $\bar{r} \to \infty$ (see Batchelor[10]). The singular part depends only upon $\bar{r}$; the finite part, A depends upon the details of the swirl distribution. For a vortex core of constant vorticity, $A = 1/4$; for a decaying line vortex, $A = -.058$ (see Appendix B).

Having found a general solution for the flow within and near a curved vortex filament, we now apply the result to two specific cases, the perturbed sinusoidal filament and the vortex ring.

2.2   Self-Induced Motion of a Sinusoidal Vortex Line

The self-induced motion of a single vortex line shaped like a sinusoid with a wavelength λ much greater than its amplitude and core diameter will now be obtained. To compute the influence of the entire line upon the region surrounding the core, the Biot-Savart law is applied to the vortex filament since to lowest order this influence does not depend upon the details of the vorticity distribution within the core.

The solution for the velocity field will be found in a coordinate system fixed on the vortex line. Since the vortex line is usually in motion with a propagation velocity $\vec{Q}_p$, this coordinate system will be in motion relative to the fluid at rest at infinity. Therefore a local velocity vector, $-\vec{Q}_p$, (non-dimensionalized with κ/R) must be added to the solution to account for this. If we wished to allow for travelling waves, it would be convenient to work in a coordinate system travelling with the wave speed in which case $\vec{Q}_p$ would include a free stream in the direction of the unperturbed vortex. However, a simple analysis, based on slender body theory indicates that in the vortex pair instability, in which rotation of the vortex is suppressed by the velocity field of the other vortex, travelling waves cannot exist to lowest order.

The self-induced motion $\vec{Q}_p$ will be determined by matching the local solution for the flow within and near the vortex core as $\bar{r} \to \infty$ with the solution for the influence of the entire vortex line as $r \to 0$. The solution outside the vortex core (outer solution) is written

$$\vec{Q} = \frac{\kappa}{2} \int_{C_v} \frac{d\vec{s} \times \vec{r}_v}{r_v^3} - \vec{Q}_p + 0[\varepsilon^2] \tag{24}$$

In order to compute the effect of the entire vortex on the self-induced motion of each part, it is necessary to find the behavior of (24) very near the vortex line. For both the sinusoidal filament and the vortex ring, the effect of curvature is to produce a self-induced motion, u_p, normal to the local plane of the vortex. Therefore, to simplify the analysis, we shall consider only u, the vertical component of (24) (See Fig. 2). The contribution of the integral in (24) to u will be called $\tilde{u}$. One cannot evaluate $\tilde{u}$ on C_v (although the integral will usually be simplest for this case) because when C_v is curved the $\tilde{u}$ diverges logarithmically. The behavior of $\tilde{u}$ in the limit $r \rightarrow 0$ will be obtained in the following way: we consider a distance, ℓ, along the vortex line on both sides of the region where the velocity is to be evaluated, and in which the shape of the vortex line can be approximated by a circular arc having a radius equal to the local radius of curvature. For all of the vortex line except this interval the denominator of the integral can be expanded for small distances from the vortex line and integrated. For the interval itself, the solution for the vortex ring can be used, provided the influence of the part of the ring not in the interval is subtracted away. This correction term for the ring is no more difficult to evaluate than the influence of the sinusoid outside the interval. All lowest order terms involving ℓ cancel when the contributions from the two regions are added.

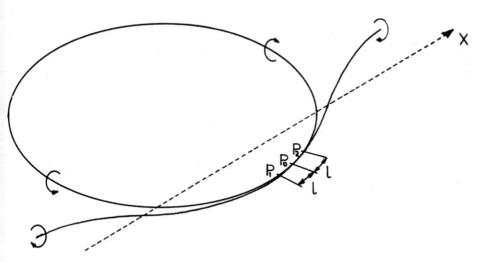

Figure 4.   Use of vortex ring to obtain local behavior.

The contribtuions of the parts of the sinusoid from minus infinity to P_1 and from P_2 to plus infinity can be evaluated for small values of ℓ in a manner similar to that of Crow[1] for a "cut-off" distance of ℓ. Retaining the terms important to the order considered, we obtain, as $r \to 0$, ($k = 2\pi/\lambda$)

$$\tilde{u} = \frac{\kappa}{2R}[\ell n \frac{1}{k\ell} - \gamma + \frac{1}{2} + \frac{(k\ell)^2}{8} + O[(k\ell)^4] - \frac{r\sin\theta}{\ell^2}R] \qquad (25)$$

The contribution of the missing segment of the sinusoid P_1P_2 is found by using the solution for a vortex ring of radius R, (the local radius of curvature of the sinusoid at P_o) with the influence of the remaining portion of the vortex ring subtracted away. The solution for a vortex ring is well known. The vertical velocity components of a vortex ring in the limit $r \to 0$ were given by Tung and Ting[6] as

$$\tilde{u} = -\frac{\kappa}{r}\sin\theta + \frac{\kappa}{2R}\ell n\frac{8R}{r} - \frac{\kappa}{2R}\cos^2\theta \qquad (26)$$

The contribution of that portion of a vortex ring of radius R which lies outside the interval P_1P_2 is obtained by a straightforward application of the Biot-Savart law. The answer written in dimensional form, expanded for small ℓ, is found to be (Bliss[8])

$$\tilde{u} = \frac{\kappa}{2R}\left(\ell n\frac{4R}{\ell} - \frac{\ell^2}{48R} + O[\frac{\ell^4}{R^4}]\right) + \frac{\kappa}{2R}\frac{r\sin\theta}{R}\left(\frac{-3}{4}\ell n\frac{4R}{\ell} - \frac{R^2}{\ell^2} + \frac{1}{24} + O[\frac{\ell^2}{R^2}]\right)$$

$$(27)$$

All the pieces to assemble the limit of the integral $\tilde{u}$ as $r \to 0$ are now available. The contribution of the sinusoid is added to the entire vortex ring and the contribution of the portion of the vortex ring which is not also part of the sinusoid is subtracted away. The result is

$$\tilde{u} = -\frac{\kappa}{r}\sin\theta + \frac{\kappa}{2R}[\ell n\frac{2}{kr} - \cos^2\theta + \frac{1}{2} - \gamma] + O[\frac{\ell^2}{R^2}] + O[\frac{r}{R}\ell n\frac{R}{\ell}] \qquad (28)$$

Terms of $O[\frac{\ell^2}{R^2}]$ are considered higher order and terms of $O[\frac{r}{R}\ell n\frac{R}{\ell}]$ vanish as $r \to 0$.

In non-dimensional variables the limit as $r \to 0$ of $\tilde{u} - u_p$ is

$$\lim_{r \to 0} u^o = -\frac{\varepsilon}{r}\sin\theta + \frac{\varepsilon}{2}[\ell n\frac{2}{kr} - \ell nR + \frac{1}{2} - \gamma - \cos^2\theta - 2\bar{u}_p] \qquad (29)$$

The behavior of the radial and swirl velocities in the neighborhood of the vortex core in the limit $\bar{r} \to \infty$ is given by (23), the vertical velocity $u^{\ell} = u\cos\theta - v\sin\theta$, becomes

$$\lim_{\bar{r}\to\infty} u^i = -\frac{\varepsilon}{r}\sin\theta + \frac{\varepsilon}{2}[\ln\frac{a}{r} - \ln R - (A - C - \frac{1}{2}) - \cos^2\theta] \quad (30)$$

The local solution valid in and near a perturbed vortex filament can now be matched to the solution valid outside the sinusoidal
vortex line to obtain the self-induced motion, $\bar{u}_p$. The inner solution and the outer solution are matched in an intermediate region
near the vortex core where both are assumed to be valid. This overlap region is considered to be "far" from the core and "very close"
to the line.

$$\lim_{r\to 0} u^o = \lim_{\bar{r}\to\infty} u^i \quad (31)$$

The value of the self-induced motion, $\bar{u}_p$, is obtained by equating these two limits as indicated in (31). The result is

$$\bar{u}_p = \frac{1}{2}[\ln\frac{2}{ka} + A - C - \gamma] \quad (32)$$

In dimensional form the result reads

$$u_p = \frac{\kappa}{2}k^2 r_o\sin kx \; [\ln\frac{1}{ka} + A - C + (\ln 2 - \gamma)] \quad (33)$$

Dividing by the local amplitude of the sinusoid, $r_o\sin kx$, we
see that the sinusoid rotates rigidly (without change of shape)
with a constant angular velocity

$$\Omega_p = \frac{\kappa}{2}k^2[\ln\frac{1}{ka} + A - C + (\ln 2 - \gamma)] \quad (34)$$

in a direction opposite to the rotation of flow in the vortex core.

The effect of axial flow upon the self-induced rotation of the
vortex enters through the value of C, which is always positive;
axial flow reduces Ω_p.

Two particular cases for the sinusoidal vortex line are of
special interest. Setting C = 0 and evaluating the constant ($\ln 2 -$
γ) in equation (34) we find

$$\Omega_p = \frac{\kappa}{2}k^2[\ln\frac{1}{ka} + A + 0.1159] \quad (35)$$

For the case of a core with constant vorticity in the uncurved
state, A = 1/4. When the vorticity is initially concentrated in
a cylindrical sheet, A = 0. These two cases are in precise agreement with results found in 1880 by Kelvin[4].

2.3 The Vortex Ring

Perhaps the simplest example of a curved rotational vortex line is the vortex ring. By matching the outer solution for the ring in the limit $r \to 0$ with the local solution for the vortex core found previously, a formula for the propagation velocity of the ring for any $O[1]$ distribution of vorticity, ζ_o, can be obtained. In the unlikely event that an axial velocity is present in the vortex core, this effect can also be included in a general manner.

Due to the symmetry of the vortex ring, the outer problem is essentially two-dimensional and it is convenient to use a stream function. The stream function for a ring is well known, see Lamb[4]; expressed in non-dimensional outer variables it is

$$\psi^o = (r + r_1)[F_1(\frac{r_1 - r}{r_1 + r}) - E_1(\frac{r_1 - r}{r_1 + r})] \tag{36}$$

where F_1 and E_1 are complete elliptic integrals of the first and second kind, respectively. The coordinate r is the same as used throughout and in this case represents the shortest distance from the ring to the point where ψ^o is evaluated. The coordinate r_1 is greatest distance from the ring to the point of evaluation.

The inner limit requires the form of equation (36) very near the ring where $r \ll 1$. The limit $r \to 0$ of the vertical velocity field for the ring (26) used previously in the development of the self-induced motion of the sinusoid vortex filament, in non-dimensional variables, is

$$\lim_{r \to 0} u^o = - \frac{\varepsilon}{r}\sin\theta + \frac{\varepsilon}{2}[\ln\frac{8R}{r} - \cos^2\theta - 2\bar{u}_p] \tag{37}$$

where the propagation velocity of the vortex ring $\bar{u}_p$ has been included.

Applying the matching principle (31) to the velocity fields in (37) and (30) we obtain the propagation velocity of the ring $\bar{u}_p$

$$\bar{u}_p = \frac{1}{2}[\ln\frac{8R}{r} + A - C - \frac{1}{2}] \tag{38}$$

In dimensional form, the propagation velocity is

$$u_p = \frac{\kappa}{2R}[\ln\frac{8R}{a} + A - C - \frac{1}{2}] \tag{39}$$

where a is the characteristic dimension chosen for the inner region. Because C is always positive, the presence of an axial velocity will reduce the speed of propagation. Although the problem of a vortex ring with axial velocity is somewhat academic, the results have a satisfying interpretation: A radial force would be necessary to support the axial momentum flux around the vortex ring. This force turns out to be exactly equal to the Kutta-Joukowski lift provided by the product of the circulation of the ring and the difference in propagation velocities of the ring with and without axial velocities.*

A formula equivalent to (39) (without axial velocity) was found by Saffman[7] by considering the energy and impulse of the ring.

The values of A for specific vorticity distributions are worked out in Appendix B. With these values, the propagation velocity can be determined for some specific cases. For the case of ζ_0 constant in the core, and no axial velocity, A = 1/4 and C = 0; the velocity of the ring is

$$u_p = \frac{\kappa}{2R}[\ln\frac{8R}{a} - \frac{1}{4}]$$ (40)

where a is the core radius. This corresponds to the classical result in Lamb[5] obtained by an energy method. It was found originally by Lord Kelvin[4] in 1867.

The propagation velocity of the decaying vortex (a = $\sqrt{4\nu t}$, A = ⁻.058) is

$$u_p = \frac{\kappa}{2R}[\ln\frac{8R}{\sqrt{4\nu t}} - 0.558]$$ (41)

This is not in agreement with Tung and Ting[6] due to an error in their calculation. The result agrees with that found by Saffman[7] using an energy method.

3.   STABILITY OF A VORTEX PAIR

Having obtained a general formula for the self-induced velocity of a perturbed sinusoidal vortex filament of arbitrary vorticity and axial velocity distribution we can complete the calculation of the stability of the perturbed vortex pair as sketched in Fig. 1. In simplified terms, the instability occurs whenever the sum of $\vec{V}_i$ and $\vec{V}_\Omega$ is parallel to $\vec{r}$, the vortex displacement. In this case, the governing equation is simply

$$\vec{V}_i + \vec{V}_\Omega = \dot{\vec{r}}(t) = Q\vec{r}(t)$$ (42)

The vortex line diverges as

$$\vec{r}(t) \sim e^{Qt}$$ (43)

* This interpretation was suggested by the remarks of Steve Crow at the Symposium.

The components of the induced velocity vector $\vec{V}_i = v_{iz}\vec{k} + v_{iy}\vec{j}$, as calculated by Crow[1], are

$$v_{iz} = \sin\frac{2\pi x}{\lambda} \frac{\Gamma r_o\cos\theta}{2\pi b^2}[1 + (\frac{2\pi b}{\lambda})\cdot K_1(\frac{2\pi b}{\lambda})]$$

$$v_{iy} = \sin\frac{2\pi x}{\lambda} \frac{\Gamma r_o\sin\theta}{2\pi b^2}[1 - (\frac{2\pi b}{\lambda})^2\cdot K_o(\frac{2\pi b}{\lambda}) - (\frac{2\pi b}{\lambda})K_1(\frac{2\pi b}{\lambda})]$$

(44)

where K_o and K_1 are Bessel functions of the second kind.

The components of the self-induced velocity $\vec{V}_\Omega = v_{\Omega z}\vec{k} + v_{\Omega y}\vec{j}$ can be obtained from the angular velocity of rotation of the vortex sinusoid given by (34), rewritten as a function of (λ/b) and a/b where b is the distance between the vortex pair.

$$v_{\Omega z} = -\sin\frac{2\pi x}{\lambda} \frac{\Gamma}{4\pi b^2} (\frac{2\pi b}{\lambda})^2 r_o\cos\theta[\ln\frac{\lambda}{\pi b} + \ln\frac{b}{a} + A - C - \gamma]$$

$$v_{\Omega y} = \sin\frac{2\pi x}{\lambda} \frac{\Gamma}{4\pi b^2} (\frac{2\pi b}{\lambda})^2 r_o\sin\theta[\ln\frac{\lambda}{\pi b} + \ln\frac{b}{a} + A - C - \gamma]$$

(45)

It is of some interest to note that the kinetic energy per unit length in the vortex wake of an aircraft modelled as a vortex pair with no net circulation is

$$T = \frac{\rho\Gamma^2}{2\pi} [\ln\frac{b}{a} + A + \frac{C}{2}]$$

(46)

where b is the vortex separation and a is the vortex core radius.

In the absence of axial velocities, $C = 0$ and the kinetic energy per unit length is equal to the induced drag. Thus, neither a nor A need be found individually to investigate wake stability since the appropriate combination is available from a knowledge of induced drag.

With axial velocities, the appropriate combination $\ln\frac{b}{a} + A - C$ is likely to have a similar relation to total drag.

Since all induced velocities are proportional to $r_o\Gamma/(2\pi b^2)$, it is convenient to introduce a non-dimensional amplification rate, α, (as was done by Crow[1])

$$Q = \frac{\Gamma}{2\pi b^2} \alpha$$

(47)

For a vortex pair with known values of a, A and C, the stability calculation is quite straightforward; choose λ, find θ, solve for α. The calculation is completely equivalent to that done by Crow, and the results would be identical for the appropriate case (constant vorticity core, c = 2a = .197b, and an axial velocity of zero in the core) except for a very slight error in his application of the Kelvin result to evaluate "cut-off" distances.

As an example, we consider a vortex core with constant vorticity and a parabolic distribution of axial velocity.

$$w_o(r) = \bar{w}[a^2 - r^2] \tag{48}$$

where $\bar{w} = w_o(0)/v_o(a)$, is the ratio of maximum axial velocity to maximum swirl velocity. For this case, $A = 1/4$ and $C = \bar{w}^2/3$.

In general, in the instability calculation for a given value of a, two groups of unstable waves are found, referred to as long and short. In each of these categories there is a "most unstable" wave with an amplification rate, α, that is a local maximum. Fig. 5 shows the most unstable wave, λ, as a function of vortex core size for several values of axial velocity $\bar{w}$. Fig. 6 shows the corresponding amplification rate.

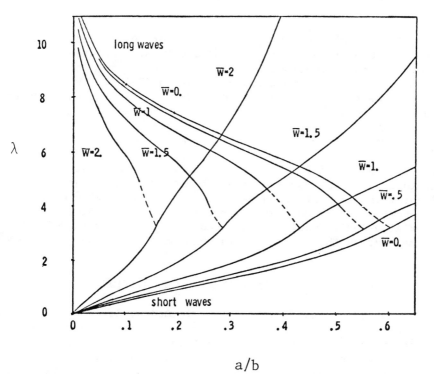

a/b

Figure 5.   Most unstable wavelength versus vortex core size.

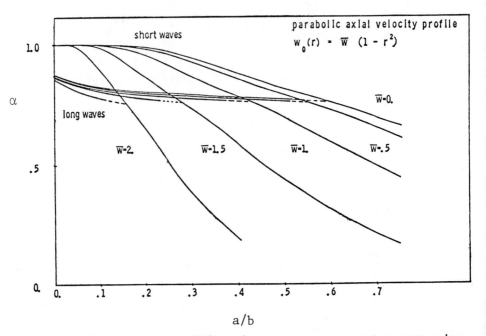

Figure 6. Maximum amplification rate versus vortex core size.

Without axial velocity the results are equivalent to those presented by Crow in a slightly different form. The effect of axial
velocities is to contract the curves to the left. This effect can
easily be understood in terms of equation (34) for the self-induced
motion; the presence of axial velocities is equivalent to an effective increase in core size

$$a_{eff} = ae^{+C} \qquad\qquad (49)$$

The effects of axial velocity are: to reduce the most unstable wavelength of the long waves most often observed in flight, to decrease
slightly the amplification rate, to cause the merging of the long
and short wave branches at lower values of core radius. (The effect
of axial velocity would be enhanced if a uniform value of $\bar{w}$ had been
assumed across the core since in this case $C = \bar{w}^2$ rather than $\bar{w}^2/3$.)

4. EXPERIMENTAL OBSERVATIONS

A laboratory experiment was performed by Zalay[11] to measure the
amplification rate of the instability of a vortex pair. Vortex line
may be impulsively generated by striking a piston which forms one
side of a chamber with a cut-out of desired shape. (e.g., a vortex
ring is produced if the cut-out is circular). The apparatus sketche
in Fig. 7

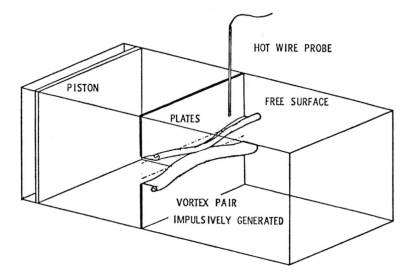

Figure 7.   Vortex tank.

was designed to produce a vortex pair. The piston chamber was sealed
except for the slit between the plates where the vortex pair was produced. The chamber beyond the piston had a free surface. The vortex
pair was produced by an impulse applied to the piston. The vortex
core size was measured with a hot wire probe. The initial core size
could be varied (by changing the impulse) over the range $.085 \leq a/b \leq$
$.2$.   Sinusoidal displacements were cut into the plate so that an
initial perturbation wavelength and displacement could be set. The
vortex pair was visualized by means of ink applied at the edges of
the plates. The instability always occurred at the preset wavelength.

     The amplification rate was obtained from motion pictures of the
instability. The observed amplification rates are compared with
theoretical predictions on the following basis: for each pre-set
wavelength λ/b, there is a theoretical maximum amplification rate
α_{max}. For all except the wave for which $\lambda/b = 12$, the vortex core
size at which $\alpha = \alpha_{max}$ is larger than the initial core size that
could be achieved in the experiment. We compare each α observed in
the experiment with its theoretical maximum for the pre-set value of
λ/b. Fig. 8 shows the theoretical values of α_{max} as a function of
λ/b. (The theoretical value of α_{max} for $\lambda/b = 12$ was taken as the
α for a core size of $a/b = .085$, the minimum that was achieved in
the experiment.) The experimental values of α as a function of λ/b
for various values of a/b also appear in Fig. 8. The scatter in the
experiment is substantial, but the general trends show some agreement
with theory.

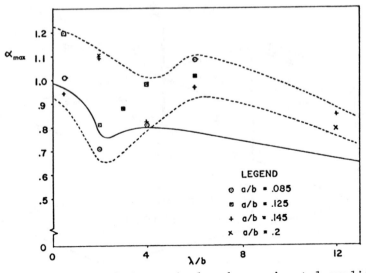

Figure 8.   Comparison of theoretical and experimental amplification
             rates.

Preliminary experiments were performed by Withycombe[12] in the
80 foot long M.I.T. towing tank. A rectangular wing of 18" span
and 2" chord was towed at a speed of 2 to 4 kts. Ink injected at
the wing tip was used to visualize the tip vortices. The presence
of axial velocity in the tip vortices was confirmed by injecting
"bubbles" of an imisible fluid into the core; these "bubbles" moved
towards the wing at about half of its speed. The angle of attack
was varied from 4° to 12°. Motion pictures of the tip vortices were
taken.

The qualitative features of the wake breakup were as follows:

1) In most cases, vortex breakdown occurred before a strong
   sinusoidal instability developed.

2) The sinusoidal instability observed was a "short" wave,
   $\lambda \simeq 1.3b$.

3) When the sinusoidal instability did develop, vortex breakdown was observed in the high pressure regions where the
   vortices are farthest apart.

This later effect is shown in Fig. 9, a sequence of stills obtained from the motion picture. The vortex breakdowns, which appear
at all four points of maximum pressure travel to the pinch, playing
quite an important role in the final breakup. (These pictures were
shot from the side of the tank; λ/b is much smaller than it appears
in this view.)

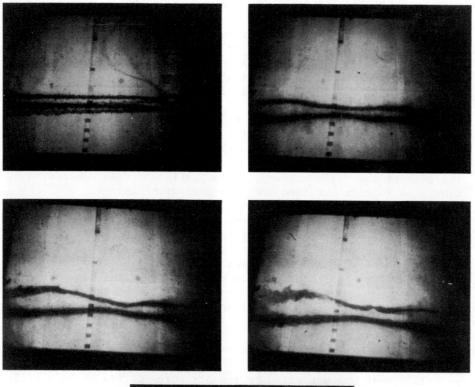

Figure 9.   Flow visualization of sinusoidal instability with induced
            vortex breakdown.

5.   CONCLUSIONS

A theory for the dynamics of curved vortex filaments of finite
size has been developed which allows the details of both swirl and
axial velocity distributions to be included in the calculation of
self-induced motion. A general solution is obtained which shows

that the self-induced motion is influenced by the details of swirl
and axial velocity only through the kinetic energy in the swirl and
the axial momentum flux. The presence of axial velocities slows
down the self-induced motion by an amount which provides the Kutta-
Joukowski lift on the filament necessary to turn the axial flow.
Specific calculations of self-induced motion for a sinusoidal vortex
filament and a vortex ring agree with earlier results.

When this model is applied to the calculation of the stability
of the vortex wake of an aircraft, general results are obtained
which suggest that it may be possible to characterize the stability
of the wake in terms of overall parameters such as total drag. When
the present theory is applied to the particular case considered by
Crow[1] the results obtained are identical with his.

The experimental observations in the vortex tank confirm in
a qualitative manner the instability predicted by theory . The
preliminary towing tank experiments give much more of an insight
into the complex mechanism of wake dissipation. Vortex breakdown
was observed either to preceed to development of the sinusoidal
instability or to be induced by the instability at the points of
maximum pressure along the deformed vortices.

6. ACKNOWLEDGMENTS

This work was supported by the Air Force Office of Scientific
Research (OSR) under contract F44620-69-C-0090.

7. REFERENCES

1. Crow, W. C., "Stability Theory for a Pair of Trailing Vortices",
    AIAA Paper No. 70-53 [January, 1970].

2. Landahl, M. and S. Widnall, "Vortex Control", Symposium on
    Aircraft Wake Turbulence, September 1970.

3. Van Dyke, M., Perturbation Methods in Fluid Mechanics, Academic
    Press, New York, [1964].

4. Thomson, Sir W. (Lord Kelvin), "Vibrations of a Columnar Vortex"
    Mathematical and Physical Papers, Volume IV, Cambridge University
    Press, Cambridge, England, [1910].

5. Lamb, Sir. H., Hydrodynamics, Dover Publications, Inc., New York
    [1945].

6. Tung, C. and Ting, L., "The Motion and Decay of a Vortex Ring", Physics of Fluids, Vol. 10, No. 5, pp. 901-10, [May, 1967].

7. Saffman, P. G., "The Velocity of Viscous Vortex Rings", Symposium on Aircraft Wake Turbulence, September 1970.

8. Bliss, D. B., "The Dynamics of Curved Rotational Vortex Lines", Massachusetts Institute of Technology Masters Thesis, September 1970.

9. Hildebrand, F. B., Advanced Calculus for Applications, Fifth Printing, Prentice-Hall, Inc., Englewood Cliffs, New Jersey, pp. 29-31, [1965].

10. Batchelor, G. K., An Introduction to Fluid Dynamics, Cambridge University Press, Cambridge, England, p. 529, [1967].

11. Zalay, A. D., "Experimental Investigation of the Decay of a Vortex Pair", S.M. Thesis, Massachusetts Institute of Technology, June, 1970.

12. Withycombe, E. II, "Wingtip Vortex Decay: An Experimental Investigation", S.B. Thesis, Massachusetts Institute of Technology, August 1970.

APPENDIX A

The Local Cylindrical Coordinate System

The purpose of this appendix is to provide a list of vector operators in the local curvilinear coordinate system of Fig. 2.

The curved cylindrical coordinates are related to the Cartesian components by

$$r = [(\sqrt{x_o + y_o} - R)^2 + z_o^2]^{\frac{1}{2}}$$

$$\theta = \arctan\left[\frac{\sqrt{x_o + y_o} - R}{z_o}\right] \tag{A.1}$$

$$s = R \arctan\left[\frac{y_o}{x_o}\right]$$

Solving (A.1) for the Cartesian components in terms of the curvilinear components yields

$$x_o = (r\sin\theta + R)\cos\frac{s}{R}$$

$$y_o = (r\sin\theta + R)\sin\frac{s}{R} \tag{A.2}$$

$$z_o = r\cos\theta$$

Vector operators in general orthogonal curvilinear coordinates are discussed in Hildebrand and Batchelor. The position vector in the Cartesian system is

$$\bar{r}_o = x_o\bar{i} + y_o\bar{j} + z_o\bar{k} \tag{A.3}$$

It is necessary to determine scale factors, h_i, such that

$$h_1\bar{i}_r = \frac{\partial \bar{r}_o}{\partial r}, \quad h_2\bar{i}_\theta = \frac{\partial \bar{r}_o}{\partial \theta}, \quad \text{and} \quad h_3\bar{i}_s = \frac{\partial \bar{r}_o}{\partial s} \tag{A.4}$$

where the derivatives can be calculated using (A.3) and (A.2). The results are

$$h_1 = 1, \quad h_2 = r, \quad h_3 = \frac{R + r\sin\theta}{R} \tag{A.5}$$

and

$$\bar{i}_r = \sin\theta \, \cos\frac{s}{R} \, \bar{i} + \sin\theta \, \sin\frac{s}{R} \, \bar{j} + \cos\theta \, \bar{k}$$

$$\bar{i}_\theta = \cos\theta \, \cos\frac{s}{R} \, \bar{i} + \cos\theta \, \sin\frac{s}{R} \, \bar{j} - \sin\theta \, \bar{k} \qquad (A.6)$$

$$\bar{i}_s = -\sin\frac{s}{R} \, \bar{i} + \cos\frac{s}{R} \, \bar{j} + 0\bar{k}$$

By using formulas from the references the vector operators can now be found.

Gradient:
$$\nabla\phi = \frac{\partial\phi}{\partial r}\bar{i}_r + \frac{1}{r}\frac{\partial\phi}{\partial\theta}\bar{i}_\theta + \frac{R}{R + r\sin\theta}\frac{\partial\phi}{\partial s}\bar{i}_s$$

Divergence:

$$\nabla\cdot\bar{A} = \frac{1}{r}\frac{\partial(rA_r)}{\partial r} + \frac{1}{r}\frac{\partial A_\theta}{\partial\theta} + \frac{R}{R + r\sin\theta}\left(\frac{A_r\sin\theta + A_\theta\cos\theta}{R} + \frac{\partial A_s}{\partial s}\right) \qquad (A.8)$$

Curl:
$$\nabla\times A = \left[\frac{1}{r}\frac{\partial A_s}{\partial\theta} + \frac{A_s\cos\theta}{R + r\sin\theta} - \frac{R}{R + r\sin\theta}\frac{\partial A_\theta}{\partial s}\right]\bar{i}_r$$

$$+ \left(\frac{R}{R + r\sin\theta}\frac{\partial A_r}{\partial s} - \frac{\partial A_s}{\partial r} - \frac{\sin\theta}{R + r\sin\theta}A_s\right)\bar{i}_\theta \qquad (A.9)$$

$$+ \left[\frac{1}{r}\frac{\partial(rA_\theta)}{\partial r} - \frac{1}{r}\frac{\partial A_r}{\partial\theta}\right]\bar{i}_s$$

Total Derivative:

$$\frac{D\bar{A}}{Dt} = \frac{\partial\bar{A}}{\partial t} + \bar{u}\cdot\nabla\bar{A} \qquad (A.10)$$

$$= \left[\frac{\partial A_r}{\partial t} + u\frac{\partial A_r}{\partial r} + \frac{v}{r}\frac{\partial A_r}{\partial\theta} + \frac{wR}{R + r\sin\theta}\frac{\partial A_r}{\partial s} - \frac{A_\theta v}{r} - \frac{A_s w\sin\theta}{R + r\sin\theta}\right]\bar{i}_r$$

$$+ \left[\frac{\partial A_\theta}{\partial t} + u\frac{\partial A_\theta}{\partial r} + \frac{v}{r}\frac{\partial A_\theta}{\partial\theta} + \frac{wR}{R + r\sin\theta}\frac{\partial A_\theta}{\partial s} + \frac{A_r v}{r} - \frac{A_s w\cos\theta}{R + r\sin\theta}\right]\bar{i}_\theta$$

$$+ \left[\frac{\partial A_s}{\partial t} + u\frac{\partial A_s}{\partial r} + \frac{v}{r}\frac{\partial A_s}{\partial\theta} + \frac{wR}{R + r\sin\theta}\frac{\partial A_s}{\partial s} + \frac{A_r w\sin\theta + A_\theta w\cos\theta}{R + r\sin\theta}\right]\bar{i}_s$$

where $\bar{u} = u\bar{i}_r + v\bar{i}_\theta + w\bar{i}_s$

Laplacian: $\quad \nabla^2\phi = \dfrac{1}{r}\dfrac{\partial\phi}{\partial r}(r\dfrac{\partial\phi}{\partial r}) + \dfrac{\sin\theta}{R + r\sin\theta}\dfrac{\partial\phi}{\partial r} + \dfrac{1}{r}\dfrac{\partial^2\phi}{\partial\theta^2} +$

$$+ \dfrac{\cos\theta}{r(R + r\sin\theta)}\dfrac{\partial\phi}{\partial\theta} + \dfrac{R^2}{(R + r\sin\theta)^2}\dfrac{\partial^2\phi}{\partial s^2} \tag{A.11}$$

APPENDIX B

Examination of the General Solution for a Perturbed Vortex Core

The general solution for the stream function $\hat{\psi}_1$ for a perturbed vortex core is (equation (21))

$$\hat{\psi}_1 = C_1\dfrac{v_0}{\bar{r}} - \dfrac{1}{2}\bar{r}^2 v_0 - v_0\int_0^{\bar{r}}\dfrac{1}{\bar{r}v_0^2}\left(\int_0^{\bar{r}}\bar{r}v_0^2 d\bar{r}\right)d\bar{r} - v_0\int_0^{\bar{r}}\dfrac{1}{\bar{r}v_0^2}\left(\int_0^{\bar{r}}\bar{r}^2\dfrac{\partial(w_0^2)}{\partial\bar{r}}d\bar{r}\right)d\bar{r}$$

$$\tag{B.1}$$

where v_0 is the swirl velocity and w_0 the axial flow in the unperturbed vortex. In this appendix we have two objectives: to examine the limiting behavior of the integrals in (B.1) as $\bar{r} \to \infty$ and to examine the solution for some simple vortex models.

For vortex flows of this type, we can identify a radius $\bar{r} = k$ outside of which the vorticity ζ_0 is effectively zero. For vortex cores having ζ_0 approaching zero asymptotically as $\bar{r} \to \infty$, we require that $\zeta_0 \sim O[\varepsilon^2]$ or less, for $\bar{r} > k$. The swirl $v_0(r)$ may then be written

$$v_0 = \begin{cases} v_0(\bar{r}), & 0 \leq \bar{r} \leq k \\ 1/\bar{r}, & k < \bar{r} < \infty \end{cases} \tag{B.2}$$

Now consider the integral

$$\int_0^{\bar{r}}\bar{r}v_0^2\, d\bar{r}$$

which appears in $\hat{\psi}_1$. Define

$$\int_0^{\bar{r}}\bar{r}v_0^2\, d\bar{r} \equiv I_1(\bar{r}), \qquad 0 \leq \bar{r} \leq k. \tag{B.3}$$

For $\bar{r} > k$,

$$\int_0^{\bar{r}} \bar{r} v_o^2 \, d\bar{r} = \int_0^k \bar{r} v_o^2 \, d\bar{r} + \int_k^{\bar{r}} \bar{r} (\frac{1}{\bar{r}})^2 \, d\bar{r} = I_1(k) - \ell n k + \ell n \bar{r} = A + \ell n \bar{r}$$

(B.4)

where

$$A \equiv I_1(k) - \ell n k$$

(B.5)

The constant A can be shown to be independent of k for all suitable values of k. Thus,

$$\int_0^{\bar{r}} \bar{r} v_o^2 \, d\bar{r} = \begin{cases} I_1(\bar{r}), & 0 \le \bar{r} \le k \\ A + \ell n \bar{r}, & k < \bar{r} < \infty \end{cases}$$

(B.6)

Now consider

$$\int_0^{\bar{r}} \frac{1}{\bar{r} v_o^2} \left(\int_0^{\bar{r}} \bar{r} v_o^2 \, d\bar{r} \right) d\bar{r}$$

(B.7)

which then becomes

$$\int_0^{\bar{r}} \frac{1}{\bar{r} v_o^2} \left(\int_0^{\bar{r}} \bar{r} v_o^2 \, d\bar{r} \right) d\bar{r} = \begin{cases} I_2(\bar{r}), & 0 \le \bar{r} \le k \\ B + \frac{\bar{r}}{2} \, n\bar{r} + \frac{\bar{r}}{2}(A - \frac{1}{2}), & k < \bar{r} < \infty \end{cases}$$

(B.8)

where

$$B = I_2(k) - \frac{k^2}{2}[I_1(k) - \frac{1}{2}]$$

(B.9)

is also independent of k.

A similar approach is taken for the two integrals involving axial velocity, which appear in the solution. We impose the further restriction on k

$$w_o \simeq 0, \quad \bar{r} > k$$

(B.10)

Therefore write

$$w_o = \begin{cases} w_o(\bar{r}), & 0 \le \bar{r} \le k \\ 0, & k < \bar{r} < \infty \end{cases}$$

(B.11)

For any valid choice of v_o and w_o a value k can be found so (B.2) and (B.10) are satisfied.

Then one of the integrals which appears in equation (B.1) can be represented as

$$\int_0^{\bar{r}} \bar{r}^2 \frac{\partial (w_o^2)}{\partial \bar{r}} \, d\bar{r} = -\bar{r}^2 w_o^2 \Big|_0^{\bar{r}} + 2 \int_0^{\bar{r}} \bar{r} w_o^2 \, d\bar{r} = I_3(\bar{r}), \quad 0 \le \bar{r} \le k \quad (B.12)$$

and

$$\int_0^{\bar{r}} \bar{r}^2 \frac{\partial (w_o^2)}{\partial \bar{r}} \, d\bar{r} = 2 \int_0^k \bar{r} w_o^2 \, d\bar{r} = -C, \quad k < \bar{r} < \infty \qquad (B.13)$$

Since $w_o = 0$ for $\bar{r} > k$, the constant C does not depend on k.

The remaining integral which contains the axial velocity can be represented in the form

$$\int_0^{\bar{r}} \frac{1}{\bar{r} v_o^2} \left[\int_0^{\bar{r}} \bar{r}^2 \frac{\partial (w_o^2)}{\partial \bar{r}} \, d\bar{r} \right] d\bar{r} = \begin{cases} I_4(\bar{r}), & 0 \leq \bar{r} \leq k \\ D - \frac{\bar{r}^2}{2} C, & k < \bar{r} < \infty \end{cases} \qquad (B.14)$$

where

$$D \equiv I_4(k) - \frac{k^2}{2} I_3(k) \qquad (B.15)$$

with I_3 defined by (B.12). D does not depend on k as long as the restrictions on k are satisfied.

The limiting behavior of the integrals in $\hat{\psi}_1$, required to complete the solution for self-induced motion of the curved vortex filament, are (B.8) and (B.13).

Although it is not required for the calculation of self-induced motion, the constant C_1 in $\hat{\psi}_1$ is needed for a complete solution, i.e. to draw streamlines within the core. In the general core, C_1 cannot be determined from the solution to this order. However, in the special case of an inviscid rotational core of finite size with the vorticity ζ_o definitely vanishing at a certain value of $\bar{r}$, the value of C_1 can be found by requiring that the pressure be continu- our in value across the boundary of the vortex core. The perturba- tion of vorticity, ζ_1, depends directly upon ζ_o so that when the latter vanishes so does the former. This means that the boundary of the vortex core remains circular when curved (at least to the order of the present solution). Choosing C_1 so that the perturbed velocities are continuous across this circular boundary will satis- fy the requirement of continuous pressure. It is found that the u is consinuous for any choice of C_1, but that v is only continu- ous when

$$C_1 = \frac{k^2}{2} + I_2(k) + I_4(k) \qquad (B.16)$$

For this value of C_1, u vanishes on the boundary of the rotational region which shows that, as expected, the requirement of continuous pressure allows no flow from the rotational to the irrotational re- gion, or vice versa. Bliss (1970) shows that this determination of C requires that $\zeta_{o\bar{r}}$ have a non-zero value at the boundary of the vortex core; otherwise C_1 cannot be found at this order of solution.

We now apply some of the results to specific distributions of vorticity and axial velocity. This should serve to illuminate some of the previous remarks and will provide solutions for particular velocity distributions.

The particular vorticity distributions to be considered are of two different forms that are often used as models for real vortices: ζ_0 equals a constant inside the vortex core, and ζ_0 equals the well known similarity solution for a decaying line vortex.

In the case of a constant vorticity core the velocity in the core corresponds to solid body rotation and is given by

$$v = \frac{\kappa r}{a^2}, \quad 0 \leq r \leq a \tag{B.17}$$

the distance a being the core diameter. In the surrounding inviscid region the velocity is

$$v = \frac{\kappa}{r}, \quad a < r < \infty \tag{B.18}$$

as it is for all cases when the core is of finite extent. The vorticity is

$$\zeta_0 = \frac{2\kappa}{a^2} \tag{B.19}$$

For this case the characteristic dimension is equal to the core radius a and therefore $k = 1$. For this case, the velocity and vorticity, in nondimensional variables, are

$$v_0 = \begin{cases} \bar{r}, & 0 \leq \bar{r} \leq 1 \\ \frac{1}{\bar{r}}, & 1 < \bar{r} < \infty \end{cases} \tag{B.20}$$

and

$$\zeta_0 = \begin{cases} 2, & 0 \leq \bar{r} \leq 1 \\ 0, & 1 < \bar{r} < \infty \end{cases} \tag{B.21}$$

Given the above very simple forms there is no difficulty in calculating the quantities that are important for the solution. From equations (B.3) and (B.7) it is found that

$$I_1(\bar{r}) = \frac{\bar{r}^4}{4} \quad \text{and} \quad I_2(\bar{r}) = \frac{\bar{r}^2}{8} \tag{B.22}$$

The constants A and B are calculated from (B.5) and (B.9):

$$A = \frac{1}{4} \quad \text{and} \quad B = \frac{1}{4} \tag{B.23}$$

And the constant C_1 from (B.16) is

$$C_1 = \frac{5}{8} \tag{B.24}$$

The great simplicity of the above case arises largely from the constancy of vorticity in the core. In the absence of axial velocity, the vorticity is found to be

$$\zeta = \zeta_0 + \varepsilon\zeta_0\bar{r}\sin\theta = 2 + \varepsilon2\bar{r}\sin\theta \tag{B.25}$$

where the $O[\varepsilon]$ term is merely a linear stretching of the constant $O[1]$ vorticity.

For the decaying line vortex the typical length is $a \sim \sqrt{4\nu t}$; the velocity and vorticity are

$$v_0 = \frac{1}{\bar{r}}(1 - e^{-\bar{r}^2}) \tag{B.26}$$

and

$$\zeta_0 = 2e^{-\bar{r}^2} \tag{B.27}$$

A more complex calculation (Bliss[8]) gives

$$A = \frac{\gamma}{2} - \frac{1}{2}\ln 2 \simeq -.058 \tag{B.28}$$

The determination of the constant B from (B.9) requires either an analytical or numerical solution for I_2, equation (B.7). A graphical integration indicates that

$$B \simeq .644 \tag{B.29}$$

There being no radius at which the vorticity definitely vanishes, the constant C_1 cannot be found without a higher order solution of the problem.

Several axial velocity distributions will now be discussed. One simple case is that of constant axial velocity in the vortex core:

$$w_0 = W[1 - H(r - a)] \tag{B.30}$$

where H is the step function and W is the magnitude of the velocity. Non-dimensionally this can be written as

$$w_0 = \bar{w}[1 - H(\bar{r} - 1)] \tag{B.31}$$

where

$$\bar{w} = \frac{aW}{\kappa} \tag{B.32}$$

with a the characteristic dimension for the vortex core.   The
parameter $\bar{w}$ then represents a ratio of typical axial and rotational
velocities associated with the vortex core.   For this case it is
found that

$$I_3(\bar{r}) = -\alpha^2 H(\bar{r} - 1) \tag{B.33}$$

The constant C from equation (B.13) is

$$C = \alpha^2 \tag{B.34}$$

Another simple case is a parabolic axial velocity distribution.
Written in non-dimensional variables this is

$$w_0 = \begin{cases} \bar{w}(1 - \bar{r}^2), & 0 \le \bar{r} \le 1 \\ 0, & 1 < \bar{r} < \infty \end{cases} \tag{B.35}$$

where $\bar{w}$ was defined in (B.32).   Straightforward calculation with
equations (B.12) and (B.13) gives

$$I_3(\bar{r}) = -2\alpha^2\bar{r}^4\left(\frac{1}{2} - \frac{\bar{r}^2}{3}\right), \quad 0 \le \bar{r} \le 1 \tag{B.36}$$

and

$$C = \frac{\alpha^2}{3} \tag{B.37}$$

An example of an axial velocity distribution that goes to zero
asymptotically as $\bar{r} \to \infty$ is expressed non-dimensionally as

$$w_0 = \alpha e^{-\bar{r}^2} \tag{B.38}$$

for which it is found that

$$I_3(\bar{r}) = -\frac{\alpha^2}{2}[1 - (2\bar{r}^2 + 1)e^{-2\bar{r}^2}], \quad 0 \le \bar{r} \le \infty \tag{B.39}$$

and

$$C = \frac{\alpha^2}{2} \tag{B.40}$$

The integral I_4, equation (B.14), will be calculated for the
case of a constant vorticity core and a parabolic axial velocity
distribution.   Using (B.36) and (B.20) it is found that

$$I_4(\bar{r}) = -\frac{\alpha^2\bar{r}^2}{2}[1 - \frac{\bar{r}^2}{3}], \quad 0 \le \bar{r} \le 1 \tag{B.41}$$

The constant D, equation (B.15), is then found to be

$$D = -\frac{\alpha^2}{6}. \tag{B.42}$$

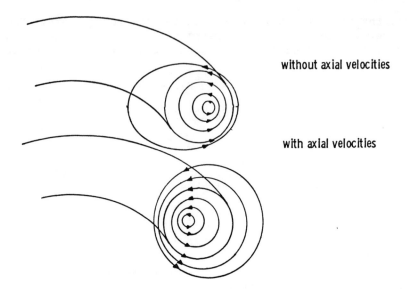

without axial velocities

with axial velocities

Figure 10. Streamlines within a perturbed vortex core.

        Typical streamline shapes in the vortex core are shown in
Figure 10. These were calculated for the case of a constant vorticity core and a parabolic axial velocity distribution. Relatively
large values of ε and α were used to exaggerate the effects.   The
boundary of the rotational region always remains circular since
the vorticity distribution is of finite extent. Without axial
velocity the streamlines in the rotational core deform into a
shape that is approximately elliptical with the centers shifted
outward. Outside the rotational core the shift is in the opposite
direction. The effect of axial velocity is to decrease the amount
the streamlines are shifted widening the distance between the outboard streamlines in order to provide the pressure gradient needed
to turn the axial flow. In the lower sketch the axial velocity is
so large that the direction of the shift is actually reversed.

STRUCTURE OF A LINE VORTEX IN AN IMPOSED STRAIN

D. W. Moore and P. G. Saffman

Imperial College, London & California Institute of

Technology, Pasadena, California   91109

The velocity of a vortex line depends on its structure, i.e. the shape of the cross-section and the detailed vorticity distribution. As a first step towards an understanding of how the structure depends on the motion and the construction of a valid approximation for the motion of vortex lines in general flow fields, we consider the structure of straight line vortices in a uniform two-dimensional straining field. Two cases are considered in detail, irrotational strain and simple shear. In the first case, it is shown that steady exact solutions of the inviscid equations exist, in which the boundary of the vortex is an ellipse with principal axes at 45° to the principal axes of strain. There are two possible axis ratios provided $e/\omega_o < 0.15$, where e is the maximum rate of extension and ω_o is the vorticity in the core. The stability of the shapes is considered, and it is shown that the more elongated shape is unstable, while the less elongated one is stable to two-dimensional deformations. There are no steady solutions of elliptical form if $e/\omega_o > 0.15$, and it is believed from some numerical work that in this case the strain field will cause the vortex to break up. For simple shear, there is one steady shape of elliptical form if the shear rotation and vorticity are in the same sense and $e' < \omega_o$, where e' is the rate of shear. The major axis is parallel to the streamlines and the shape is stable to two-dimensional deformations. For shear rotation and vorticity in opposite senses, there are two steady elliptical shapes if $e'/\omega_o < 0.21$, with major axes perpendicular to the streamlines. The more elongated form is unstable, and the less elongated one is stable. Disturbances of three-dimensional form are also considered in the limit of extremely large axial wavelength.

i.   INTRODUCTION

Recently Crow (1970) has analyzed the stability of the trailing vortices behind a lifting wing. The effect on their stability of the flow near the wing itself is neglected and the trailing vortex system is represented by a pair of infinite parallel line vortices. Each vortex is subjected to a small sinusoidal deformation and the stability of the system elucidated in terms of the growth rate of these deformations. However, once the vortices are allowed to deform it is necessary to drop the idealisation to a line vortex, because the line-integral (Biot-Savart law) expressing the self-induced velocity of the deformed vortex is logarithmically infinite. To remove the singularity, the vortices must be regarded as having a small but definite radius.

Once the vortices are regarded as having a finite cross section it is possible to raise questions about their structure and stability not answered by Crow (or by Rosenhead (1930) in his very similar treatment of the stability of the Karman vortex street). To see what these are, it will be helpful to describe Crow's work in a little more detail.

Let us look in the direction of translation and consider the effect of the left (or port tip) vortex on the right (or starboard tip) vortex. We refer positions and velocities to rectangular axes Oxyz with Oz in the undisturbed position of the right vortex and Ox in the plane of the two undisturbed vortices. The left vortex produces a downward velocity $\Gamma/2\pi b$ at the right vortex, where b is the distance between the undisturbed vortices and $\Gamma (>0)$ the circulation around the left vortex. Thus the undisturbed vortex pair (and the coordinate axes) descent with velocity $\Gamma/2\pi b$. However, suppose the right vortex is slightly deformed so that it no longer occupies the z axis. The velocity produced by the left vortex is not uniform and is equal to zero (in the moving axes) only at x = y = 0. For small x/b and y/b the relative velocity field is $(\Gamma y/2\pi b^2, \Gamma x/2\pi b^2, 0)$, that is an irrotational plane strain. If no other effects were present we would have, since vortex lines move with the fluid

$$\frac{dx}{dt} = \left(\frac{\Gamma}{2\pi b^2}\right) y$$

$$\frac{dy}{dt} = \left(\frac{\Gamma}{2\pi b^2}\right) x$$

(1.1)

which indicates a growing disturbance of growth rate $\Gamma/2\pi b^2$. For large times, x and y become equal which means that the disturbed vortex lies in a plane inclined at 45° to the plane of the wing.

Evidently it is the strain field produced by one vortex in the neighborhood of the other which is the cause of the instability. The self induction effect stabilizes the vortices. This is clear from Kelvin's (1880) investigation of the vibrations of a line vortex. A band or loop in a line vortex rotates about the vortex and if this rate of rotation is sufficiently rapid it can dominate the destabilizing mechanism. However, in order to calculate the self induction, the vortex must be regarded as having a small but definite cross section and further it must be recognized that the self induction effect depends on the shape of the cross-section and the distribution of vorticity inside it. (The speed of vortex rings with arbitrary distributions of core vorticity have recently been calculated by Fraenkel (1970) and Saffman (1970).)

The first question we propose to discuss is the effect of the imposed irrotational plane strain on the structure of the vortex. In §2 we show that for the case when the vorticity is uniform inside the cross section there is an exact solution in which the cross section is an ellipse whose axes are inclined at $\pi/4$ to the principal axes of the plane-strain. That such a solution might be anticipated is suggested by Kirchoff's (1876) exact solution for an elliptical vortex rotating steadily in a fluid at rest at infinity. However, our solution cannot be derived from this and has rather different properties.

A striking feature of the results is that there are two possible ellipses for a given strain rate e* if

$$\left| \frac{e}{\omega} \right| < 0.15 \ldots$$

and no solution of elliptical form when

$$\left| \frac{e}{\omega} \right| > 0.15 \ldots$$

where ω is the vorticity. While we have not been able analytically to exclude the possibility of a steady (or periodic) solution of non-elliptical form in this latter case, a numerical study in which the core was represented by 49 point vortices showed that the vorticity is rapidly dispersed by the imposed strain. Thus we suggest that a cylindrical vortex has only a finite "strength" to resist deformation by an imposed strain. This raises the question of how close two vortices can be before the strain field of one disrupts the other. A crude calculation indicates that they must overlap before this happens; and a numerical study, with the cores again

---

* $|e|$ is the maximum rate of extension.

represented by point vortices, showed that even when the vortices
are initially touching they do not break up. Crow (1965) obtained the criterion for two line vortices to remain in equilibrium
in an irrotational plane strain.

In §3 we examine the stability of the elliptical vortex.
Unfortunately, we have not been able to solve the stability problem
in general and have, for the most part, restricted our discussion
to the stability against perturbations independent of the axial
coordinate. A similar analysis of Kirchoff's rotating vortex was
given by Love (1893). There are good reasons for believing that
these disturbances are the most unstable, and this is confirmed in
§4, where distrubances whose wavelength in the axial direction is
very large compared to the core radius are considered.

The two dimensional stability analysis for irrotational plane
strain shows that the more elongated of the two steady elliptical
solutions is always unstable to perturbations involving a change of
shape, while the less elongated is stable to such perturbations,
though it is unstable to a translational disturbance of exactly the
form considered above. This shows that the structure of the vortex
remains stable in the presence of imposed strain and thus supports
the work of Rosenhead and Crow.

The long wave analysis of §4 confirms that, as asserted by
Crow, bending of the vortex has a stabilizing influence and we
show that an approximate treatment of the self-induced velocities
introduced by this bending reproduces the results of the long wave
analysis. In principle, this analysis determines how the cut-off
radius of Rosenhead and Crow is modified by the distortion of the
core, but the lengthy calculations have not been completed.

2.   THE STEADY SOLUTION

We choose rectangular axes Oxy perpendicular to the axis of
the vortex with 0 at the center of the ellipse and with Ox lying
along the major axis of the ellipse. Then the equation of the
ellipse E is

$$\frac{x^2}{a^2} + \frac{y^2}{b^2} = 1 \qquad\qquad , \qquad\qquad (2.1)$$

where a > b.

If (u,v) are the velocity components it is convenient to form-
ulate the problem in terms of a stream function ψ where $u = \partial\psi/\partial y$
and $v = -\partial\psi/\partial x$. Then we want to show that there is a ψ such that

$$\nabla^2_1 \psi = 0 \qquad \text{outside } E;$$
$$= -\omega_0 \qquad \text{inside } E, \qquad (2.2)$$

$$\psi \sim \frac{1}{2} e (x^2 - y^2) \text{ as } |x^2 + y^2| \to \infty, \qquad (2.3)$$

$$\frac{\partial \psi}{\partial n} \text{ is continuous across } E, \qquad (2.4)$$

and

$$\psi = \text{constant on } E. \qquad (2.5)$$

In these equations ω_0 is the uniform vorticity inside the ellipse, e is the strain rate and $\partial/\partial n$ denotes differentiation along the normal to E. The boundary condition (2.3) says that the flow at large distances from E is an irrotational plane strain with principal axes inclined at $\pi/4$ to the axes of the ellipse. The boundary condition (2.4) makes the tangential velocity at E the same on both sides thus, by virtue of Bernoulli's theorem, securing continuity of the pressure.

The flow inside E is uniquely determined by (2.2) and (2.5) and is

$$\psi = -\frac{1}{2} \Omega ab\left(\frac{x^2}{a^2} + \frac{y^2}{b^2} - 1\right) \qquad (2.6)$$

where

$$\Omega = \frac{\omega_0 ab}{a^2 + b^2} . \qquad (2.7)$$

The interior streamlines are a family of similar ellipses and for any one of these it can be shown that the eccentric angle of a fluid particle increases at a constant rate Ω.

The flow outside E is most conveniently expressed in elliptic coordinates defined by

$$x = c \cosh \xi \cos \eta ,$$
$$y = c \sinh \xi \sin \eta . \qquad (2.8)$$

The curves ξ = constant are a family of confocal ellipses and provided that

$$c = (a^2 - b^2)^{1/2} \tag{2.9}$$

one of these, $\xi = \xi_o$ say, will coincide with the given ellipse E. Then

$$a = c \cosh \xi_o$$
$$b = c \sinh \xi_o \tag{2.10}$$

and the region $\xi > \xi_o$ corresponds to the outside of E.

Now we observe that both the interior flow (2.6) and the imposed plane strain (2.3) have η dependence like 1 or $\cos 2\eta$. Thus we try to find ψ in the region outside E in the form

$$\psi = \frac{1}{4} ec^2 \cosh 2\xi \cos 2\eta +$$

$$Se^{-2(\xi - \xi_o)} \cos 2\eta - \frac{\Gamma}{2\pi} (\xi - \xi_o) \tag{2.11}$$

where S and Γ are constants. The second term represents a noncirculatory disturbance to the imposed strain which decays as $\xi \to \infty$ and the last term allows for the circulation in the potential flow outside the vortex, the value of the circulation being Γ.

It is now easy to verify that (2.4) and (2.5) are satisfied poovided that

$$\Gamma = \pi ab\omega_o \quad ,$$

$$S + \frac{1}{4} ec^2 \cosh 2\xi_o = 0 \quad , \tag{2.12}$$

$$-2S + \frac{1}{2} ec^2 \sinh 2\xi_o = \frac{1}{2} \Omega(a^2 - b^2) \quad ,$$

where (2.10) has been used to simplify the expressions.

If we eliminate S from the last two equations of (2.12) we find on simplifying that

$$\frac{e}{\omega_o} = \frac{\theta(\theta-1)}{(\theta^2+1)(\theta+1)} \quad , \tag{2.13}$$

where $\theta = a/b$ is the axis ratio of the ellipse. This function has a single maximum at $\theta = \theta_{cr} \doteq 2.9$ where θ_{cr} is the unique root of

$$\theta^4 - 2\theta^3 - 2\theta^2 - 2\theta + 1 = 0$$

in the interval $(1,\infty)$ and the maximum value itself is about 0.15.

Thus for any value of the imposed strain such that $|e/\omega_0| < 0.15 \ldots$ there are two possible steady elliptical vortices. In the next section we study the stability of the flow and show that the more distorted of the two possible configurations is unstable.

Since in steady inviscid motion Bernoulli's theorem holds along streamlines the pressure on the boundary of the ellipse is greatest where the speed is least, that is, on the extremities of the major axis. Now the highest pressures in the external flow are where the speeds are lowest, that is, where the imposed strain together with the disturbance to it caused by the ellipse oppose the circulatory motion. Thus we can anticipate that the major axis of the ellipse is at $\pi/4$ to the strain axes and lies in the quadrants where the strain and the circulation are in opposite directions. We can confirm this from the analytical solution. Thus for the case of two parallel vortices of small cross section,* the cross sections are ellipses whose major axes are parallel to the direction of the motion.

The analysis can be generalized to give the shape of an elliptical vortex in an arbitrary uniform strain with uniform vorticity outside the core. A special case will be simple shear. We replace (2.3) by

$$\psi \sim \alpha xy + \frac{1}{2} e(x^2-y^2) - \frac{1}{2}\gamma(x^2-y^2). \tag{2.14}$$

This stream function gives a uniform rotation with vorticity 2γ plus a pure strain with principal rates of strain $\pm\sqrt{e^2+\alpha^2}$ at angle $1/2 \tan^{-1}(e/\alpha)$ and $\pi/2 + 1/2 \tan^{-1}(e/\alpha)$ to the x-axis. The flow will be a simple shear if $\gamma = \pm\sqrt{\alpha^2+e^2}$. We now take ψ outside E in the form

* This problem was discussed in a general way by Lichtenstein (1929, p. 456).

$$\psi = \frac{\alpha c^2}{4} \sin 2\eta \left\{ \sinh 2\xi - \sinh 2\xi_o \; e^{2(\xi_o - \xi)} \right\} + \frac{ec^2}{4} \cos 2\eta$$

$$\left\{ \cosh 2\xi - \cosh 2\xi_o \; e^{2(\xi_o - \xi)} \right\} - \frac{\gamma c^2}{4} \left\{ \cosh 2\xi - \cosh 2\xi_o \; + \right.$$

$$\left. \cos 2\eta \left(1 - 3 \; e^{2(\xi_o - \xi)} \right) \right\} - \frac{\Gamma}{2\pi} (\xi - \xi_o) \; . \tag{2.15}$$

(for $\xi > \xi_o$)

The boundary condition (2.5) is automatically satisfied, as well as (2.14) and the condition of uniform vorticity outside E. Application of the remaining boundary condition (2.4) gives after a little reduction

$$\alpha = 0$$

$$\Gamma = \pi a b (\omega_o - 2\gamma) \qquad , \tag{2.16}$$

$$e \frac{(\theta + 1)}{(\theta - 1)} - \gamma = \omega_o \frac{\theta}{\theta^2 + 1}$$

For given external flow, we know the value of γ (either positive or negative) and the value of $|e|$. We do not know *a priori* the sign of e, because this gives the direction of maximum extension relative to the principal axes of the elliptical core. If e > 0, the direction of maximum extension is at angle $- 1/4 \; \pi$ to the x-axis.

If γ and ω_o have the same sign, there are again two solutions or none depending on the magnitude of e. But if ω_o and γ have opposite signs, there may be none, one, two, or three solutions depending on the relative magnitudes of e, γ and ω_o.

For the case of simple shear, e = $\pm\gamma$. First take $\gamma > 0$, i.e. shear rotation and vortex in the same sense. Then e = γ is necessary for a solution and the values of θ are the roots of

$$\frac{2\gamma}{\omega_o} (\theta^2 + 1) = \theta^2 - \theta \qquad , \tag{2.17}$$

which has one root (with $\theta > 1$) provided $\gamma < 1/2 \; \omega_o$. As $\gamma \to 1/2 \; \omega_o$, $\theta \to \infty$. The flow at infinity is u = $-2\gamma y$, v = 0 and the major axis is parallel to the streamlines.

Next, take $\gamma < 0$, i.e. shear rotation and vortex in opposite senses. Then if $e > 0$, the value of θ must satisfy

$$\frac{2|\gamma|}{\omega_o} (\theta^2+1) = \theta - 1 , \qquad (2.18)$$

which has two roots with $\theta > 1$ if

$$\frac{2|\gamma|}{\omega_o} < \frac{\sqrt{2} - 1}{2} , \qquad (2.19)$$

and none otherwise. One root is larger than $1 + \sqrt{2}$ and the other is smaller. The flow at infinity is $u = 0$, $v = 2\gamma x$ and the major axis is perpendicular to the streamlines. On the other hand, with $e < 0$, it is easy to see that there is no solution.

The Kirchoff solution, referred to the rotating frame in which the motion is steady, is the special case of (2.16) with $e = 0$ and $\gamma < 0$.

3.   STABILITY ANALYSIS

We have not been able to give a discussion of the stability of the elliptical vortex against general disturbances. The diffi- culty is that the linearized vorticity equation inside the vortex is awkward to handle. However, if we restrict our discussion to deformations of the boundary which are independent of the axial co- ordinate, the resulting perturbations will merely translate the vortex lines without distortion. Thus, since the vorticity is uni- form, the vorticity at any point inside the vortex is unchanged, so that the perturbation velocity field is irrotational everywhere.

Thus we assume that the boundary of the ellipse is distorted so that

$$\xi = \xi_o + e^{\sigma t}F(\eta) . \qquad (3.1)$$

The disturbance to the flow caused by this change in the boundary shape must satisfy the exact boundary conditions

$$\frac{D}{Dt} (\xi - \xi_o - e^{\sigma t}F) = 0 \qquad (3.2)$$

and

$$\underset{\sim}{u} \cdot \hat{\underset{\sim}{t}} \quad \text{continuous} \qquad (3.3)$$

where in both conditions the velocity field is to be evaluated at the disturbed surface, and where in (3.3), u is the velocity and $\hat{t}$ is a unit vector tangential to the disturbed surface. The second boundary condition ensures that the disturbances do not cause a vortex sheet to appear at the vortex boundary, as this would violate the vorticity equation. We could equally have insisted on continuity of pressure at the disturbed boundary and one can show that (3.3) secures this. However, we prefer to avoid calculating the pressure field and (3.3) is more convenient.

Some care is needed in dealing with these boundary conditions. We write

$$\underset{\sim}{u} = \bar{\underset{\sim}{u}}(\xi,\eta) + e^{\sigma t}\tilde{\underset{\sim}{u}}(\xi,\eta) \tag{3.4}$$

where $\bar{u}$ is the velocity field in the undisturbed state and $\tilde{\underset{\sim}{u}}$ is the perturbation, which proves to be $O(F)$.

Retaining only terms of the first order in F in (3.2) gives $[\underset{\sim}{u} = (u_\xi,u_\eta)]$

$$u_\xi = e^{\sigma t} h(\xi_0,\eta)\sigma F + \bar{u}_\eta(\xi_0,\eta) \frac{\partial F}{\partial \eta} \quad , \tag{3.5}$$

where $h(\xi,\eta)$ is the line element of the elliptic coordinates. But

$$u_\xi = \bar{u}_\xi(\xi_0,\eta) + (\xi-\xi_0) \frac{\partial \bar{u}_\xi}{\partial \xi}(\xi_0,\eta) + \ldots + \tilde{u}_\xi(\xi_0,\eta) + \ldots$$

so that to order F

$$u_\xi = \bar{u}_\xi + e^{\sigma t} \left(F \frac{\partial \bar{u}_\xi}{\partial \xi} + \tilde{u}_\xi \right) \quad , \tag{3.6}$$

where all the quantities on the right hand side of (3.6) are evaluated at $\xi = \xi_0$. If we substitute into (3.4) and use the fact that

$$\bar{u}_\xi(\xi_0,\eta) = 0$$

we find

$$\tilde{u}_\xi = h\sigma F + \bar{u}_\eta \frac{\partial F}{\partial \eta} - F \frac{\partial \bar{u}_\xi}{\partial \xi} \quad , \tag{3.7}$$

where now all the field quantities entering the equation are evaluated on $\xi = \xi_0$. This equation can be simplified by use of the equation of continuity

$$\frac{\partial}{\partial \xi} (h\bar{u}_\xi) + \frac{\partial}{\partial \eta} (hu_\eta) = 0$$

and we get finally

$$\tilde{u}_\xi = h\sigma F + \frac{1}{h} \frac{\partial}{\partial \eta} (h \bar{F} \bar{u}_\eta) \tag{3.8}$$

Next we consider (3.3). We have, again to order F,

$$\underset{\sim}{u} \cdot \hat{\underset{\sim}{t}} = \bar{\underset{\sim}{u}} \cdot \hat{\underset{\sim}{t}} + e^{\sigma t} \left(F \frac{\partial \bar{\underset{\sim}{u}}}{\partial \xi} \cdot \hat{\underset{\sim}{t}} + \bar{F}\underset{\sim}{u} \cdot \frac{\partial \hat{\underset{\sim}{t}}}{\partial \xi} + \tilde{\underset{\sim}{u}} \cdot \hat{\underset{\sim}{t}} \right) \tag{3.9}$$

where all the quantities on the right, including t, are evaluated on $\xi = \xi_0$. Now u is continuous across $\xi = \xi_0$ so that, in view of (3.9), the boundary condition (3.3), implies the continuity of

$$F \frac{\partial \bar{u}_\eta}{\partial \xi} + \tilde{u}_\eta$$

across $\xi = \xi_0$. But if ω_0 is the vorticity inside the vortex

$$\frac{1}{h} \frac{\partial \bar{u}_\eta}{\partial \xi} + \frac{1}{h^2} \frac{\partial h}{\partial \xi} \bar{u}_\eta = \begin{array}{l} 2\gamma ; \xi = \xi_0 + 0 \quad ; \\[2mm] \omega_0 ; \xi = \xi_0 - 0 \quad , \end{array} \tag{3.10}$$

so that, since the second term on the left is continuous at $\xi = \xi_0$, we have finally

$$\tilde{u}_\eta(\xi_0+0,\eta) - \tilde{u}_\eta(\xi_0-0,\eta) = Fh(\omega_0-2\gamma) \tag{3.11}$$

We can express these boundary conditions in terms of the velocity potential ϕ of the disturbance and, on remarking that the steady solution of §2 shows that

$$\bar{u}_\eta(\xi_0,\eta) = \Omega h(\xi_0,\eta) \quad ,$$

where Ω is the quantity defined in (2.7), we get

$$\frac{\partial \tilde{\phi}}{\partial \xi}(\xi_0 \pm 0,\eta) = \sigma h^2 F + \Omega \frac{\partial}{\partial \eta}(h^2 F) \tag{3.12}$$

and

$$\frac{\partial \tilde{\phi}}{\partial \eta}(\xi_0+0,\eta) - \frac{\partial \tilde{\phi}}{\partial \eta}(\xi_0-0,\eta) = h^2(\omega_0-2\gamma)F \quad . \tag{3.13}$$

We assume that the disturbance to the shape is of the form (which keeps the core area constant)

$$h^2F = A \cos m\eta + B \sin m\eta \tag{3.14}$$

where m > 0 is an integer whose values label the different modes,
while for the velocity potential we take

$$\tilde{\phi} = e^{-m\xi}(C \cos m\eta + D \sin m\eta), \quad \xi \geq \xi_o \tag{3.15}$$

and

$$\tilde{\phi} = E \cosh m\xi \cos m\eta + F \sinh m\xi \sin m\eta, \quad \xi \leq \xi_o \tag{3.16}$$

The combination of hyperbolic and trigonometric functions in (3.16)
is dictated by the requirement that the velocity field be regular
everywhere inside the ellipse.

It is now a straightforward matter to show that

$$\frac{\sigma^2}{\omega_o^2} = -\frac{1}{4}\left\{\left(\frac{2m\theta}{\theta^2+1} - 1 + \frac{2\gamma}{\omega_o}\right)^2 - \left(\frac{1-2\gamma}{\omega_o}\right)^2\left(\frac{\theta-1}{\theta+1}\right)^{2m}\right\}, \tag{3.17}$$

where, as in §2, $\theta(>1)$ is the axis ratio of the undisturbed vortex.

We examine now the case $\gamma = 0$, i.e. a vortex in a pure strain.
Consider first m = 1. Then

$$\frac{\sigma^2}{\omega_o^2} = \frac{\theta^2(\theta-1)^2}{(\theta+1)^2(\theta^2+1)^2} = \frac{e^2}{\omega_o^2}$$

in view of (2.13). Thus the m = 1 mode is always unstable with
growth rate e. Now we can show that the motion corresponding to
$\sigma = e$ is a translation of the ellipse without change of shape
parallel to the outward principal axis of strain. Thus the results
are so far in complete agreement with those obtained in §1. The
disturbances with m = 2 represents a combination of extension and
contraction of the axes and changes of orientation of the axes.
Thus an instability to this mode of disturbance would suggest a
disintegration of the vortex, rendering the theory based on a line
vortex invalid.

On putting m = 2 in (3.17) we find that $\sigma^2 < 0$ if $1 \leq \theta < \theta_{cr}$
and $\sigma^2 > 0$ if $\theta > \theta_{cr}$, where θ_{cr} is the critical axis ratio defined
in §2. This is a satisfying result, because it shows that the more
distorted of the two possible ellipses for a given strain is always
unstable to distortions of shape and thus only one of the two solutions is physically significant. Moreover, since it can easily be
shown that $\sigma^2 < 0$ for $1 \leq \theta < \theta_{cr}$ for the modes $m \geq 3$, the vortex
is structurally stable to two dimensional disturbances involving
distortion and an analysis of stability which neglects the possibility of the instability of the vortex structure will be correct.

In the next section we examine three-dimensional effects, and suggest that they do not affect these conclusions.

For the case of a vortex in a simple shear we find that $\sigma = 0$ for the m = 1 mode of disturbance. This corresponds to the fact that a small displacement of the vortex as a whole will cause it to translate with a constant velocity. When the shear rotation and core rotation are in the same sense, the unique steady configuration is stable to disturbances with m $\geq$ 2. When the shear rotation and core rotation are in opposite senses, the less elongated of the two possible configurations is stable to such disturbances, whereas the more elongated is unstable.

4. DISTURBANCES OF LARGE AXIAL WAVELENGTH

If the disturbances are allowed to vary in the direction parallel to the axis of the vortex, bending of the vortex lines constituting the core is inevitable. Moreover, such bending as is produced initially is compounded by the relatively complicated velocity field in the core itself. Kelvin (1880) solved the problem for a circular vortex, but our problem is harder. We will not attempt a full discussion of this complicated situation, but content ourselves with an examination of disturbances which vary only slowly in the axial direction. Also, we shall restrict the analysis to the case of a vortex in an irrotational strain.

We assume that the disturbances depend on the axial coordinate z through a factor e^{ikz} and we consider only the case ka << 1. Then the tilting of the vortex lines of the core produces a component of vorticity parallel to the xy-plane of magnitude, k. Associated with this vorticity is a velocity field whose components in the xy plane are $O(k^2)$. This velocity field induced by the slightly tilted vortex lines is not the only contribution to the disturbance velocity field. There is also the irrotational disturbance caused by the displacement of the vortex boundary. This consists of two parts, one of $O(1)$ identical with that of §3, and one due to three-dimensional effects which is generally of $O(k^2)$. However, for the m = 1 disturbances, which, as we have seen, correspond to a drift of the vortex away from its equilibrium position, this latter effect proves to be $O(k^2 \log k)$.

Suppose that the disturbed vortex is

$$\xi = \xi_o + e^{\sigma t + ikz}(A \cos\eta + B \sin\eta) \qquad (4.1)$$

Inside the disturbed vortex the motion is rotational, but outside it is irrotational and we have as an appropriate solution

$$\tilde{\phi}_E = e^{ikz} \Bigg\{ \hat{A}_1 \, Ke_1(\xi,-q) ce_1(\eta,q) + \hat{B}_1 \, Ko_1(\xi,-q) Se_1(\eta,q)$$

$$+ \, \hat{A}_3 \, Ke_3(\xi,-q) ce_3(\eta,q) + \hat{B}_3 \, Ko_3(\xi,-q) Se_3(\eta,q) + \dots \Bigg\} \qquad (4.2)$$

where the Mathieu functions are written in the notation of Abramovitz and Stegun (1964) and where

$$q = -\frac{1}{4} c^2 k^2 \qquad . \qquad (4.3)$$

The choice of angular functions is dictated by the symmetry inherent in (4.1) and the choice of the functions of ξ is forced by the requirement that $\tilde{\phi} \to 0$ as $\xi \to \infty$.

Now if $c\gamma \ll 1$, we have

$$ce_{2m+1}\left(\eta, \, -\frac{1}{4} c^2 k^2\right) = \cos(2m+1)\eta + 0(k^2) \qquad (4.4)$$

and

$$se_{2m+1}\left(\eta, \, -\frac{1}{4} c^2 k^2\right) = \sin(2m+1)\eta + 0(k^2) \quad . \qquad (4.5)$$

Thus if (4.2) is to reduce to (3.15) as $k \to 0$, all the terms in the series must vanish as $k \to 0$ except the first. Now as $ck \to 0$, it can be shown* that

$$Ke_1(\xi,Q) = Q^{-1/2} e^{-\xi} + Q^{1/2} \log Q^{1/2} \cosh\xi + 0(Q^{1/2}) \qquad (4.6)$$

and

$$Ko_1(\xi,Q) = Q^{-1/2} e^{-\xi} + Q^{1/2} \log Q^{1/2} \sinh\xi + 0(Q^{1/2}) \, , \qquad (4.7)$$

where

$$Q = -q = \frac{1}{4} c^2 k^2 \qquad . \qquad (4.8)$$

In view of (4.4) to (4.7) the condition that (4.2) reduces to (3.15) as $k \to 0$ implies that $\hat{A}_1 Q^{-1/2} = A$ and $\hat{B}_1 Q^{-1/2} = C$. Thus we can assert that

---

* The calculation uses the infinite series representation of Ke_1 and Ko_1 in terms of products of Bessel functions.

$$\tilde{\phi}_E = A(e^{-\xi} + \frac{1}{4} c^2k^2\log(ak) \cosh\xi) \cos\eta$$
$$+ B(e^{-\xi} + \frac{1}{4} c^2k^2\log(ak) \sin\xi) \sin\eta + 0(k^2) \qquad (4.9)$$

We can also see that, in view of the estimates of the vorticity given above, the disturbance velocity field inside the vortex is $\tilde{u}_I$, where

$$\tilde{u}_I = \text{grad}\tilde{\phi}_I + 0(k^2) \qquad (4.10)$$

An examination of the appropriate Mathieu functions shows that no logarithmic terms arise in the limit $k \to 0$, and that

$$\tilde{\phi}_I = E \cosh m\xi \cos\eta + F \sinh m\xi \sin\eta + 0(k^2) \qquad (4.11)$$

We can now repeat the calculation of the growth rate, replacing (3.15) by (4.9) and imposing the same boundary conditions.* We find that

$$\sigma^2 = e^2 - [\frac{\Gamma}{4\pi} k^2 \log(\frac{1}{ak})]^2, \qquad (4.12)$$

where $\Gamma = \pi ab\omega_o$ is the circulation around the vortex.

Thus, three-dimensional effects reduce the growth rate of the instability. Crow (1970) has traced this stabilizing effect to the large self induced velocities introduced when a rectilinear vortex of small cross section is deformed. Precisely, if a portion of the vortex of length λ is moved off the original axis, it experiences at every point a velocity q of magnitude

$$|\tilde{q}| = \frac{\Gamma}{4\pi R} \{\log (\frac{\lambda}{\sigma}) + 0(1)\} , \qquad (4.13)$$

where R is the local radius of curvature, and σ is the core radius. This velocity is directed along the binormal in such a sense as to give always a contra-rotation of the portion in question.**

We can use this result to extend the physical discussion of a vortex in a plane strain given in §1. If we suppose that the vortex has as its equation

$$x = \hat{x}(t)e^{ikz}$$
$$y = \hat{y}(t)e^{ikz}$$

---

* The axial velocity is linked to the $0(k^2)$ terms in the xy velocity field and thus we need not consider its continuity at the vortex boundary to the present order of approximation.

** Batchelor (1968), p. 510, has given a careful derivation of this formula.

then a short calculation shows that approximately

$$q = \frac{\Gamma}{4\pi} \log \left(\frac{1}{ak}\right) (-k^2\hat{y}, \; k^2\hat{x}, \; 0)$$

where, in applying (4.13) we have taken a as a measure of the core size and k^{-1} as a measure of the horizontal extent of the disturbance - the leading term in (4.13) is, of course, indifferent to the precise choice of σ and λ.

Thus equations (1.1) become

$$\frac{d\hat{x}}{dt} = \hat{y} \left(\frac{\Gamma}{2\pi b^2} - \frac{\Gamma k^2}{4\pi} \log \left(\frac{1}{ak}\right)\right)$$

$$\frac{dy}{dt} = x \left(\frac{\Gamma}{2\pi b^2} + \frac{\Gamma k^2}{4\pi} \log \left(\frac{1}{ak}\right)\right)$$

and, since $e = \frac{\Gamma}{2\pi b^2}$, we recover the result (4.12) for the growth rate of the disturbance.

However, the calculation is incomplete in that determination of a "cut-off radius" for the Biot-Savart law requires the $O(k^2)$ term in (4.12).

This work was supported by Air Force contract AFOSR 69-1804.

REFERENCES

Abramovitz, M. and Stegun, I. A. (1964), "Handbook of Mathematical Functions", Washington, D. C., Nat'l. Bureau of Standards.

Batchelor, G. K. (1968), "Fluid Dynamics", Cambridge.

Crow, S., (1965), Ph.D. Thesis, California Institute of Technology.

Crow, S. (1970), to appear in AIAA J.

Fraenkel, L. E. (1970), Proc. Roy. Soc., A316, 29.

Kelvin, Lord (1880), Phil. Mag. 10, 155. Mathematical and Physical Papers, Vol. 4, p. 152.

Kirchoff, G. (1876), "Mechanik", Leipzig.

Lichtenstein, L., (1929), "Grundlagen der Hydromechanik", Springer Verlag, Berlin.

Love, A. E. (1893), Proc. London Math. Soc. (1) xxv.18.

Rosenhead, L. (1930), Proc. Roy. Soc., A 127, 590.

Saffman, P. G. (1970), "Studies in Applied Math", To appear December 1970.

A NEW LOOK AT THE DYNAMICS OF VORTICES WITH FINITE CORES

P. C. Parks

NASA Langley Research Center and University of

Warwick

ABSTRACT

The stability theory for wave-like disturbances in a pair of
trailing line vortices developed by S. C. Crow has been modified
to take account of finite core radii and appropriate distributions
of vorticity within these cores. The difficulties encountered by
Crow in calculating the self-induction effects of each vortex are
avoided and, for a uniform distribution of vorticity within the
cores, the self-induction function is expressible in terms of
modified Bessel functions of the second kind. The essential
features of Crow's theory are, however, confirmed with small
numerical changes - for example, the most unstable long waves which
develop in the trail have a wavelength of 7.2b, where b is the
separation distance of the two vortices (compared with Crow's
result of 8.4b).

The effect of axial flow within the cores may be considered
by wrapping the core in a sheath of vortex rings. This model
leads to a new stability criterion for a single "jet-vortex". The
stability diagrams for the trailing vortex pair are also modified
by axial flows, the stable regions being reduced in size and the
growth rate of the most unstable mode being increased.

The theory, which assumes the vortices extend from infinity
to infinity, cannot be used to calculate the growth of perturbations deliberately introduced at the wing of the aircraft, but a
modified discretised theory amenable to digital computation has
been developed to investigate initial growth of these excited
waves. Some results from digital computations are presented.

NOMENCLATURE

A	Subscript for anti-symmetric mode
a	Growth rate of unstable waves, or complex exponent
a'	$M c k /4$
a_{ij}	Matrix elements, defined after (9.2)
B_1	$1-(\beta^2[\psi(\gamma)-1]/\gamma^2)$
B_2	$-1-(\beta^2[\psi(\gamma)-1]/\gamma^2)$
b	Separation distance of undistrubed vortices
C	Subscript for center line
C_L	Lift coefficient of wing
c	Radius of circular core
E	β^2/γ^2
e	Exponential function
H	$-\psi(\beta)$
I_1	Modified Bessel function
i	$\sqrt{-1}$
$\underline{i}',\underline{j}',\underline{k}'$	Units vectors, defined in Section 3
J	$-\chi(\beta)$
K_o,K_1	Modified Bessel functions of the second kind
k	Wave number
L	Subscript for axial vortex lines
M	$2\pi cU/\Gamma_o$
M'	$2\pi bU/\Gamma_o$
M*	M'bk
n	Circumferential mode number of Batchelor and Gill[11]
q	Distance defined in Fig. 3
R	Subscript for vortex rings
r	Radial distance from center line
S	Subscript for symmetrical mode
s	Dummy variable in integrals in 8.1, also eigenvalues $2\pi c^2 a/\Gamma_o$ in (8.23) and $2\pi b^2 a/\Gamma_o$ in (9.2)
t	Time
U	Axial velocity within core
$\underline{u}_1,\underline{u}_2$	Induced velocities, defined in (2.1)

$u*, \hat{u}*$	Defined in and after (8.1.2)
V_o	Speed of aircraft
$w*, \hat{w}*$	Defined in and after (8.1.2)
x, y, z	Coordinate system, defined in Fig. 1
x', y', z'	Coordinate system, defined in Fig. 3
y_1, z_1, y_2, z_2	Displacements of vortex center lines, 1 for starboard, 2 for port
$\hat{y}_1, \hat{z}_1, \hat{y}_2, \hat{z}_2$	Amplitude of displacements, when $y_1 = \hat{y}_1 \exp(ikx+at)$, etc.
α	Non-dimensional growth rate, $2\pi b^2 a / \Gamma_o$
β	bk
Γ_o	Total strength of circulation of one vortex
γ	Non-dimensional core radius, ck
$\gamma(\ )$	Strength of axial vorticity per unit area
$\gamma(y', z')$	Strength of axial vorticity per unit area
$\gamma*$	Euler's constant, 0.5772157.....
$\Delta x, \Delta z$	Distances, defined after (8.1.2)
δ	Non-dimensional cut off distance used by Crow[1]
$\Theta(x,t)$	Angular perturbation of vortex ring, defined in Fig. 7
$\hat{\Theta}$	Defined after (8.1.2)
κ	Strength of vortex ring, defined in (8.1.1)
μ	$s + \frac{1}{2} ckMi$, defined after (8.2.4)
μ'	$(-s/M'bk) - i/2$, defined after (9.4)
ξ	x/b
π	Pi, 3.1415926....
$\phi(x,t)$	Angular perturbation of vortex ring, defined in Fig. 7
$\chi(\beta)$	$\beta K_1(\beta)$
$\psi(\beta)$	$\beta^2 K_0(\beta) + \beta K_1(\beta)$
$\psi(r,z)$	Stokes' stream function defined in (8.1.1)
$\omega(\beta, \delta)$	Integral defined in (2.3)
$\tilde{\omega}_o$	Radius of vortex ring in (6.3)

1. INTRODUCTION

In a recent paper[1] S. C. Crow has developed an interesting and penetrating stability analysis showing how sinusoidal standing waves can develop in the parallel vortices that trail behind an aircraft. Crow considers the two vortices to be concentrated filaments, and this assumption gives rise to mathematical difficulty when the self-induction function of each vortex is to be calculated. Crow overcomes this difficulty by truncating the integral involved, which otherwise would be infinite, in such a way that correct numerical solutions are obtained to two classical vortex problems treated by Kelvin[2] and Lamb[3].

In this paper, however, the vortices are assumed to have finite cores ab initio. This assumption avoids the difficulty encountered by Crow as the self-induction integral is now finite and, for a uniform vorticity distribution within the core, may be expressed in terms of modified Bessel functions of the second kind. The essential results of Crow's theory are, however, confirmed.

The basic theory assumes the vortex pair extends from infinity to infinity. This theory is inadequate to calculate initial transient growth of perturbations deliberately introduced by the aircraft, and so a digital computer program has been written. Some preliminary results are given.

This paper also considers the effect of axial flow within the core. Experimental evidence suggests that quite high axial velocities exist within aircraft trailing vortices. A uniform axial velocity is modeled by wrapping the circular bundle of vortex lines in a sheath of vortex rings. These rings move along in an axial direction and also undergo small angular perturbations. A new stability criterion for a single "jet-vortex" is deduced from this model, and the instability of a straight circular jet as calculated by Batchelor and Gill[11] is recovered as a special case.

Finally, a vortex-jet pair is considered. Here, the axial velocity modifies the stability boundaries calculated for zero axial flow as well as introducing a new stability boundary which is essentially the criterion for the single jet-vortex. The effect of the axial flow is destabilizing both in the sense of reducing the extent of stable regions of the stability diagrams, and in the sense of increasing the growth rate in the unstable region.

In theory the models described here could be extended to cover nonuniform vorticity and axial velocity distributions within the core, although in practice the evaluation of resulting integrals in closed form would present considerable difficulties.

2.   THEORY OF S. C. CROW[1]

As an introduction and for completeness the theory developed by S. C. Crow[1] is briefly described. The geometry is shown in Fig. 1. The induced velocities at the points   A   and   B   are calculated using the Biot-Savart law as

$$
\left.
\begin{array}{l}
\underset{\sim}{u}_1 = \dfrac{\Gamma_0}{4\pi} \displaystyle\int_1 \dfrac{d\underset{\sim}{s}_1 \times \underset{\sim}{r}_{11}}{\left|\underset{\sim}{r}_{11}\right|^3} - \dfrac{\Gamma_0}{4\pi} \displaystyle\int_2 \dfrac{d\underset{\sim}{s}_2 \times \underset{\sim}{r}_{21}}{\left|\underset{\sim}{r}_{21}\right|^3} \\[4mm]
\underset{\sim}{u}_2 = -\dfrac{\Gamma_0}{4\pi} \displaystyle\int_2 \dfrac{d\underset{\sim}{s}_2 \times \underset{\sim}{r}_{22}}{\left|\underset{\sim}{r}_{22}\right|^3} + \dfrac{\Gamma_0}{4\pi} \displaystyle\int_1 \dfrac{d\underset{\sim}{s}_1 \times \underset{\sim}{r}_{12}}{\left|\underset{\sim}{r}_{12}\right|^3}
\end{array}
\right\}
\tag{2.1}
$$

By using axes that are moving downward with a velocity $\dfrac{\Gamma_0}{2\pi b}$, by considering small perturbations in   y   and   z   of each vortex and by assuming wave-like solutions in which $y_1 = \hat{y}_1\, e^{ikx+at}$ and so forth, Crow arrives at the following characteristic equation in α for the motion:

$$
\det
\begin{bmatrix}
-\alpha & 1+\omega & 0 & -\psi \\
1-\omega & -\alpha & -\chi & 0 \\
0 & \psi & -\alpha & -1-\omega \\
\chi & 0 & -1+\omega & -\alpha
\end{bmatrix}
= 0 ,
\tag{2.2}
$$

with eigenvector $[\hat{y}_1 \quad \hat{z}_1 \quad \hat{y}_2 \quad \hat{z}_2]'$ in which $\alpha = \dfrac{2\pi b^2}{\Gamma_0}\, a$, $\beta = bk$,

$$
\left.
\begin{array}{l}
\chi(\beta) = \displaystyle\int_0^\infty \dfrac{\cos \beta\xi \; d\xi}{(1+\xi^2)^{3/2}} = \beta K_1(\beta) \\[5mm]
\psi(\beta) = \displaystyle\int_0^\infty \dfrac{\cos \beta\xi + \beta\xi \sin \beta\xi}{(1+\xi^2)^{3/2}} \, d\xi = \beta^2 K_0(\beta) + \beta K_1(\beta) \\[5mm]
\omega(\beta,\delta) = \displaystyle\int_\delta^\infty \dfrac{\cos \beta\xi + \beta\xi \sin \beta\xi - 1}{\xi^3} \, d\xi
\end{array}
\right\}
\tag{2.3}
$$

The integral $\omega(\beta,\delta)$ diverges as $\delta \to 0$ and so Crow truncates this so that $\delta = 0.642 \dfrac{c}{b}$ where   c   is the core radius and   b   the

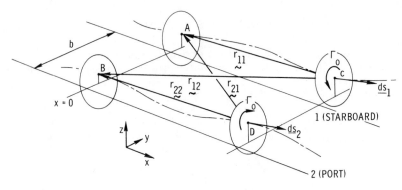

Figure 1.- Geometry of perturbed vortex pair.

separation distance of the vortices. It turns out that numerically
correct solutions to two classical vortex problems - low-frequency
wave motions of a columnar vortex (Kelvin[2]) and the translational
velocity of a circular vortex ring (Lamb[3]) may be obtained using
concentrated vortex line theory provided that the self-induction
integrals, corresponding to $\omega(\beta,\delta)$ above, are truncated at a
distance 0.642c from their singularities. This apparently unique
cut-off distance is itself rather remarkable, and so far
unexplained.

3.   SELF-INDUCTION FUNCTION FOR A VORTEX WITH FINITE CORE

Let us assume that the vortex has a finite core. Fig. 2 shows
the geometry of such a core. For a distribution of vorticity that
is uniform or one that is a function of radial distance from the
center line, it is convenient to consider the geometry of Fig. 3.
If the strength of the vorticity is $\gamma(y',z')$ per unit area of
cross section then the velocity induced at A is

$$\frac{1}{4\pi} \int\int_{\text{Disc}} \frac{\gamma(y',z')\ (\underline{i}'\ dx') \times (-x'\underline{i}' - y'\underline{j}' + (q - z')\underline{k}')}{\left\{x'^2 + y'^2 + (q - z')^2\right\}^{3/2}}\ dy'\ dz' \quad (3.1)$$

where $q = CF = EA$ in Fig. 3, $\underline{i}'$, $\underline{j}'$, $\underline{k}'$, are unit vectors in the x',
y', z' directions, and the double integral is taken over the circu-
lar cross section of radius c and center C. We now linearize the
expression (3.1) in terms of the displacement q which is assumed
small compared with $(x'^2 + y'^2 + z'^2)^{1/2}$. Usually this will be
true as $x' \gg q$. As $x' \to 0$ however, we must consider the geometry of
the center line, and the relationship

$$q \doteq \frac{1}{2}\ x'^2/\rho \quad (3.2)$$

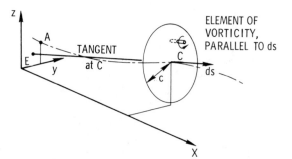

Figure 2.- Finite core geometry for self-induction.

where ρ is the radius of curvature of the center line at A. For finite ρ $q/x' \to 0$ as $x' \to 0$ so once again $x' \gg q$.

The integral (3.1) on expanding the denominator and assuming γ is a function of $r = (y'^2 + z'^2)^{1/2}$ only becomes, the first order in q,

$$-\frac{q}{2} \underset{\sim}{j}' \, dx' \int_{r=0}^{c} \frac{\gamma(r)r \, dr}{(x'^2 + r^2)^{3/2}} + \frac{q\underset{\sim}{j}' \, dx'}{4} \int_{r=0}^{c} \frac{3\gamma(r)r^3 \, dr}{(x'^2 + r'^2)^{5/2}} \quad (3.3)$$

If we assume further that $\gamma(r)$ is uniform and equal to $\Gamma_0/\pi c^2$ then (3.3) reduces on integration by parts to

$$-\frac{q\Gamma_0}{4\pi} \frac{1}{(x'^2 + c^2)^{3/2}} \underset{\sim}{j}' \, dx' \quad (3.4)$$

Had we considered a concentrated vortex at C with strength Γ_0 the induced velocity at A would have been

$$-\frac{q\Gamma_0}{4\pi} \frac{1}{(x'^2 + q^2)^{3/2}} \underset{\sim}{j}' \, dx' \quad (3.5)$$

or, to first order in q,

$$-\frac{q\Gamma_0}{4\pi} \frac{1}{x'^3} \underset{\sim}{j}' \, dx' \quad (3.6)$$

Comparing (3.4) and (3.6) we observe that the effect of the finite core may be accounted for by replacing x'^3 by $(x'^2 + c^2)^{3/2}$ in the denominator of the self-induction function integral

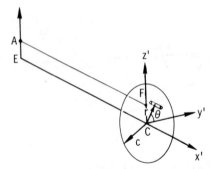

Figure 3.- Rearranged geometry.

The effect of this change is to replace $\omega(\beta,\delta)$ appearing in (2.3) by

$$\omega(\beta,\gamma) = \frac{\beta^2}{\gamma^2} \int_0^\infty \frac{\cos\gamma\xi + \gamma\xi\,\sin\gamma\xi - 1}{(1 + \xi^2)^{3/2}}\,d\xi$$

or

$$\omega(\beta,\gamma) = \frac{\beta^2}{\gamma^2}\,(\psi(\gamma) - 1) \tag{3.7}$$

where $\gamma = ck$ and the function $\psi(\gamma) = \gamma^2 K_0(\gamma) + \gamma K_1(\gamma)$ as in (2.3). The modes of oscillation of a single vortex are shown in Fig. 13.

4. NEW STABILITY BOUNDARIES

This simple modification to the self-induction function allows us to deduce stability boundaries in the $(\gamma/\beta, \beta)$ plane. We should first note that the effect of the finite cores on the mutual induction functions is of order $(c/b)^2$, an effect that will be neglected in this analysis. The characteristic equation (2.2) may be solved to yield roots

$$\alpha_s = \pm\left\{(1 - \psi(\beta) + \frac{\beta^2}{\gamma^2}\,[\psi(\gamma) - 1])\,(1 + \chi(\beta) - \frac{\beta^2}{\gamma^2}\,[\psi(\gamma) - 1])\right\}^{1/2}$$

and

$$\alpha_A = \pm\left\{(1 + \psi(\beta) + \frac{\beta^2}{\gamma^2}\,[\psi(\gamma) - 1])\,(1 - \chi(\beta) - \frac{\beta^2}{\gamma^2}\,[\psi(\gamma) - 1])\right\}^{1/2} \tag{4.1}$$

where the roots α_s correspond to symmetric modes in which

$\hat{y}_1 = -\hat{y}_2$, $\hat{z}_1 = \hat{z}_2$, and the roots α_A correspond to antisymmetric modes in which $\hat{y}_1 = \hat{y}_2$, $\hat{z}_1 = -\hat{z}_2$.

The stability of these modes will depend on the sign of α^2 in each case, negative α^2 yielding stability and positive α^2 yielding instability with a growth rate α. The behavior of the functions which form the two factors of α_s^2 and α_A^2 is such that the maximum growth rate occurs approximately when the two factors are equal in magnitude. It follows from an examination of the eigenvector $(\hat{y}_1 \quad \hat{z}_1 \quad \hat{y}_2 \quad \hat{z}_2)'$ that $\hat{y}_1/\hat{z}_1 \doteq 1$ and $\hat{y}_2/\hat{z}_2 \doteq -1$ in both the symmetric and antisymmetric modes of maximum growth. Thus, initially small perturbations will grow as standing waves in planes 45° to the plane formed by the unperturbed trailing vortices, which usually will be the horizontal plane.

The stability boundaries, maximum growth rates and the curve on which the self-induction function is zero are shown in Figs. 4 and 5 which should be compared with Figs. 6 and 7 of Crow's paper[1]. The self-induction function is zero when $\gamma = 1.11$ thus yielding the rectangular hyperbola in Figs. 4 and 5 given by

$$\left(\frac{\gamma}{\beta}\right) \cdot \beta = 1.11 \qquad (4.2)$$

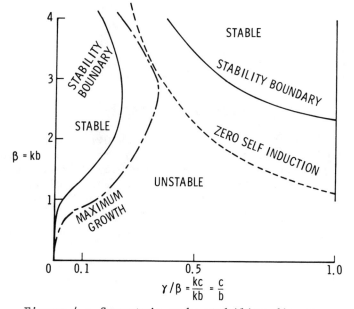

Figure 4.- Symmetric mode stability diagram.

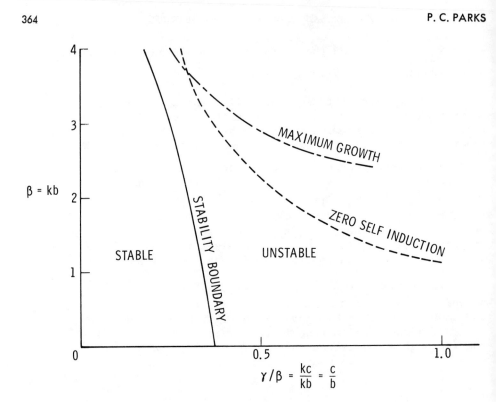

Figure 5.- Antisymmetric mode stability diagram.

5.   COMPARISON OF RESULTS

Taking the value of c/b or γ/β as 0.1 as predicted by Spreiter and Sacks[8] we may deduce that the most unstable long waves occur in a symmetric mode with $\beta = kb = 0.875$ or at a wavelength $2\pi/k = 7.18b$. The growth rate $\alpha = 0.81$. For an elliptically loaded wing $\Gamma_0 = \dfrac{8}{\pi^2} \dfrac{V_0 \, bC_L}{A_R}$ (Thwaites[9], pp. 309, 310) so that the instability grows by a factor e in a time $a^{-1} = 1.23 \dfrac{2\pi b^2}{\Gamma_0} = 9.53 \dfrac{A_R \, b}{C_L \, V_0}$.

The results of Crow[1] may be compared and confirmed in the following table:

Crow[1]	Result	Parks [This paper]
8.4b	Most unstable long wavelength:	7.2b
$9.4 \dfrac{A_R \, b}{C_L \, V_0}$	Grows by factor e in time:	$9.5 \dfrac{A_R \, b}{C_L \, V_0}$
0.42b	Most unstable short wavelength:	0.57b
$7.75 \dfrac{A_R \, b}{C_L \, V_0}$	Grows by factor e in time:	$7.75 \dfrac{A_R \, b}{C_L \, V_0}$

For comparison, vortex roll-up times are calculated as

$$0.36 \frac{A_R \, b}{C_L \, V_0}, \quad \text{Spreiter and Sacks}[8],$$

or

$$0.93 \frac{A_R \, b}{C_L \, V_0}, \quad \text{Crow}[9].$$

The form of the most unstable symmetrical mode is shown in Fig. 16.

6.   FINITE CORE THEORY APPLIED TO SPECIAL PROBLEMS

It is interesting to apply the finite core theory developed in section 3 to the two special problems treated by Kelvin[2] and Lamb[3]. Kelvin[2], Eq. 61, obtains the frequency of vibration of a columnar vortex as

$$\frac{\Gamma_0}{4\pi} k^2 \left\{ \log_e \frac{1}{ck} + \frac{1}{4} + \log_e 2 - \gamma^* \right\} \tag{6.1}$$

in our notation, where $\gamma^* = 0.5772157\ldots$ (Euler's constant). We obtain

$$\frac{\Gamma_0}{4\pi} k^2 \left\{ \log_e \frac{1}{ck} - \frac{1}{2} + \log_e 2 - \gamma^* + 0(\gamma^2 \log \gamma) \right\} \tag{6.2}$$

where $\gamma = ck$.

The second problem concerns the translational speed of a circular vortex ring. Lamb[3], p. 241, Eq. 7, gives the speed as

$$\frac{\Gamma_0}{4\pi\tilde{\omega}_0} \left\{ \log_e \frac{8\tilde{\omega}_0}{c} - \frac{1}{4} \right\} \tag{6.3}$$

where $\tilde{\omega}_0$ is the mean radius of the ring and c the cross-section radius.

We obtain an expression in terms of elliptic integrals which reduces to

$$\frac{\Gamma_0}{4\pi\tilde{\omega}_0} \left\{ \log_e \frac{8\tilde{\omega}_0}{c} - 1 + 0 \left(\left(\frac{c}{\tilde{\omega}_0}\right)^2 \log_e \frac{c}{\tilde{\omega}_0} \right) \right\} \tag{6.4}$$

It is interesting to note that (6.2) is the same result as that obtained by Rosenhead[10], p. 598, Eq. 8, using a different argument, and that the -1 in (6.4) was the number "favored by J. J. Thomson and others," as explained by Fraenkel[6], p. 38. Of course, (6.1) and (6.3) are correct, the uniform distribution of vorticity assumed in deriving (6.2) and (6.4) being only a first approximation to the true vorticity distribution. However, for $ck \ll 1$ the error involved in this approximation is small.

7.   DIGITAL COMPUTATION OF INITIAL GROWTH OF EXCITED PERTURBATIONS

It is natural to suggest making use of the unstable standing wave phenomenon, predicted by Crow[1] and in section 4, to break up the aircraft vortex wake. One method might be to excite the most unstable mode by oscillation of pairs of ailerons or flaps on the wings of the aircraft in a symmetrical fashion which leaves the total lift unchanged but affects the initial y coordinate of the rolled up vortices. Unfortunately the theory of section 2 assumes the vortices extend from infinity to infinity and cannot be used to calculate the growth of aircraft excited perturbations proposed here. However, a discretised semi-infinite linear theory has been set up as a programme on the Langley CDC 6600 computer, and prelimi-nary results are shown in Fig. 6. This shows an aircraft (approxi-mately a Boeing 747) flying from right to left and starting a sinusoidal excitation of the short wavelength motion (wavelength 0.57b) at $t = 0$. The slow growth and diffusion to the right of the perturbation are shown at four subsequent times.

Such a programme requires optimization to reduce excessive running times. In this problem with $c/b = 0.1$ satisfactory discretization along the vortices was achieved with a spacing of 0.015b.

Even the few wavelengths depicted in Fig. 6 required approximately 1 hour's machine time to compute!

8. THE EFFECT OF AXIAL FLOW WITHIN THE CORE - THE ISOLATED "JET-VORTEX"

Observations of vortices made visible by smoke, as well as theoretical calculations, such as those of Cone[5], suggest axial flow within the vortex core may be an important factor in discussing the vortex dynamics.

We first investigate the effect of a uniform axial flow along the vortex core of a single "jet-vortex." Following the vortex line technique used earlier in the paper, we model the axial flow U by wrapping the core in a sheath of vortex rings with an unperturbed strength U per unit distance in the x-direction. These rings will be convected at a speed U/2 in the x-direction under their own self-induction; the rings will also be capable of

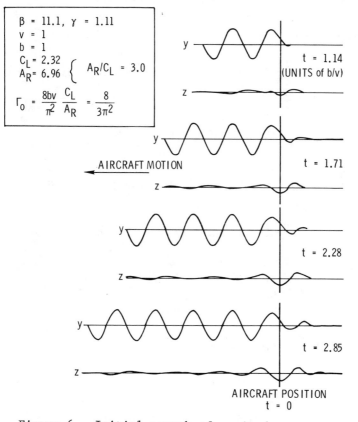

Figure 6.- Initial growth of excited waves.

small y- and z-displacements of their centers following
perturbations of the core as well as small angular perturbations
$\theta(x,t)$ and $\phi(x,t)$ as shown in Fig. 7. The angular perturbations
are an essential part of the operation of the model, in which we
first consider the stability of a straight circular jet $(\Gamma_0 = 0)$.

8.1. The Stability of a Circular Jet

Consider a straight circular jet of fluid of radius c having
a velocity U relative to similar fluid at rest. Let us consider
the motion of the vortex rings described in the preceding paragraph
restricting the motion to $z(x,t)$ and $\theta(x,t)$ defined in Fig. 7,
with $y(x,t)$ and $\phi(x,t)$ zero for all x and t.

We calculate the velocity (u^*,w^*) induced at the point G of
the ring whose center is instantaneously at the point $(0, 0, z(0,t))$
Using Fig. 8 and the Stokes stream function ψ for a single vortex
ring of strength K given by Lamb[3], p. 239, as

$$\psi(x,z) = - \frac{1}{2} Kcz \int_0^\infty e^{-sx} J_1(sz) J_1(sc) \, ds \qquad (8.1.1)$$

for x > 0, we obtain

$$u^*(t) = \int_{x=0}^\infty \frac{\partial u}{\partial x}(-x,c) \, \Delta x(x,t) + \frac{\partial u}{\partial z}(-x,c) \, \Delta z(x,t) + u(-x,c)$$

$$+ w(-x,c) \, \theta(x,t) + \int_{x=0}^\infty \frac{\partial u}{\partial x}(x,c) \, \Delta x(-x,t)$$

$$+ \frac{\partial u}{\partial z}(x,c) \, \Delta z(-x,t) + u(x,c) + w(x,c) \, \theta(-x,t)$$

$$w^*(t) = \int_{x=0}^\infty \frac{\partial w}{\partial x}(-x,c) \, \Delta x(x,t) + \frac{\partial w}{\partial z}(-x,c) \, \Delta z(x,t) + w(-x,c)$$

$$- u(-x,c) \, \theta(x,t) + \int_{x=0}^\infty \frac{\partial w}{\partial x}(x,c) \, \Delta x(-x,t)$$

$$+ \frac{\partial w}{\partial z}(x,c) \, \Delta z(-x,t) + w(x,c) - u(x,c) \, \theta(-x,t)$$

$$(8.1.2)$$

where $u(x,z) = - \frac{1}{z} \frac{\partial \psi}{\partial z}$, $w(x,z) = \frac{1}{z} \frac{\partial \psi}{\partial x}$, $K = U \, dx$,

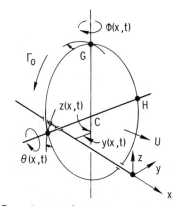

Figure 7.- General vortex ring geometry.

$\Delta x(x,t) = c\left[\theta(0,t) - \theta(x,t)\right]$, $\Delta z(x,t) = z(0,t) - z(x,t) - x\theta(x,t)$,

and where $\dfrac{\partial u}{\partial x}$ $(-x,c)$, and so forth, are calculated from $\dfrac{\partial u}{\partial x}$ (x,c),

and so forth, remembering that $u(x,c)$, $\dfrac{\partial w}{\partial z}$ (x,c), $\dfrac{\partial w}{\partial x}$ (x,c) are

even functions of x and $w(x,c)$, $\dfrac{\partial w}{\partial z}$ (x,c), $\dfrac{\partial u}{\partial x}$ (x,c) are odd

functions of x. Putting $z(x,t) = \hat{z}e^{ikx+at}$, $\theta(x,t) = \hat{\theta}e^{ikx+at}$

$u^*(t) = \hat{u}^* e^{at}$, $w^*(t) = \hat{w}^* e^{at}$ and using symmetry considerations
we obtain

$$\frac{\hat{u}^*}{U} = \int_{x=0}^{\infty}\left[-c\int_{s=0}^{\infty} 2se^{-sx} J_1^{\,2}(sc)\ ds\right.$$

$$\left.- c^2\int_{s=0}^{\infty} s^2 e^{-sx} J_1(sc)\,J_1'(sc)\ ds\right]\ i\ \sin kx\ \hat{\theta}\ dx$$

$$+ \int_{x=0}^{\infty}\left[-\frac{1}{c}\int_{s=0}^{\infty} e^{-sx} J_1^{\,2}(sc)\ ds + \int_{s=0}^{\infty} se^{-sx} J_1(sc)\,J_1'(sc)\ ds\right.$$

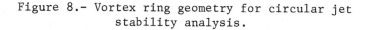

Figure 8.- Vortex ring geometry for circular jet
stability analysis.

$$+ c \int_{s=0}^{\infty} s^2 e^{-sx} J_1(sc) J_1''(sc) ds \Bigg] \times \Bigg[(1 - \cos kx) \, \hat{z}$$

$$- ix \sin kx \, \hat{\theta} \Bigg] dx + \frac{1}{2} e^{-at}$$

$$\frac{\hat{w}^*}{U} = \int_{x=0}^{\infty} \Bigg[- c^2 \int_{s=0}^{\infty} s^2 e^{-sx} J_1^2(sc) \; ds \Bigg] (1 - \cos kx) \, \hat{\theta} \, dx$$

$$+ \int_{x=0}^{\infty} \Bigg[c \int_{s=0}^{\infty} s^2 e^{-sx} J_1(sc) J_1'(sc) ds \Bigg] (i \sin kx \, \hat{z}$$

$$+ x \cos kx \, \hat{\theta}) \, dx + \int_{x=0}^{\infty} \Bigg[- \int_{s=0}^{\infty} e^{-sx} J_1^2(sc) \; ds$$

$$- c \int_{s=0}^{\infty} s e^{-sx} J_1(sc) J_1'(sc) ds \Bigg] \cos kx \, \hat{\theta} \, dx$$

$$(8.1.3)$$

These integrals may be evaluated following the procedures of Watson[7], Chapter 13 in particular. First we reverse the order of integration and integrate with respect to   x   which yields

$$\frac{\hat{u}^*}{U} = i\hat{\theta} \Bigg\{ -2ck \int_{s=0}^{\infty} \frac{s}{s^2 + k^2} J_1^2(sc) \; ds$$

$$- c^2 k \int_{s=0}^{\infty} \frac{s^2}{s^2 + k^2} J_1(sc) J_1'(sc) \; ds + \frac{2k}{c} \int_{s=0}^{\infty} \frac{s}{(s^2 + k^2)^2} J_1^2(sc)$$

$$- 2k \int_{s=0}^{\infty} \frac{s^2}{(s^2 + k^2)^2} J_1(sc) J_1'(sc) \; ds$$

$$- 2kc \int_{s=0}^{\infty} \frac{s^3}{(s^2 + k^2)^2} J_1(sc) J_1''(sc) \; ds \Bigg\}$$

$$+ \hat{z} \left\{ - \frac{k^2}{c} \int_{s=0}^{\infty} \frac{1}{s(s^2 + k^2)} J_1^{\;2}(sc) \; ds \right.$$

$$+ k^2 \int_{s=0}^{\infty} \frac{1}{s^2 + k^2} J_1(sc) \; J_1'(sc) \; ds$$

$$\left. + ck^2 \int_{s=0}^{\infty} \frac{s}{s^2 + k^2} J_1(sc) \; J_1''(sc) \; ds \right\} + \frac{1}{2} \, e^{-at}$$

$$\frac{\hat{w}^*}{U} = \hat{\theta} \left\{ - c^2 k^2 \int_{s=0}^{\infty} \frac{s}{s^2 + k^2} J_1^{\;2}(sc) \; ds \right.$$

$$+ c \int_{s=0}^{\infty} \frac{s^2(s^2 - k^2)}{(s^2 + k^2)^2} J_1(sc) \; J_1'(sc) \; ds$$

$$\left. - \int_{s=0}^{\infty} \frac{sJ_1^{\;2}(sc) \; ds}{s^2 + k^2} - c \int_{s=0}^{\infty} \frac{s^2}{s^2 + k^2} J_1(sc) \; J_1'(sc) \; ds \right\}$$

$$+ i\hat{z} \left\{ ck \int_{s=0}^{\infty} \frac{s^2}{s^2 + k^2} J_1(sc) \; J_1'(sc) \; ds \right\} \tag{8.1.4}$$

The integrals appearing in (8.1.4) are evaluated in terms of Bessel functions as follows:

$$\int_{s=0}^{\infty} \frac{s}{s^2 + k^2} J_1^{\;2}(sc) \; ds = I_1(ck) \; K_1(ck)$$

$$\int_{s=0}^{\infty} \frac{s^2}{s^2 + k^2} J_1(sc) \; J_1'(sc) \; ds = \frac{k}{2} \left[I_1'(ck) \; K_1(ck) + I_1(ck) \; K_1'(ck) \right]$$

$$\int_{s=0}^{\infty} \frac{s}{(s^2 + k^2)^2} J_1^{\;2}(sc) \; ds = - \frac{c}{2k} \left[I_1'(ck) \; K_1(ck) + I_1(ck) \; K_1'(ck) \right]$$

$$\int_{s=0}^{\infty} \frac{s^2}{(s^2 + k^2)^2} J_1(sc) \, J_1{}'(sc) \, ds = -\frac{1}{4k} \Big[I_1{}'(ck) \, K_1(ck)$$

$$+ \, I_1(ck) \, K_1{}'(ck) \Big]$$

$$- \, \frac{c}{4} \Big[I_1{}''(ck) \, K_1(ck)$$

$$+ \, 2I_1{}'(ck) \, K_1{}'(ck)$$

$$+ \, I_1(ck) \, K_1{}''(ck) \Big]$$

$$\int_{s=0}^{\infty} \frac{s^3}{(s^2 + k^2)^2} J_1(sc) \, J_1{}''(sc) \, ds = \Big[-k + \frac{1}{c^2 k} \Big] I_1(ck) \, K_1(ck)$$

$$+ \, I_1{}'(ck) \, K_1{}'(ck)$$

$$- \, \Big[ck^2 + \frac{1}{c} \Big] \Big[I_1{}'(ck) \, K_1(ck)$$

$$+ \, I_1(ck) \, K_1{}'(ck) \Big]$$

$$\int_{s=0}^{\infty} \frac{1}{s(s^2 + k^2)} J_1{}^2(sc) \, ds = \frac{1}{k^2} \Big[\frac{1}{2} - I_1(ck) \, K_1(ck) \Big]$$

$$\int_{s=0}^{\infty} \frac{1}{(s^2 + k^2)} J_1(sc) \, J_1{}'(sc) \, ds = -\frac{1}{2k} \Big[I_1{}'(ck) \, K_1(ck)$$

$$+ \, I_1(ck) \, K_1{}'(ck) \Big]$$

$$\int_{s=0}^{\infty} \frac{s}{s^2 + k^2} J_1(sc) \, J_1{}''(sc) \, ds = -I_1(ck) \, K_1(ck) \times \Big[1 + \frac{1}{c^2 k^2} \Big]$$

$$+ \, \frac{1}{ck} K_1(ck) \, I_1{}'(ck)$$

$$\int_{s=0}^{\infty} \frac{s^2(s^2 - k^2)}{(s^2 + k^2)^2} J_1(sc) J_1'(sc) \, ds = k\left[I_1'(ck) K_1(ck) + I_1(ck) K_1'(ck)\right]$$

$$+ \frac{ck^2}{2}\left[I_1''(ck) K_1(ck)\right]$$

$$+ 2I_1'(ck) K_1'(ck)$$

$$+ I_1(ck) K_1''(ck)\right]$$

On substitution into (8.1.4) considerable simplification takes place leaving the expressions

$$\frac{\hat{u}^*}{U} = -c^2k^2 I_1(ck) K_1(ck) \frac{\hat{z}}{c} + \frac{c^2k^2}{2}\left[I_1'(ck) K_1(ck) + I_1(ck) K_1'(ck)\right] i\hat{\theta}$$

$$+ \frac{1}{2} e^{-at}$$

$$\frac{\hat{w}^*}{U} = \frac{c^2k^2}{2}\left[I_1'(ck) K_1(ck) + I_1(ck) K_1'(ck)\right] i\frac{\hat{z}}{c} + c^2k^2 I_1'(ck) K_1'(ck) \hat{\theta}$$

$$(8.1.5)$$

Now the unperturbed rings travel in the x-direction with a steady velocity $\frac{U}{2}$ so that

$$u^*(t) = c \frac{\partial\theta}{\partial t} (0,t) + \frac{1}{2} Uc \frac{\partial\theta}{\partial x} (0,t) + \frac{1}{2} U$$

$$w^*(t) = \frac{\partial z}{\partial t} (0,t) + \frac{1}{2} Uc \frac{\partial z}{\partial x} (0,t)$$

or

$$\hat{u}^* = ca\hat{\theta} + \frac{1}{2} Ucki\hat{z} + \frac{1}{2} Ue^{-at}$$

$$\hat{w}^* = a\hat{z} + \frac{1}{2} Uki\hat{z}$$

$$\left.\right\} \qquad (8.1.6)$$

Substituting for $\hat{u}^*$ and $\hat{w}^*$ from (8.1.6) in (8.1.5) we obtain the characteristic equation

$$
\det \begin{bmatrix}
\dfrac{ck}{2}\left[I_1{}'(ck)\,K_1(ck) \right. & c^2k^2\,I_1{}'(ck)\,K_1{}'(ck) \\[6pt]
\left. +\,I_1(ck)\,K_1{}'(ck)\right]i & \\[6pt]
-\dfrac{1}{2}\,i - s & \\[12pt]
-I_1(ck)\,K_1(ck) & \dfrac{ck}{2}\left[I_1{}'(ck)\,K_1(ck) \right. \\[6pt]
& \left. +\,I_1{}'(ck)\,K_1{}'(ck)\right]i \\[6pt]
& -\dfrac{1}{2}\,i - s
\end{bmatrix} = 0
$$

$$(8.1.7)$$

where $s = \dfrac{a}{Uk}$ and the column eigenvector is $[k\hat{z}\ \ \hat{\theta}\,]'$. Solving the resulting quadratic equation we obtain after some reduction, using the identity

$$K_1{}'(ck)\,I_1(ck) - I_1{}'(ck)\,K_1(ck) = -\,1/ck, \qquad (8.1.8)$$

the roots

$$s = \frac{a}{Uk} = ick\,K_1{}'(ck)\,I_1(ck) \pm \left[\sqrt{-I_1(ck)\,K_1(ck)\,I_1{}'(ck)\,K_1'(ck)}\right]ck$$

$$(8.1.9)$$

This agrees exactly with the result of Batchelor and Gill[11] for their $n = 1$ mode in which a circular cross section is maintained. Moreover the unstable mode in which the plus sign is taken in (8.1.9) leads to a relationship $\dfrac{k\hat{z}}{\hat{\theta}} \doteq -1$ for small ck so that the motion of the rings appears as shown in Fig. 9, which agrees qualitatively with similar diagrams for motions of the vortex sheets bounding a two-dimensional jet as given by Abernathy and Kronauer.[12]

For small ck (8.1.9) yields

$$a = \left(-\frac{1}{2} \pm \frac{1}{2}\right) Uk \qquad (8.1.10)$$

The two modes corresponding to these roots are shown in Fig. 14.

8.2. The Stability of a Circular Jet-Vortex

We now extend the model established in section 8.1 to the case of a circular jet-vortex. We assume a uniform axial velocity U within the core and a uniform axial vorticity distribution of strength $\Gamma_0/\pi c^2$ per unit area within the core. We model this with a circular bundle of vortex line filaments surrounded by a sheath of circular vortex rings. We consider perturbations in y and z of the center line and small angular perturbations θ and ϕ of the rings, as in Fig. 7.

By a procedure similar to that employed in section 8.1 we calculate the induced velocity of the rings at the center line, C, and replace the second equation of (8.1.5) by

$$\frac{\hat{w}_{RC}^*}{U} = \frac{c^2 k^2}{2} K_1(ck)\, i\,\frac{\hat{z}}{c} + \frac{c^2 k^2}{2} K_1'(ck)\hat{\theta} \,, \qquad (8.2.1)$$

retaining the first equation of (8.1.5) to calculate $\hat{u}_{RG}^*$ at the point G. We also replace the second equation of (8.1.6) by $\hat{w}_C^* = a\hat{z} + Uki\hat{z}$. To $\hat{w}_{RC}^*$ we have to add the induced velocity $\hat{w}_{LC}^*$ of the axial vortex filaments; this has been calculated as the self-induction function in section 3 and is given by

$$\hat{w}_{LC}^* = -\frac{\Gamma_0}{2\pi c^2}\left[\psi(ck) - 1\right]\hat{y} \qquad (8.2.2)$$

We now obtain the equations

$$a\hat{z} + Uki\hat{z} = U\frac{c^2 k^2}{2} K_1(ck)\, i\,\frac{\hat{z}}{c} + U\frac{c^2 k^2}{2} K_1'(ck)\hat{\theta} - \frac{\Gamma_0}{2\pi c^2}\left[\psi(ck) - 1\right]\hat{y}$$

and

$$ca\hat{\theta} + \frac{1}{2} Ucki\hat{\theta} = -Uc^2 k^2 I_1(ck)\, K_1(ck)\,\frac{\hat{z}}{c}$$

$$+ U\frac{c^2 k^2}{2}\left[I_1'(ck)\, K_1(ck) + I_1(ck)\, K_1'(ck)\right] i\hat{\theta}$$

$$+ \frac{\Gamma_0}{2\pi c}\hat{\phi}.$$

The two corresponding equations in $\hat{y}$ and $\hat{\phi}_2$ lead to the following characteristic equation in s, where $s = \dfrac{2\pi c^2}{\Gamma_0}\, a$ and $M = 2\pi\, cU/\Gamma_0$

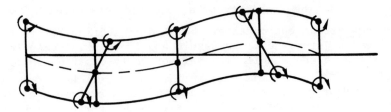

Figure 9.- Vortex ring motion in the unstable mode
of a circular jet.

$$\det \begin{bmatrix} s + ckMi \left[1 - \frac{ck}{2} K_1(ck)\right] & -\frac{Mc^3k^3}{2} K_1{}'(ck) & -\psi(ck) + 1 & 0 \\ MckI_1(ck)\, K_1(ck) & s + Mi\left\{\frac{ck}{2} - \frac{c^2k^2}{2}\left[I_1{}'(ck)\, K_1(ck) + I_1(ck)\, K_1{}'(ck)\right]\right\} & 0 & 1 \\ \psi(ck) - 1 & 0 & s + ckMi\left[1 - \frac{ck}{2} K_1(ck)\right] & -\frac{Mc^3k^3}{2} K_1{}'(ck) \\ 0 & -1 & MckI_1(ck)\, K_1(ck) & s + Mi\left\{\frac{ck}{2} - \frac{c^2k^2}{2}\left[I_1{}'(ck)\, K_1(ck) + I_1(ck)\, K_1{}'(ck)\right]\right\} \end{bmatrix} = 0$$

(8.2.3)

where the column eigenvector is $[k\hat{y} \quad \hat{\phi} \quad k\hat{z} \quad \hat{\theta}]'$.

For $ck \ll 1$ this reduces to

$$\det \begin{bmatrix} s + \frac{1}{2} ckMi & \frac{Mck}{2} & -\psi(ck) + 1 & 0 \\ \frac{Mck}{2} & s + \frac{1}{2} ckMi & 0 & 1 \\ \psi(ck) - 1 & 0 & s + \frac{1}{2} ckMi & \frac{Mck}{2} \\ 0 & -1 & \frac{Mck}{2} & s + \frac{1}{2} ckMi \end{bmatrix} = 0$$

(8.2.4)

Putting $s + \frac{1}{2} ckMi = \mu$ we obtain the quadratic equation in μ^2

$$\mu^4 + (1 + b'^2 - 2a') \mu^2 + (a' - b')^2 = 0 \qquad (8.2.5)$$

where $a' = \dfrac{M^2 c^2 k^2}{4}$ and $b' = \psi(ck) - 1$. For stability of the jet vortex we require the quadratic equation in μ^2 to have two real roots less than or equal to zero. The conditions are

$$1 + b'^2 - 2a' > 0$$

and
$$\qquad (8.2.6)$$

$$(1 + b')^2 > 4a'$$

The second condition is the more critical and yields

$$\frac{2\pi cU}{\Gamma_0} = M < \frac{\psi(ck)}{ck} \doteq \frac{1}{ck} \qquad \text{as} \quad ck \ll 1. \qquad (8.2.7)$$

This is believed to be a new criterion for the stability of a jet vortex. For given U, c, and Γ_0 there is a critical wave number k_{crit} given by

$$k_{crit} = \frac{\Gamma_0}{2\pi c^2 U} \qquad (8.2.8)$$

so that a perturbation with a wavelength shorter than $2\pi/k_{crit}$ is unstable.

If Γ_0 is small the roots of (8.2.4) are the repeated root

$$s = \left(-\frac{i}{2} \pm \frac{1}{2}\right) Mck$$

or

$$a = \left(-\frac{i}{2} \pm \frac{1}{2}\right) Uk \qquad (8.2.9)$$

agreeing with the solution for circular jets for small ck given in equation (8.1.10) and by Batchelor and Gill[11].

For zero U or M the equation (8.2.5) factorises as

$$[\mu^2 + (\psi(ck) - 1)] [\mu^2 + 1] = 0 \qquad (8.2.10)$$

The first factor corresponds to stable spiral motions of the isolated vortex under its own self-induction (Crow[1]), while the second factor represents simple harmonic motions in θ and ϕ of the vortex rings. These rings have zero strength however, so the second factor has a mathematical, rather than physical, significance. The modes and ring motions for the critical case are shown in Fig. 15.

9. STABILITY OF A TRAILING JET VORTEX PAIR WITH AXIAL FLOWS

We now reconsider the trailing vortex pair, each vortex having a uniform axial flow U within its core. It is convenient in this problem to define

$$M' = \frac{2\pi b U}{\Gamma_0} \tag{9.1}$$

to replace M as defined in (8.2.3).

The interaction of the vortex rings of one vortex on the behavior of the other is of order $(c/b)^2$ and will be neglected. The characteristic equation of the equations describing the motion now becomes

$$
\det
\begin{vmatrix}
a_{11} - s & a_{12} & -B_2 & 0 & 0 & 0 & H & 0 \\
a_{21} & a_{22} - s & 0 & -E & 0 & 0 & 0 & 0 \\
B_1 & 0 & a_{11} - s & a_{12} & J & 0 & 0 & 0 \\
0 & E & a_{21} & a_{22} - s & 0 & 0 & 0 & 0 \\
0 & 0 & -H & 0 & a_{11} - s & a_{12} & B_2 & 0 \\
0 & 0 & 0 & 0 & a_{21} & a_{22} - s & 0 & E \\
-J & 0 & 0 & 0 & -B_1 & 0 & a_{11} - s & a_{12} \\
0 & 0 & 0 & 0 & 0 & -E & a_{21} & a_{22} - s
\end{vmatrix}
= 0
\tag{9.2}
$$

with a column eigenvector $[k\hat{y}_1 \quad \hat{\phi}_1 \quad k\hat{z}_1 \quad \hat{\theta}_1 \quad k\hat{y}_2 \quad \hat{\phi}_2 \quad k\hat{z}_2 \quad \hat{\theta}_2]'$ and where

$$s = \frac{2\pi b^2}{\Gamma_0} a, \quad a_{11} = M'bki \, [-1 + \tfrac{1}{2} ck \, K_1(ck)] \quad a_{12} = M'bk \left[\tfrac{1}{2} c^2 k^2 K_1{}'(ck) \right]$$

$$a_{21} = -M'bk \, K_1(ck) \, I_1(ck) \quad a_{22} = M'bki \left\{ \frac{ck}{2} \left[I'(ck) \, K_1(ck) \right.\right.$$

$$\left.\left. + I_1(ck) \, K_1{}'(ck) \right] - \tfrac{1}{2} \right\}$$

$$B_1 = 1 - \frac{\beta^2}{\gamma^2} [\psi(\gamma) - 1], \; B_2 = -1 - \frac{\beta^2}{\gamma^2} [\psi(\gamma) - 1], \; E = \frac{\beta^2}{\gamma^2} , \; J = -\chi(\beta),$$

$$H = -\psi(\beta), \; \gamma = ck, \; \beta = bk, \; \chi(\beta) = \beta K_1(\beta), \; \psi(\beta) = \beta^2 K_0(\beta) + \beta K_1(\beta).$$

$$\tag{9.3}$$

The determinant (9.2) breaks into two determinants for the symmetric and antisymmetric modes:

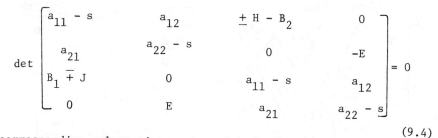

$$\det \begin{bmatrix} a_{11} - s & a_{12} & \pm H - B_2 & 0 \\ a_{21} & a_{22} - s & 0 & -E \\ B_1 \mp J & 0 & a_{11} - s & a_{12} \\ 0 & E & a_{21} & a_{22} - s \end{bmatrix} = 0$$

(9.4)

with corresponding column eigenvectors for (9.2)

$$[k\hat{y}_1 \quad \hat{\phi}_1 \quad k\hat{z}_1 \quad \hat{\theta}_1 \quad -k\hat{y}_1 \quad -\hat{\phi}_1 \quad k\hat{z}_1 \quad \hat{\theta}_1]'$$

and

$$[k\hat{y}_1 \quad \hat{\phi}_1 \quad k\hat{z}_1 \quad \hat{\theta}_1 \quad k\hat{y}_1 \quad \hat{\phi}_1 \quad -k\hat{z}_1 \quad -\hat{\theta}_1]'$$

Assuming $ck \ll 1$ so that $a_{11} = a_{22} = -\dfrac{M'bki}{2}$, $a_{12} = a_{21} = -\dfrac{M'bk}{2}$ and putting $\mu' = -\dfrac{i}{2} - \dfrac{s}{M'bk}$ we obtain, in the symmetric case,

$$\mu'^4 + \mu'^2 \left(-\frac{1}{2} + \frac{(B_1 - J)(B_2 - H)}{M'^2 b^2 k^2} + \frac{E^2}{M'^2 b^2 k^2} \right) + \frac{1}{16} + \frac{1}{4} \frac{(B_1 - J)E}{M'^2 b^2 k^2}$$

$$+ \frac{1}{4} \frac{(B_2 - H)E}{M'^2 b^2 k^2} + \frac{(B_1 - J)(B_2 - H)E^2}{M'^4 b^4 k^4} = 0$$

(9.5)

Once again, for stability we require this quadratic equation in μ'^2 to have two real nonpositive roots. The first condition to be satisfied is

$$\left(\frac{1}{4} + \frac{(B_1 - J)E}{M'^2 b^2 k^2} \right) \left(\frac{1}{4} + \frac{(B_2 - H)E}{M'^2 b^2 k^2} \right) > 0$$

(9.6)

which modifies the stability boundaries obtained in section 4 and Fig. 4 as shown in Fig. 10. Additionally, a second condition is

$$M'^2 b^2 k^2 < \frac{[E^2 - (B_1 - J)(B_2 - H)]^2}{(E + B_1 - J)(E + B_2 - H)}$$

which in the region of Fig. 10 of interest is essentially

or

$$M'bk < E$$

$$\frac{U \, 2\pi b^2 k}{\Gamma_0} < \frac{b^2}{c^2}$$

or

$$\frac{2\pi cU}{\Gamma_0} < \frac{1}{ck}$$

(9.7)

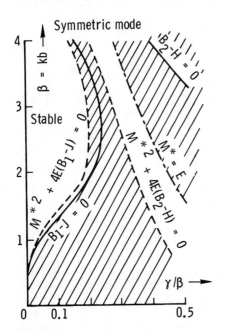

Figure 10.- Stability diagram for symmetric mode with axial flow.

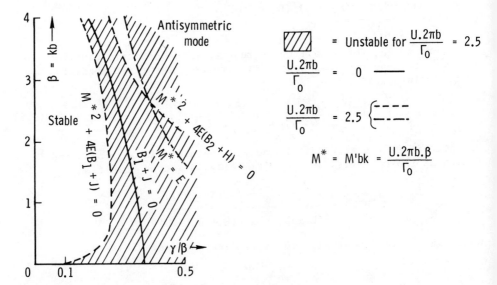

Figure 11.- Stability diagram for antisymmetric with axial flow.

This is the stability criterion for the single jet vortex given in (8.2.7), and is shown also in Fig. 10. Similar considerations apply to the antisymmetric case which has the stability boundaries shown in Fig. 11.

A third condition for stability that the coefficient of μ'^2 in (9.5) should be positive is already satisfied if (9.7) holds.

Figs. 10 and 11 should be compared with Figs. 4 and 5, respectively. It is seen that the stability boundaries have moved to the left. In Fig. 10 for $\beta = 1$, for example, we encounter stability, instability, stability and instability as we move in the direction of increasing γ/β crossing, in turn, the curves given by

$$\frac{M'^2 b^2 k^2}{4} + (B_1 - J)E = 0, \quad \frac{M'^2 b^2 k^2}{4} + (B_2 - H)E = 0 \quad \text{and} \quad M'bk = E,$$

respectively.

The effect of axial flow on the maximum growth rate is very small as is shown in Fig. 12, which is plotted for a traverse of the line $\beta = 1$ of Fig. 9. The calculation of the growth rates is a quite delicate computation.

On the line $c/b = 0.1$ the effect will be that the maximum growth curve shown in Fig. 4 that is moved to the left intersects the line $c/b = 0.1$ at a slightly larger value of β for slightly higher values of the growth factor α. Thus the growth rate of the most unstable mode is increased.

The unstable symmetric and antisymmetric motions will still be in planes at approximately $45°$ to the horizontal until the $M'bk = E$ curve is crossed when unstable spiral motions will be encountered. The symmetric modes are shown in Fig. 16.

10.   CONCLUSIONS

A stability theory for a trailing vortex pair has been presented which takes account of the finite core size of the vortices. The findings of S. C. Crow[1] are essentially confirmed.

The work has been extended to account for axial flows with the vortex cores and stability criteria for the single jet vortex and a pair of trailing jet vortices have been presented.

There is a need to confirm the theoretical results obtained here by alternative theoretical methods and also by suitably devised experiments. Further theoretical developments are

required to cover other core deformation modes, to take account
of nonuniform vorticity and axial flow distributions within the
core and to relate the present theory to other developments such
as vortex breakdown or bursting explanations.

    In particular experimental measurements of axial flows
encountered in practice are required in order that more realistic
mathematical models may be built.

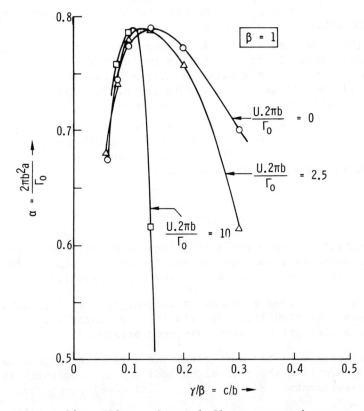

Figure 12.- Effect of axial flow on growth rates.

The theoretical results are summarised in the following table:

SUMMARY OF RESULTS

Result	Section	Equation	Figures	Modes in Fig.
Self induction function for a single vortex of circular cross-section	3	3.7	2	13
Stability boundaries for a trailing vortex pair	4,5	4.1	4,5	16
Computed growth of excited perturbations	7	–	6	–
Instability of a circular jet	8.1	8.1.9	7	9,14
Stability criterion for a single jet-vortex	8.2	8.2.7	–	15
Stability boundaries for the trailing jet-vortex pair	9	9.6,9.7	10,11	16

11. ACKNOWLEDGMENTS

This work was undertaken while the author held a National Research Council Postdoctoral Resident Research Associateship supported by the National Aeronautics and Space Administration at NASA Langley Research Center, Hampton, Virginia. It is a pleasure to acknowledge valuable discussions in connection with this work with NASA colleagues, particularly Mr. A. A. Schy and Mr. R. E. Dunham. The computer programme referred to in section 7 was written by Mr. R. B. Scher, Senior Student in Physics at Harvard University as part of his work as a summer trainee at Langley Research Center. The author also gratefully acknowledges his sabbatical leave granted by the University of Warwick, Coventry, England.

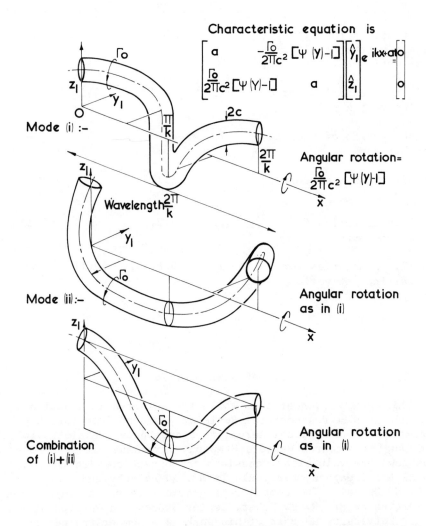

Characteristic equation is

$$\begin{bmatrix} a & -\dfrac{\Gamma_0}{2\pi c^2}\left[\Psi(Y)-1\right] \\ \dfrac{\Gamma_0}{2\pi c^2}\left[\Psi(Y)-1\right] & a \end{bmatrix}\begin{bmatrix}\hat{y}_1 \\ \hat{z}_1\end{bmatrix}e^{ikx+at}=\begin{bmatrix}0 \\ 0\end{bmatrix}$$

Mode (i) :-

Angular rotation=
$$\dfrac{\Gamma_0}{2\pi c^2}\left[\Psi(Y)-1\right]$$

Wavelength $\dfrac{2\pi}{k}$

Mode (ii):-

Angular rotation
as in (i)

Combination
of (i)+(ii)

Angular rotation
as in (i)

FIG. 13. MODES OF SINGLE VORTEX (NEUTRAL STABILITY)

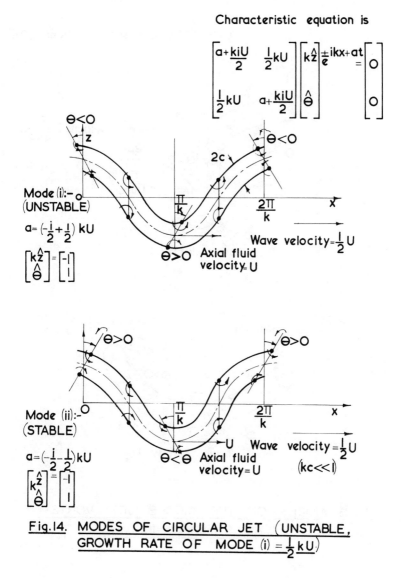

Characteristic equation is

$$\begin{bmatrix} a+\dfrac{kiU}{2} & \dfrac{1}{2}kU \\[2mm] \dfrac{1}{2}kU & a+\dfrac{kiU}{2} \end{bmatrix} \begin{bmatrix} k\hat{z} \\[2mm] \hat{\Theta} \end{bmatrix} e^{\pm ikx+at} = \begin{bmatrix} 0 \\[2mm] 0 \end{bmatrix}$$

Mode (i):-
(UNSTABLE)

$$a=\left(-\dfrac{i}{2}+\dfrac{1}{2}\right)kU$$

$$\begin{bmatrix} k\hat{z} \\ \hat{\Theta} \end{bmatrix} = \begin{bmatrix} -1 \\ 1 \end{bmatrix}$$

$\Theta<0$

$\Theta<0$

$2c$

Wave velocity$=\dfrac{1}{2}U$

$\Theta>0$ Axial fluid velocity$=U$

Mode (ii):-
(STABLE)

$$a=\left(-\dfrac{i}{2}-\dfrac{1}{2}\right)kU$$

$$\begin{bmatrix} k\hat{z} \\ \hat{\Theta} \end{bmatrix} = \begin{bmatrix} -1 \\ 1 \end{bmatrix}$$

$\Theta>0$

$\Theta>0$

Wave velocity$=\dfrac{1}{2}U$

$\Theta<\Theta$ Axial fluid velocity$=U$ $(kc\ll1)$

Fig.14. MODES OF CIRCULAR JET (UNSTABLE, GROWTH RATE OF MODE (i) $=\dfrac{1}{2}kU$)

Characteristic equation: (8.2.4.)

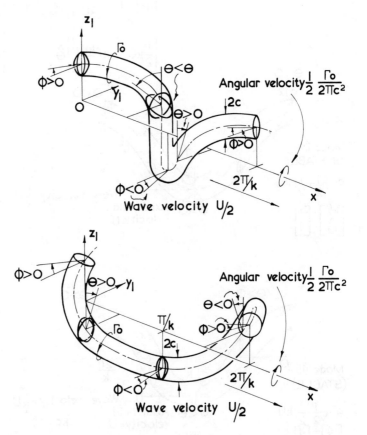

FIG. 15 <u>MODES OF THE SINGLE JET VORTEX</u>
(<u>CRITICAL CASE</u>: $Uc/\Gamma_0 = 1/2\pi ck$)

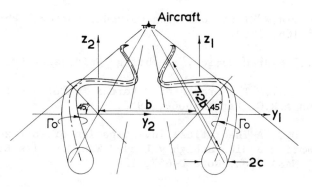

U=O, Growth rate   0·81 $\Gamma_0/2\pi b^2$, for c/b = 0·1.

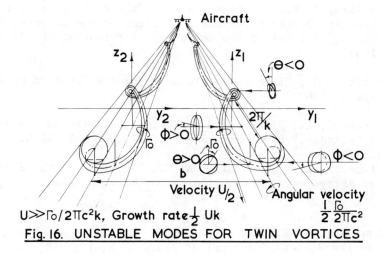

$U \gg \Gamma_0/2\pi c^2 k$, Growth rate $\frac{1}{2}$ Uk

Velocity $U_{/2}$ Angular velocity $\frac{1}{2}\frac{\Gamma_0}{2\pi c^2}$

Fig. 16. UNSTABLE MODES FOR TWIN VORTICES

12.   REFERENCES

1. Crow, S. C., "Stability Theory for a Pair of Trailing
    Vortices," AIAA Paper 70-53, Eighth AIAA Aerospace Sciences
    Meeting, New York, January 19-21, 1970. AIAA Journal (to
    appear)

2. Lord Kelvin, "Vibrations of a Columnar Vortex," Phil. Mag.,
    X, 155-168, 1880.

3. Lamb, H., "Hydrodynamics," Cambridge University Press, 1932.

4. Squire, H. B., "The Growth of a Vortex in Turbulent Flow,"
    British A.R.C. Paper 16666, 1955.

5. Cone, C. D., "A Theoretical Investigation of Vortex-Sheet
    Deformation Behind a Highly Loaded Wing and its Effect on
    Lift," NASA Tech. Note D-657, 1961.

6. Fraenkel, L. E., "On Steady Vortex Rings of Small Cross
    Section in an Ideal Fluid," Proc. Royal Soc. A, 316,
    29-62, 1970.

7. Watson, G. N., "A Treatise on the Theory of Bessel Functions,"
    Cambridge University Press, 1952.

8. Spreiter, J. R., and Sacks, A. H., "The Rolling Up of the
    Trailing Vortex Sheet and its Effect on the Downwash Behind
    Wings," Journal of the Aeronautical Sciences, 18, 21-32,
    1951.

9. Thwaites, B., "Incompressible Aerodynamics," Oxford University
    Press, 1960.

10. Rosenhead, L., "The Spread of Vorticity in the Wake Behind
    a Cylinder," Proc. Roy. Soc. A, 127, 590-612, 1930.

11. Batchelor, G. K., and Gill, A. E., "Analysis of the Stability
    of Axisymmetric Jets," Journal of Fluid Mechanics, 14,
    529-551, 1962.

12. Abernathy, F. H., and Kronauer, R. E., "The Formation of
    Vortex Streets," Journal of Fluid Mechanics, 13, 1-20,
    1962.

DECAY OF AN ISOLATED VORTEX

Coleman duP. Donaldson and Roger D. Sullivan

Aeronautical Research Associates of Princeton, Inc.

INTRODUCTION

This paper presents a short review of a program in progress at A.R.A.P. which is aimed at increasing our understanding of the decay of the trailing vortex systems behind aircraft. Although a number of mechanisms have been identified by which a vortex system may be dissipated, i.e., by shear, by the growth of disturbances due to mutual interference, and by vortex breakdown, we have confined ourselves in developing the present program to a study of the effect of turbulent shear on the decay of a single vortex. The ultimate aim of the program is the development of a method by which the effects of turbulence, whether introduced initially in the center of a vortex or contained in the fluid in which the vortex is formed, could be calculated.

The program consists of two parts. The first part is directed towards the development of analytical tools for the prediction of the effects of turbulence. The second part is an experimental program designed to obtain data against which the computational techniques may be checked.

Since the program has just now reached the point where we are beginning to obtain results, this paper should be considered a progress report on the work that is presently being sponsored at A.R.A.P. by the Air Force Office of Scientific Research under Contract F44620-69-C-0089.

THEORETICAL STUDIES

During the past few years, an attempt has been made at A.R.A.P. to develop a method for computing turbulent shear flows that followed the generation of turbulence and turbulent shear from the introduction of a small amount of turbulent energy into a laminar shear layer to the development of a fully turbulent layer[1,2,3]. The basis of this method is a modeling of certain terms in the fundamental equation for the double velocity correlation $\overline{u_i'u_j'}$ (see Eq. (2) below) so as to form a closed set of equations. In view of the success that had been realized in applying this method to atmospheric flows[3] and the new insights into the generation of atmospheric turbulence and shear that resulted, it was thought desirable to apply the method to the case of an isolated vortex for which $\overline{u} = \overline{w} = 0$ and $\overline{v} = \overline{v}(r,t)$ where $\overline{u}$, $\overline{v}$, and $\overline{w}$ are components of the mean velocity in the radial, tangential, and axial directions, respectively.

The basic equation which governs the mean velocity of such a flow is easily obtained. It is

$$\frac{\partial \overline{v}}{\partial t} = \nu\left(\frac{\partial^2 \overline{v}}{\partial r^2} + \frac{1}{r}\frac{\partial \overline{v}}{\partial r} - \frac{\overline{v}}{r^2}\right) - \frac{\partial}{\partial r}(\overline{u'v'}) - \frac{2\overline{u'v'}}{r} \tag{1}$$

The boundary conditions on v are

$$r \to \infty \qquad r\overline{v} \to \Gamma_\infty/2\pi$$

$$r \to 0 \qquad \overline{v} = \alpha r$$

where α is a function of time to be determined and Γ_∞ is the circulation at large r. One might proceed to try to solve this equation by making an assumption concerning the relationship between the correlation $\overline{u'v'}$ and the mean tangential velocity $\overline{v}$. This is the usual technique that has been employed in eddy viscosity models.

If we follow the method of invariant modeling that is under development at A.R.A.P., we will compute the value of $\overline{u'v'}$ in (1) by computing the development through the use of the basic equation for the second-order correlations in an incompressible medium. This equation is[4]

$$\frac{\partial}{\partial t}\,\overline{u_i'u_k'} + \bar{u}^j\,\overline{(u_i'u_k')}_{,j} = -\,\overline{u'^j u_k'}\,\bar{u}_{i,j} - \overline{u'^j u_i'}\,\bar{u}_{k,j}$$

$$-\,\overline{(u'^j u_i'u_k')}_{,j} - \frac{1}{\rho}\,\overline{(p'u_i')}_{,k} - \frac{1}{\rho}\,\overline{(p'u_k')}_{,i}$$

$$+\,\frac{1}{\rho}\,\overline{p'(u_{i,k}' + u_{k,i}')} + \nu g^{mn}\,\overline{(u_i'u_k')}_{,mn}$$

$$-\,2\nu g^{mn}\,\overline{u_{i,m}'u_{k,n}'} \tag{2}$$

This equation alone is not sufficient to define the problem. We must express the higher order and unknown correlations in this equation with physical models of these terms which contain, at most, the known mean velocity and second-order correlations. A modeling that has been used successfully in the past[3] is

$$\overline{u_i'u_j'u_k'} = -M\left[\overline{(u_i'u_j')}_{,k} + \overline{(u_i'u_k')}_{,j} + \overline{(u_j'u_k')}_{,i} \right] \tag{3}$$

$$\overline{p'(u_{i,k}' + u_{k,i}')} = \rho\,\frac{M}{\Lambda^2}\left(\frac{g_{ik}\overline{u'^m u_m'}}{3} - \overline{u_i'u_k'} \right) \tag{4}$$

$$\overline{p'u_i'} = -\rho M \overline{(u'^\ell u_i')}_{,\ell} \tag{5}$$

$$g^{mn}\,\overline{u_{i,m}'u_{k,n}'} = \frac{\overline{u_i'u_k'}}{\lambda^2} \tag{6}$$

where

$$M = \Lambda\sqrt{\overline{u'^m u_m'}} \tag{7}$$

$$\lambda = \Lambda/\sqrt{a + bRe_\Lambda} \tag{8}$$

$$Re_\Lambda = \frac{\rho M}{\mu} \tag{9}$$

Here a and b are dimensionless constants and Λ is a length related to the scale of the turbulence. In the model it is assumed that Λ is proportional to the scale of the mean motion.

This model is almost identical to that used for the boundary layer and clear air turbulent calculations which have been made previously.[3] We take here the same values for a and b, namely 2.5 and 0.125, respectively. In the previous work, Λ was chosen equal to 0.064 times the layer thickness δ. In a vortex, a measure of the scale of the mean motion is the "core radius", $r_1(t)$; that is, the radius at which $\bar{v}$ reaches its maximum or "core velocity", $\bar{v}_1(t)$. However, the relationship of δ as a measure of the breadth of the shear dimension in a boundary layer and r_1 as a similar measure in a vortex is not clear. Feeling that the characteristic length for a vortex should be taken as somewhat larger than the core radius, we have, in the present calculation, arbitrarily assumed that the scale Λ is given by $0.16r_1(t)$.

The numerical experiment was an attempt to apply the invariant model defined by Eqs. (3) through (9) to the simple vortex decay problem. If the correlations are functions of r and t only, the result of substituting Eqs. (3) through (6) into (2) is given by the following set of equations for the correlations that are necessary to describe the motion.

$$\frac{\partial \overline{u'u'}}{t} = \frac{4\bar{v}\ \overline{u'v'}}{r} + \frac{\partial M}{\partial r}\left[5\frac{\partial \overline{u'u'}}{\partial r} + \frac{2(\overline{u'u'} - \overline{v'v'})}{r}\right]$$

$$+ M\left[5\frac{\partial^2 \overline{u'u'}}{\partial r^2} + \frac{5}{r}\frac{\partial \overline{u'u'}}{\partial r} - \frac{4}{r}\frac{\partial \overline{v'v'}}{\partial r} - \frac{6(\overline{u'u'}-\overline{v'v'})}{r^2}\right]$$

$$+ \frac{M}{\Lambda^2}\left(\frac{\overline{v'v'} + \overline{w'w'} - 2\overline{u'u'}}{3}\right) - 2\nu\frac{\overline{u'u'}}{\lambda^2}$$

$$+ \nu\left[\frac{\partial^2 \overline{u'u'}}{\partial r^2} + \frac{1}{r}\frac{\partial \overline{u'u'}}{\partial r} - \frac{2(\overline{u'u'}-\overline{v'v'})}{r^2}\right] \qquad (10)$$

$$\frac{\partial \overline{v'v'}}{\partial t} = -2\overline{u'v'}\left(\frac{\partial \overline{v}}{\partial r} + \frac{\overline{v}}{r}\right) + \frac{\partial M}{\partial r}\left[\frac{\partial \overline{v'v'}}{\partial r} + \frac{2}{r}\,(\overline{u'u'}-\overline{v'v'})\right]$$

$$+ M\left[\frac{\partial^2 \overline{v'v'}}{\partial r^2} + \frac{1}{r}\frac{\partial \overline{v'v'}}{\partial r} + \frac{4}{r}\frac{\partial \overline{u'u'}}{\partial r} + \frac{6\,(\overline{u'u'}-\overline{v'v'})}{r^2}\right]$$

$$+ \frac{M}{\Lambda^2}\left(\frac{\overline{u'u'} + \overline{w'w'} - 2\overline{v'v'}}{3}\right) - 2\nu\,\frac{\overline{v'v'}}{\lambda^2}$$

$$+ \nu\left[\frac{\partial^2 \overline{v'v'}}{\partial r^2} + \frac{1}{r}\frac{\partial \overline{v'v'}}{\partial r} + \frac{2\,(\overline{u'u'}-\overline{v'v'})}{r^2}\right] \tag{11}$$

$$\frac{\partial \overline{w'w'}}{\partial t} = \frac{\partial M}{\partial r}\frac{\partial \overline{w'w'}}{\partial r} + M\left(\frac{\partial^2 \overline{w'w'}}{\partial r^2} + \frac{1}{r}\frac{\partial \overline{w'w'}}{\partial r}\right)$$

$$+ \frac{M}{\Lambda^2}\left(\frac{\overline{u'u'} + \overline{v'v'} - 2\overline{w'w'}}{3}\right) - 2\nu\,\frac{\overline{w'w'}}{\lambda^2}$$

$$+ \nu\left(\frac{\partial^2 \overline{w'w'}}{\partial r^2} + \frac{1}{r}\frac{\partial \overline{w'w'}}{\partial r}\right) \tag{12}$$

$$\frac{\partial \overline{u'v'}}{\partial t} = -\overline{u'v'}\left(\frac{\partial \overline{v}}{\partial r} + \frac{\overline{v}}{r}\right) + 2\overline{v'v'}\,\frac{\overline{v}}{r} + 3\,\frac{\partial M}{\partial r}\frac{\partial \overline{u'v'}}{\partial r}$$

$$+ M\left(3\,\frac{\partial^2 \overline{u'v'}}{\partial r^2} + \frac{3}{r}\frac{\partial \overline{u'v'}}{\partial r} - 12\,\frac{\overline{u'v'}}{r^2}\right)$$

$$- \frac{M}{\Lambda^2}\,\overline{u'v'} - 2\nu\,\frac{\overline{u'v'}}{\lambda^2}$$

$$+ \nu\left(\frac{\partial^2 \overline{u'v'}}{\partial r^2} + \frac{1}{r}\frac{\partial \overline{u'v'}}{\partial r} - 4\,\frac{\overline{u'v'}}{r^2}\right) \tag{13}$$

The boundary conditions we will apply to these equations are as follows:

(a) at $r \rightarrow \infty$

$$\overline{u'u'} = \overline{v'v'} = \overline{w'w'} = \overline{u'v'} = 0$$

(b) at $r \rightarrow 0$

$$\overline{u'u'} = a_1 + b_1 r^2$$

$$\overline{v'v'} = a_1 + b_2 r^2$$

$$\overline{w'w'} = a_3 + b_3 r^2$$

$$\overline{u'v'} = \beta r^2$$

where the a's, the b's, and β are functions of time to be determined. These boundary conditions (as $r \rightarrow 0$) are consequences of the symmetry of the problem. It is not obvious that $\overline{u'u'}$ must equal $\overline{v'v'}$ at $r = 0$ from a first look at the equations for these quantities (Eqs. (10) and (11)); nevertheless, it is a fact that the two equations become identical for $r \rightarrow 0$.

To perform a calculation of vortex decay using Eqs. (1) and (10 through (13), an initial distribution of turbulent energy was assume to exist in a vortex whose initial velocity $\overline{v}(r,t_0)$ was given. This velocity distribution was assumed to be the well-known laminar solution for the diffusion of a line vortex[5], namely

$$\overline{v}(r,t_0) = \frac{\Gamma_\infty}{2\pi r}\left(1 - e^{-r^2/4\nu t_0}\right) \tag{14}$$

The computer program is, of course, written in nondimensional form. The velocity $\overline{v}$ (along with other velocities) is made nondimensiona by means of the initial core velocity $\overline{v}_{10}$. The radius r (along with other lengths) is made nondimensional by means of the initial core radius r_{10}. The time t is made nondimensional by means of the characteristic time $r_{10}/\overline{v}_{10}$. The values of r_{10} and v_{10} are related to the parameters Γ_∞ and t_0 of Eq. (14) by the following equations:

$$r_{10}^2 = 4x_1 \nu t_0 \tag{15}$$

$$r_{10}\overline{v}_{10} = \frac{\Gamma_\infty}{2\pi} \frac{2x_1}{1 + 2x_1} \tag{16}$$

where x_1 is the nonzero solution of

$$e^{x_1} = 1 + 2x_1 \tag{17}$$

Its approximate value is 1.25643.

When this procedure is followed, the effects of viscosity in the computation are defined by the initial core Reynolds number

$$Re = \frac{\rho v_{10} r_{10}}{\mu} \tag{18}$$

Its value was assumed to be 10,000.

The initial conditions that were chosen for the turbulence distributions at the start of the calculation need some discussion. First of all, it should be observed that the set of equations used here to define a turbulent vortex permits a similarity solution. The local mean velocity is of the form

$$\bar{v} = \bar{v}_1(t) f(r/r_1) \tag{19}$$

and the correlations are of the form

$$\overline{u_i' u_k'} = \bar{v}_1^2 g_{ik}(r/r_1) \tag{20}$$

where

$$\bar{v}_1(t) \sim t^{-1/2} \tag{21}$$

and

$$r_1(t) \sim t^{1/2} \tag{22}$$

This behavior is the same as in a laminar vortex. The ordinary differential equations which govern the similarity solution are given in the appendix.

Prior to obtaining any solutions of the equations being used here, it was not known what the distributions $g_{ik}(r/r_1)$ looked like or how fast the solution of the problem would seek the similarity solution if indeed it did so. In previous experience using the method, it had been found that no matter what initial conditions were put on these distributions (provided it was not $\overline{u_i' u_k'} \equiv 0$) the outcome of the computations became, in time, independent of the initial conditions. It was felt that the vortex should behave in the same manner, so that it was not really important, from the long time point of view, what initial conditions were applied to the problem. What we did try to do, however, was to put in a distribution of turbulent kinetic energy that was essentially confined to the region of the core and to assume that the initial shear correlation $\overline{u'v'}$ was zero. This condition was thought to be in some way a partial representation of the turbulence rolled up in the core of a trailing vortex as a result of wing drag. As will be pointed out in the next section, the turbulence production terms in Eqs. (10) through (13), when considered alone, permit oscillatory solutions when $\partial/\partial r (\overline{v}r) > 0$. When initial conditions on the turbulence distributions are chosen that are far from an equilibrium state, these oscillations are excited. Although they are damped, the first steps in a computation must be taken in small increments of time in order to track these oscillations properly so that negative values of $\overline{u'}^2$, $\overline{v'}^2$, and $\overline{w'}^2$ do not result. In order to ease this starting problem somewhat, the following artifice was used. The initial conditions on the turbulence distributions were taken to be of the form

$$\overline{u'v'} = 0 \tag{23}$$

$$\overline{u'u'} = 2 \, \frac{\overline{v}}{r} \, h(r) \tag{24}$$

$$\overline{v'v'} = \left(\frac{\overline{v}}{r} + \frac{\partial \overline{v}}{\partial r} \right) h(r) \tag{25}$$

where $h(r)$ is arbitrary. This choice makes the initial time derivatives due to only the production terms in Eqs. (10) through (13) equal to zero.

The actual initial conditions used in the computations reported here were (in nondimensional form)

$$\overline{v} = \frac{1+2x_1}{2x_1 r} \left(1 - e^{-x_1 r^2} \right) \tag{26}$$

$$\overline{u'u'} = \frac{0.01}{x_1 r^2} \, e^{-x_1 r^2} \left(1 - e^{-x_1 r^2} \right) \tag{27}$$

$$\overline{v'v'} = 0.01e^{-2x_1 r^2} \tag{28}$$

$$\overline{w'w'} = \frac{1}{2}(\overline{u'u'} + \overline{v'v'}) \tag{29}$$

$$\overline{u'v'} = 0 \tag{30}$$

with x_1 given by Eq. (17). In terms of the arbitrary function $h(r)$ given in Eqs. (24) and (25), the above-noted distributions are equivalent to

$$h(r) = \frac{0.01}{1+2x_1} e^{-x_1 r^2} \tag{31}$$

RESULT OF A TYPICAL COMPUTATION

After much difficulty with small errors in the program for carrying out the computation of a turbulent vortex according to the scheme just outlined, results are now being obtained which are considered reliable.* The results for the case described in the previous section may be discussed in terms of Figures 1 through 4 where we have plotted the behavior of the nondimensional maximum velocity $\overline{v}_1/\overline{v}_{10}$, the nondimensional core radius r_1/r_{10}, the nondimensional circulation ratio $r_1\overline{v}_1/r_{10}\overline{v}_{10}$, and the maximum of the sum of the correlations, $K = \overline{u'u'} + \overline{v'v'} + \overline{w'w'}$, divided by the instantaneous value of the maximum velocity squared $\overline{v}_1^2$.

These results show that after an initial time of the order of fifty to one hundred times $r_{10}/\overline{v}_{10}$ the vortex core starts to grow and the maximum velocity begins to decrease in a manner suggesting that a similarity solution is being approached. If a similarity solution were to be achieved, the product $\overline{v}_1 r_1$ should approach a constant and the ratio of K_{max} to $\overline{v}_1^2$ should also approach a constant value. It is seen in Figures 3 and 4 that this behavior is emerging.

---

*In the oral presentation of this paper, it was stated that no similarity solution was approached and the turbulence in the vortex died away faster than the maximum velocity $\overline{v}_1^2$. It was pointed out at that time that this conclusion was tentative. It will be seen in what follows that, following a correction and rewrite of the program, this conclusion has been shown to be in error and a similarity solution for a single vortex does appear to be approached at large times.

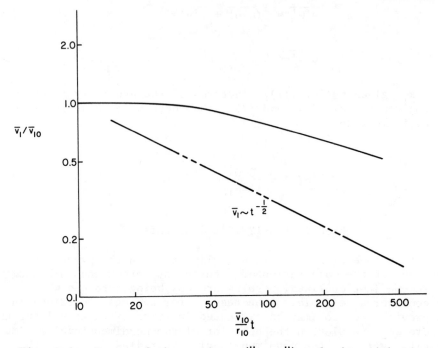

Figure 1. Decay of the maximum ("core") velocity with time

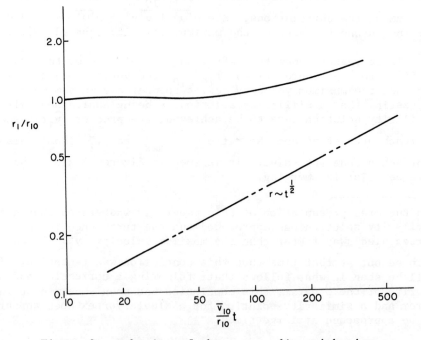

Figure 2. Behavior of the core radius with time

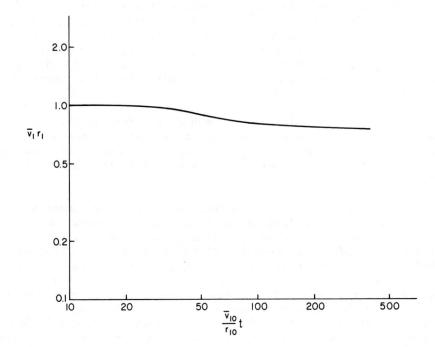

Figure 3.   Behavior of the circulation at the core radius with time

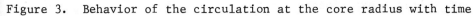

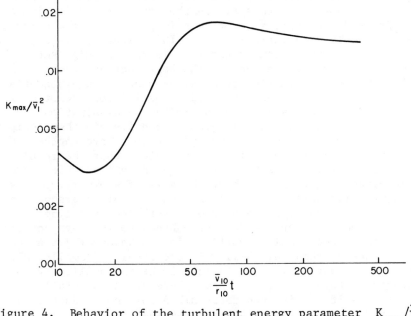

Figure 4.   Behavior of the turbulent energy parameter $K_{max}/\bar{v}_1^{2}$
with time

We may also test for the development of a similarity solution by comparing the nondimensional velocity, circulation, turbulent stress, and turbulent energy profiles as time increases. Such comparisons are shown in Figures 5, 6, 7, and 8. From these figures it appears that the profiles are not changing significantly with time at the later times. This is indicative of an approach to the similarity solution.

Of great interest is a comparison of the results that we have obtained with experimental results. In Figure 5 we show a comparison of our computed mean velocity profile with some measurements of Hoffmann and Joubert.[6] It is seen that the agreement is quite good.

Two other results of our computations are in agreement with their results. At the largest nondimensional distances downstream of vortex formation observed by Hoffmann and Joubert, they observed that the ratio of the circulation at the core radius to that at infinity was 0.53 and that the ratio of the core radius r_1 to the radius where the circulation first became equal to the far field circulation, r_0, was 0.34. These numbers are in general agreement with those we have obtained, namely, $\Gamma_1/\Gamma_\infty = 0.55$ and $r_1/r_0 = 0.42$. It would appear that the vortex we have computed is closely related to the equilibrium vortex state being approached in Hoffmann and Joubert's experiments.

There is an interesting point to be made in conjunction with Figure 6. We note from this figure that the circulation increases from $r/r_1 = 0$ to $r/r_1 \simeq 3$. From this point out, the circulation decreases slowly with increasing r. This situation is such that this portion of the vortex is unstable in the Rayleigh sense. In our equations, as will be seen in the next section, when $\partial(vr)\partial r < 0$, the production terms alone exhibit an exponential instability where they had exhibited an oscillatory solution for $\partial(vr)/\partial r > 0$. This instability is limited by dissipation and diffusion at the small scale lengths we consider in these computations. The flow is, however, unstable to long wavelengths and it is well known from the experiments of Taylor[7] that the result of this instability is a breakdown to a secondary flow which forms doughnut-like ring vortices about the axis of symmetry. It is, perhaps, the instability that is inherent in this overshoot in the circulation well outside the core of a turbulent trailing vortex that is responsible for the doughnut-shaped smoke rings sometimes observed about the trailing vortices in the wakes of jet aircraft. The existence of this instability is the result of small-scale turbulent transport. The excitation of the instability by long wavelength disturbances in the atmosphere or in the trailing vortex system itself is then responsible for the production of the individual smoke rings that are observed.

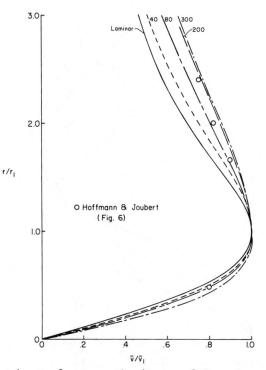

Figure 5.   Comparison of mean velocity profiles normalized to the point of maximum velocity for values of $(\bar{v}_{10}/r_{10})t$

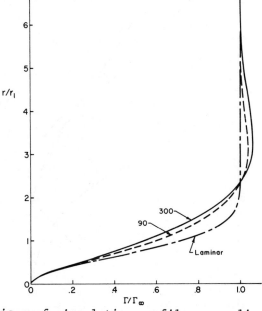

Figure 6.   Comparison of circulation profiles normalized to Γ_∞ for several values of $(\bar{v}_{10}/r_{10})t$

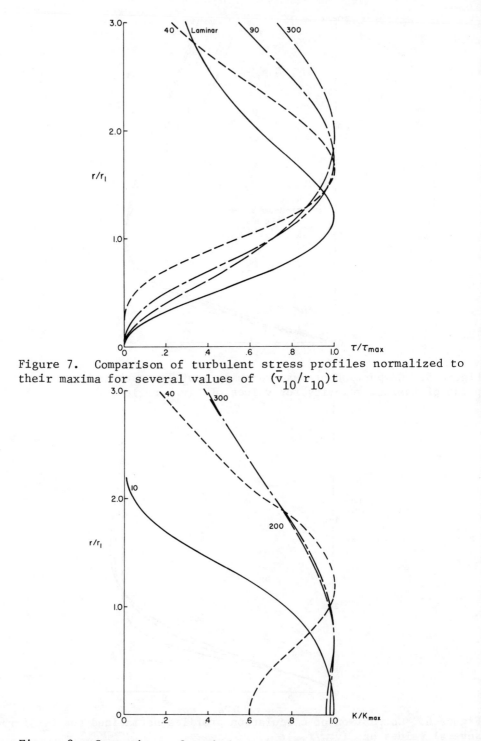

Figure 7.   Comparison of turbulent stress profiles normalized to their maxima for several values of $(\bar{v}_{10}/r_{10})t$

Figure 8.   Comparison of turbulent energy profiles normalized to their maxima for several values of $(\bar{v}_{10}/r_{10})t$

The results that have just been presented are most encouraging, and we are pressing forward with our program for the calculation of turbulent vortices. Our attention in the near future will be directed to two problems. The first is an investigation of the effect of a continuously maintained external level of turbulence on the decay of vortices. Only a slight modification of the way in which the program is presently run is needed to study this effect. The second part of our theoretical effort is the development of a program which will enable us to compute the development in space of a turbulent vortex that has an initial distribution of both tangential and axial mean velocity.

EXAMINATION OF THE PRODUCTION TERMS

In the previous section we mentioned the possibility of the production terms in the equations for the components of $\overline{u_i'u_k'}$ yielding oscillatory solutions which one had to be careful to track in the early stages of a computation so as not to produce negative values of the energies $\overline{u'u'}$ or $\overline{v'v'}$. We will now explore this behavior of the equations a little more closely.

If only the production terms in the equations for the components of $\overline{u_i'u_k'}$ are considered, the equations we obtain for the relevant correlations for the case we are investigating are

$$\frac{\partial \overline{u'v'}}{\partial t} = -\overline{u'u'} \frac{1}{r} \frac{\partial (r\bar{v})}{\partial r} + 2\overline{v'v'} \frac{\bar{v}}{r} \tag{32}$$

$$\frac{\partial \overline{u'u'}}{\partial t} = 4\overline{u'v'} \frac{\bar{v}}{r} \tag{33}$$

$$\frac{\partial \overline{v'v'}}{\partial t} = -2\overline{u'v'} \frac{1}{r} \frac{\partial (r\bar{v})}{\partial r} \tag{34}$$

$$\frac{\partial \overline{w'w'}}{\partial t} = 0 \tag{35}$$

If we make use of the fact that, in some circumstances, the time scale for changes in $\bar{v}$ is much longer than for the changes in the correlations, we may seek an approximate solution by assuming that $\bar{v}$ is independent of t. Differentiating Eq. (32) with this assumption and substituting from Eqs. (33) and (34), we find

$$\frac{\partial^2 \overline{u'v'}}{\partial t^2} = -q^2 \overline{u'v'} \tag{36}$$

where

$$q^2 = 8 \frac{\overline{v}}{r^2} \frac{\partial (r\overline{v})}{\partial r} \tag{37}$$

It is obvious from this equation that oscillatory solutions will be found if $q^2 > 0$. The solution to the set of equations for the case when

$$\overline{u'v'} \equiv 0 \quad \text{at} \quad t = 0 \tag{38}$$

and

$$\overline{u'u'} = \overline{v'v'} = E_0 \quad \text{at} \quad t = 0 \tag{39}$$

where E_0 is a constant, is easily written down. It is

$$\frac{\overline{u'v'}}{E_0} = -\frac{1}{q} \left(\frac{\partial \overline{v}}{\partial r} - \frac{\overline{v}}{r} \right) \sin qt \tag{40}$$

$$\frac{\overline{u'u'}}{E_0} = \frac{\frac{\partial \overline{v}}{\partial r} (1 + \cos qt) + \frac{\overline{v}}{r} (3 - \cos qt)}{\frac{2}{r} \frac{\partial (r\overline{v})}{\partial r}} \tag{41}$$

$$\frac{\overline{v'v'}}{E_0} = \frac{\frac{\partial \overline{v}}{\partial r} (1 - \cos qt) + \frac{\overline{v}}{r} (3 + \cos qt)}{4 \frac{\overline{v}}{r}} \tag{42}$$

It will be seen that there is no possibility of negative values of $\overline{u'u'}$ or $\overline{v'v'}$ unless

$$\frac{\overline{v}}{r} - \frac{\partial \overline{v}}{\partial r} > 3 \frac{\overline{v}}{r} + \frac{\partial \overline{v}}{\partial r} \tag{43}$$

or

$$\frac{\partial}{\partial r} (r\overline{v}) < 0 \tag{44}$$

Eq. (44) is only true if $q^2 < 0$ in which case the solutions are not oscillatory but grow exponentially. In general, the oscillatory solutions cannot produce any negative energies.

However, in making computations, it is necessary, when examining
solutions for which the initial conditions are far from those of
similarity, to take small steps in time so as to accurately track
the oscillations described above so that no negative energies are
found as the solution is developed.

EXPERIMENTAL PROGRAM

In addition to the theoretical studies which we are under-
taking at A.R.A.P., an experimental program has just been started
by means of which we hope to obtain answers to certain questions
that must be considered before any understanding of vortex motion
can be deemed adequate. In our experimental program we are again
concentrating our efforts on the behavior of a single isolated
vortex in what we hope will be a well-documented environment.

The experimental apparatus that has been built in which
these studies are being carried out is shown schematically in
Figure 9. A photograph of the completed test setup is shown
in Figure 10.

The apparatus consists of a long, 18-inch diameter,
cylindrical tube through which air can be sucked by means of a
fan. Thé entrance to the tube is carefully baffled and screened
ahead of a carefully constructed contraction section so that a
steady flow of air having zero angular momentum can be drawn
through the tube. Velocities up to 120 ft/sec can be achieved
with this apparatus.

After a smooth flow has been established in the tube, two
airfoils are introduced from opposite sides of the tube so that
each produces a vortex of the same sign and strength on the center
of the tube. By virtue of the symmetry of the device, the vortex,
once located at the center of the tube, stays there. This method
of producing a single isolated vortex was used by Hoffmann and
Joubert in their interesting study which we have discussed
previously.

Downstream of the vortex production section, an observation
and surveying station is provided as shown in Figure 11.
Provision is made at this section for both directionally sensitive
pitot tube and hot-wire surveys, as well as for observing the
vortex visually when tracers such as smoke are used to probe the
vortex.

The tube between the vortex production section and the
observation section can be made almost any length, from approximately 5 feet to 55 feet. In this manner we plan to study the
downstream development of a vortex in a known environment. This

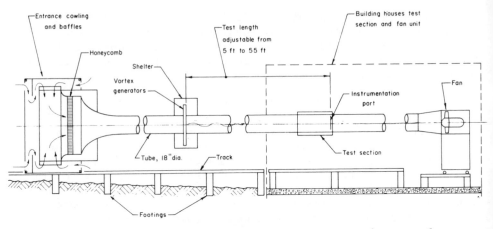

Figure 9.   Schematic of A.R.A.P. vortex tube tunnel

Figure 10.   View of A.R.A.P. vortex tube

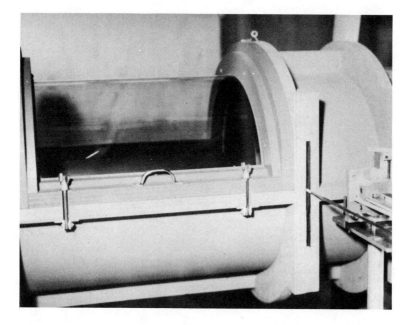

Figure 11. Close-up of observation and surveying station

environment is to be determined from detailed measurements of the flow in the tube when no vortex is introduced. During the course of our investigation, we plan to vary this environment by systematically producing turbulence along the walls of the tube so as to study the effects of externally generated turbulence on the development of a vortex. With our present apparatus which has generating airfoils of 3.5 in. chord, we are able to study the vortex at any age from $z/c = 17.6$ to $z/c = 189$ over a wide range of environmental conditions.

For certain vortices, the axial momentum profile may have a very pronounced effect on the behavior of the vortex. We plan to investigate this effect by building the apparatus shown in Figure 12 with which to form a vortex. The figure shows the normal generating airfoils that were discussed previously with a small central body separating them. This central body is connected to a supply of high pressure air through a tube which passes through one of the airfoils. By changing the length of the central body upstream of the airfoils or by introducing a jet of air on the centerline through the port in the rear of the central body, the axial momentum profile for a given vortex strength may be varied.

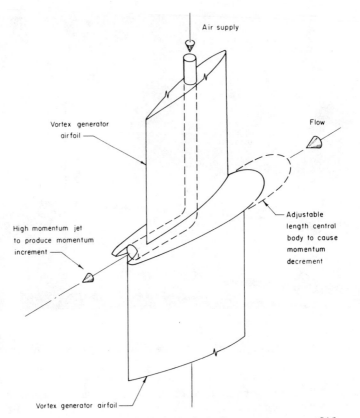

Figure 12.   Method of controlling initial momentum profile of vortex

As of this writing we have just finished the construction of
the apparatus described above and are making initial calibration
runs and refining our experimental techniques so that no significant
experimental results are available. We are hopeful that we will be
able to obtain answers to the following basic questions through the
use of the experimental facility described here.

(a) How does the vortex decay in the basic environment
provided by the simple tube?

(b) What are the levels of turbulence in the vortex at
various strengths and how are they related to the mean rotational
velocities?

(c) How does a change in the turbulence level of the
environment affect the decay of a vortex, if at all?

(d) What role does the axial velocity distribution play in
the development of a vortex of a given strength?

CONCLUSIONS

We have given a very brief outline of a program which is designed to study the decay of an isolated vortex. Results are available at the present time only from a theoretical portion of the overall program. These results are most encouraging. The method of invariant modeling, when applied to the problem of turbulent vortex decay, has yielded results that are in good agreement with experimental measurements. From these results, it appears that an isolated turbulent vortex approaches a self-similar form well downstream of its point of formation. This self-similar form is such that there is a circulation overshoot at a radius of some three times the core radius. It is tentatively conjectured that this circulation overshoot, which is unstable in the Rayleigh sense, is responsible for the production of the vortex rings that have been observed surrounding the trailing vortices of large jet aircraft.

Finally it is believed that the method that has been developed for calculating vortex decay can be adapted to the study of the effects of external turbulence level and axial momentum profile on the decay rate of isolated vortices.

REFERENCES

[1]Donaldson, Coleman duP.: "A Computer Study of an Analytical Model of Boundary-Layer Transition," AIAA Journal, Vol. 7, no. 2, 1969, pp. 271-278.

[2]Donaldson, Coleman duP. and Rosenbaum, Harold: "Calculation of Turbulent Shear Flows Through Closure of the Reynolds Equations by Invariant Modeling," A.R.A.P. Report No. 127, December 1968.

[3]Donaldson, Coleman duP., Sullivan, Roger D., and Rosenbaum, Harold: "Theoretical Study of the Generation of Atmospheric Clear Air Turbulence," AIAA Paper No. 70-55, presented at the Eighth Aerospace Sciences Meeting, New York, January 1970.

[4]Hinze, J. O.: Turbulence, McGraw-Hill, New York, 1959, p. 251.

[5]Lamb, Horace: Hydrodynamics (6th edition), Dover Press, New York, 1945, p. 592.

[6]Hoffmann, E. R. and Joubert, P.N.: "Turbulent Line Vortices," J. Fluid Mech., Vol. 16, no. 3, 1963, pp. 395-411.

[7]Taylor, G. I.: "Stability of a Viscous Liquid Contained Between Two Rotating Cylinders," Phil. Trans. A, ccxxiii, 1922, p. 289.

APPENDIX

The similarity conditions of Eqs. (21) and (22) may be used to reduce Eqs. (1) and (10) through (13) to ordinary differential equations by means of the following substitions:

$$\xi = \frac{r^2}{t} \tag{A.1}$$

$$\bar{v} = \frac{1}{r} C(\xi) \tag{A.2}$$

$$\overline{u'u'} = \frac{1}{t} k_1(\xi) \tag{A.3}$$

$$\overline{v'v'} = \frac{1}{t} k_2(\xi) \tag{A.4}$$

$$\overline{w'w'} = \frac{1}{t} k_3(\xi) \tag{A.5}$$

$$\overline{u'v'} = \frac{1}{t} \tau(\xi) \tag{A.6}$$

$$\Lambda = t^{1/2}\bar{\Lambda} \tag{A.7}$$

$$\lambda = t^{1/2}\bar{\lambda} \tag{A.8}$$

We see that M, by its definition, Eq. (7), is automatically a function of ξ alone

$$M = \bar{\Lambda}\sqrt{k_1 + k_2 + k_3} \tag{A.9}$$

and that the relation between $\bar{\lambda}$ and $\bar{\Lambda}$ is the same as that between λ and Λ as given by Eq. (8).

Performing the substititions results in the following set of equations where primes denote differentiation with respect to ξ.

$$\frac{4}{R} \xi C'' + \xi C' - 2(\xi\tau)' = 0 \tag{A.10}$$

$$4\left[\xi\left(5M + \frac{1}{R}\right)k_1'\right]' + 2\left[2M' - \frac{1}{\xi}\left(3M + \frac{1}{R}\right)\right](k_1 - k_2)$$

$$-8Mk_2' + (\xi k_1)' + \frac{M}{3\bar{\Lambda}^2}(k_2 + k_3 - 2k_1) - \frac{2}{R}\frac{k_1}{\bar{\lambda}^2} + 4\frac{C}{\xi}\tau = 0 \tag{A.11}$$

$$4\left[\xi\left(M + \frac{1}{R}\right)k_2'\right]' + 2\left[2M' + \frac{1}{\xi}\left(3M + \frac{1}{R}\right)\right](k_1 - k_2)$$

$$+ 8Mk_1' + (\xi k_2)' + \frac{M}{3\bar{\Lambda}^2}(k_1 + k_3 - 2k_2) - \frac{2}{R}\frac{k_2}{\bar{\lambda}^2} - 4C'\tau = 0 \quad (A.12)$$

$$4\left[\xi\left(M + \frac{1}{R}\right)k_3'\right]' + (\xi k_3)' + \frac{M}{3\bar{\Lambda}^2}(k_1 + k_2 - 2k_3) - \frac{2}{R}\frac{k_3}{\bar{\lambda}^2} = 0 \quad (A.13)$$

$$4\left[\xi\left(3M + \frac{1}{R}\right)\tau'\right]' - 4\left(3M + \frac{1}{R}\right)\frac{\tau}{\xi} + (\xi\tau)' - \frac{M}{\bar{\Lambda}^2}\tau - \frac{2}{R}\frac{\tau}{\bar{\lambda}^2}$$

$$-2k_1C' + 2k_2\frac{C}{\xi} = 0 \quad (A.14)$$

DECAY OF A VORTEX PAIR BEHIND AN AIRCRAFT[*]

J. N. Nielsen and Richard G. Schwind

Nielsen Engineering & Research, Inc.

ABSTRACT

A model of a trailing vortex pair behind an aircraft is presented which is thought to represent a case of extreme vortex persistency and which therefore is relevant from the safety point of view. Three stages are considered in the analysis: a rolling-up stage directly behind the aircraft, a second stage in which the vortices act independently as constant strength equilibrium turbulent vortices, and a third stage where the vortices physically interact and decay in strength. An overall theory is presented encompassing all three stages and aimed at obtaining equilibrium solutions. Calculative examples are presented for all stages.

SYMBOLS

a	lateral spacing between a pair of trailing vortices
c	wing chord
C_1, C_2	constants of integration
C_θ, C_z	nondimensional parameters, equation (59)

[*]The work reported herein was sponsored by the Air Force Office of Scientific Research, Office of Aerospace Research, United States Air Force, under Contract F44620-70-C-0052.

$d\vec{r}$ $dx + i\,dy$

D   drag of single vortex of a trailing pair or drag of generator forming vortex

Ei   exponential function

f,g,h   characteristic functions depending only on r_1/r_0, equation (49)

K $\Gamma/2\pi$, circulation parameter

K_c value of K corresponding to vorticity rolled up in vortex core

K_i value of K at $r = r_i$

K_0 $\Gamma_0/2\pi$

K_1 value of K for $v = v_1$

L   torque on outer edge of cross section of vortex of unit thickness

m_2 mass flow per unit time across area S_2 of control volume enclosing vortex

p   static pressure

p_∞ static pressure of free stream

$\vec{q}$ $v_x + i\,v_y$

r,θ,z cylindrical coordinates with positive z along downstream axis of vortex

r_i value of r where the eye of the vortex joins the logarithmic region

r_j value of r where the logarithmic region of the vortex joins the outer region

r_0 outer radius of vortex where $\Gamma = 0.99\,\Gamma_0$

r_1 value of r where $K = K_1$ and $v = v_1$

s   wing semispan

s_v semispan of trailing vortices

t	time measured behind wing trailing edge
u, v, w	velocity components along r, θ, z directions, respectively
u', v', w'	turbulent fluctuating values of u, v, and w
v_x, v_y	components of velocity along x, y axes
v_1	maximum value of v
W	free-stream velocity
x, y	Cartesian axes in crossflow plane; $x = r \sin \theta$, $y = r \cos \theta$
Γ	circulation around any contour enclosing entire trailing vortex of radius r_0
Γ_0	initial value of Γ near wing for no decay
Δ	axial velocity defect at vortex centerline
Λ	constant in equation (3)
μ	absolute viscosity of air
ν	laminar kinematic viscosity of air
ν_t	turbulent kinematic viscosity, eddy viscosity
ν_{t_0}	eddy viscosity for turbulent shear at edge of vortex, equation (30)
ν_{t_1}	eddy viscosity for turbulent shear near axis of vortex, equation (45)
ξ	vorticity in crossflow plane of a turbulent vortex
ρ	mass density of air
τ_{rz}	shear in z direction lying in plane of r and z
$\tau_{r\theta}$	shear in θ direction lying in plane of r and θ
$(\tau_0)_\ell$	value of $\tau_{r\theta}$ at edge of vortex for laminar flow
$(\tau_0)_t$	value of $\tau_{r\theta}$ at edge of vortex for turbulent flow

ψ stream function

ψ_T stream function for pair of trailing vortices and their
                        induced crossflow

ω angular velocity of eye of vortex

1. INTRODUCTION

Trailing vortices behind an aircraft can persist for many
minutes after their formation. Other aircraft encountering these
vortices can undergo severe motions which are sometimes fatal.
The persistency of a vortex pair behind an aircraft is thus
important, as it may affect the safety of other aircraft the pair
may encounter. The introduction of heavier transports and the
appearance of V/STOL aircraft both tend to aggravate the safety
problems associated with trailing vortices. It is of interest to
determine under what conditions trailing vortices are most persistent. For this reason, the analysis of the equilibrium characteristics of a trailing vortex pair presented herein assumes a quiet
atmosphere with no buoyancy. Several stages occur in the life of
a trailing vortex pair behind an aircraft, and analytical solutions
for all stages are investigated. The paper critically reviews
existing theory and experiment, and points out areas requiring
further attention.

2. GENERAL CONSIDERATIONS

2.1 Preliminary Remarks

The asymptotic model of a vortex pair behind an aircraft can
assume many forms. One form of a spectacular nature is composed
of a number of vortex rings into which the trailing line vortices
form because of instability (ref. 1). However, it appears from
observation that the vortex wake does not always form such ring
vortices, and that other configurations are possible. Aircraft
operating conditions, atmospheric conditions, and engine parameters
among other things can affect the wake characteristics. The
conditions which produce vortex wakes of greatest persistency are
not clear at the present state of knowledge. Vortex ring formation, buoyancy, inhomogeneities of atmospheric air movements, and
lateral aircraft maneuvers all appear to reduce persistency.

A prime mechanism which seems to be important in reducing the
danger of concentrated vortices to aircraft in their proximity is
vortex breakdown (ref. 2) or, as it is sometimes called, vortex
bursting. When a vortex is closely concentrated, it manifests

large tangential velocities which can induce large forces on a
neighboring aircraft. If the vorticity is less concentrated, the
induced forces are reduced. Since vortex breakdown tends to move
the vorticity from its area of high concentration near the axis
radially outward, it tends to enhance safety. If ring vortices
form, there is no a priori reason to expect the induced velocities
to be less than for the trailing pair from which they were formed.
However, there exists evidence that, at least in some cases,
vortex breakdown occurs in the ring vortices because of distortion
of the rings.

It appears that a model which purports to represent the most
persistent trailing vortex pair should neglect the above four
factors known to reduce persistency. Nevertheless, a means of
reduction of vortex strengths must be included in such a model.
For a single vortex of given circulation, the total vorticity
cannot be changed within the fluid itself, since no means for
destroying the vorticity exists. However, for a vortex pair where
the net circulation about both vortices is zero, the total vorticity
is zero, and it is possible to annihilate vorticity of one vortex
by that of opposite sign associated with the other vortex. A
model for vorticity annihilation based on this notion is subsequently discussed which may be a limiting case from the safety
point of view.

2.2 Vortex Model

Three stages are envisioned in the history of the trailing
vortex pair as depicted in figure 1. In the first stage, the flat
vortex sheet shed from the trailing edge of the aircraft rolls up
into two spiral vortices. At some distance downstream, each vortex
becomes essentially axisymmetric because of convection and diffusion, and can be approximated by an equilibrium turbulent vortex.
In this second stage, the vortices grow in radius until they make
contact, marking the beginning of the third stage. The flow
pattern shown for the third stage is that seen by a viewer moving
downward with the vortices. (This pattern also exists in the
stage II flow.) A stagnation streamline in the crossflow plane
encloses the vortices with their recirculating flows. Vorticity
is diffused across the vertical plane of symmetry within the
recirculating flow and annihilated there. Vorticity is also
diffused across the streamline enclosing the recirculating flow,
convected upward by the outer flow, and annihilated on the plane
of symmetry above the recirculating flow.

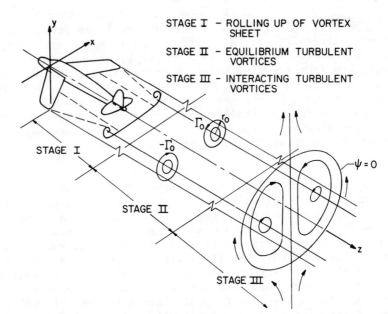

Figure 1.- Stages in development of trailing vortex pair behind an aircraft.

2.3 Dimensional Analysis

A dimensional analysis has been made to determine the non-dimensional groups important in the decay of a vortex pair behind an aircraft. Eight independent variables which can influence the persistency of a vortex pair in a quiet atmosphere with no buoyancy are the following ones:

(1) Γ_0 = initial value of Γ

(2) r_0 = initial characteristic radius of vortex

(3) a = vortex lateral spacing

(4) D = drag of either vortex

(5) W = airplane velocity

(6) ρ = air density

(7) ν = kinematic viscosity

(8) z = distance behind airplane

If Γ is the value of the circulation about one vortex at any time, we can write

$$\Gamma = \Gamma(\Gamma_0, r_0, a, D, W, \rho, \nu, z) \tag{1}$$

With eight independent variables, we find that Γ/Γ_0 is a function of five dimensionless groups.

$$\frac{\Gamma}{\Gamma_0} = F\left(\frac{Wz}{\Gamma_0}, \frac{\Gamma_0}{\nu}, \frac{D}{\rho\Gamma_0^2}, \frac{r_0}{a}, \frac{z}{a}\right) \tag{2}$$

The first parameter has been used by Hoffmann and Joubert (ref. 3) to correlate their data on circulation distribution through turbulent vortices. The second parameter is a nondimensional measure of the turbulent vortex strength. While the product of the first and second parameters yields the usual Reynolds number based on distance, it is felt that the first parameter is a more appropriate one than Reynolds number because it scales for Γ_0 which is subject to more variability than ν. The second group is a measure of the ratio of the eddy viscosity ν_t to the molecular viscosity ν through the relationship of Owen (ref. 4)

$$\frac{\nu_t}{\nu} = \Lambda\sqrt{\frac{\Gamma}{\nu}} \tag{3}$$

in which Λ is a constant somewhat less than unity. The remaining three groups are self-evident.

3. TURBULENT VORTEX DESCRIPTION

The present description of a turbulent vortex is based on an analogy between a turbulent vortex and a turbulent boundary layer. Consider a turbulent vortex divided into three parts which are to be compared with three parts of a turbulent boundary layer.

Vortex	Boundary Layer
Eye of the vortex	Laminar sublayer
Logarithmic region	"Law-of-Wall" region
Outer region	Wake region

In a turbulent vortex, the experiments of Hoffmann and Joubert exhibit a region close to the axis of rotation which is in solid body rotation. This region is termed the "eye" of the vortex and is analogous to the laminar sublayer. The next region in the turbulent vortex is a region wherein a logarithmic law of the circulation distribution exists analogous to the logarithmic

velocity distribution (law-of-the-wall) for a turbulent boundary
layer. Outside the logarithmic region, the torque must drop off
to zero at the boundary between the turbulent jet and the outer
inviscid fluid in the same way the shear in a boundary layer drops
off through the wake from its uniform value in the law-of-the-wall
region to zero at the edge of the boundary layer.

Characteristic profiles of tangential velocity, v, and circulation parameter, K = vr, through a turbulent vortex are shown in
figure 2. To obtain a reference value of K, we note that a convenient distinguished radius is the radius r_1 at which v attains
its maximum value. The value of K_1 at this point is taken as the
reference value of K, analogous to the reference value of friction
velocity used in turbulent boundary-layer analysis. It is also
useful for analytical purposes to define an outer radius of the
vortex similar to the thickness of a boundary layer. For this
purpose, we define the radius of the vortex, r_0, as the value of
r for which the value of K attains 99 percent of its asymptotic
value, K_0. At some radius r_i the eye is joined to the logarithmic
region.

3.1 Logarithmic Law

The data of Hoffmann and Joubert (ref. 3) shows that in the
logarithmic region the universal circulation distribution is given
by

$$\frac{K}{K_1} = 2.14 \log_{10}\left(\frac{r}{r_1}\right) + 1 = 0.928 \log_e\left(\frac{r}{r_1}\right) + 1 \qquad (4)$$

We adopt the foregoing expression for the present analysis.

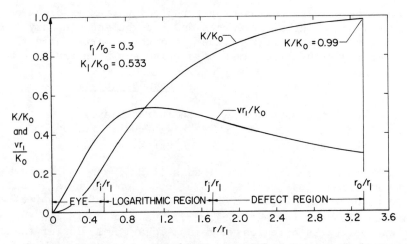

Figure 2.- Velocity and circulation profiles in a turbulent vortex.

A condition which will be used is the continuity of the vorticity across the vortex. The vorticity in cylindrical coordinates is given by

$$\xi = \frac{1}{r} \frac{\partial (vr)}{\partial r} = \frac{1}{r} \frac{\partial K}{\partial r} \tag{5}$$

For the logarithmic region we then find from equation (4) that

$$\xi = \frac{0.928 \ K_1}{r^2} \tag{6}$$

The vorticity of the logarithmic region thus increases rapidly as the axis is approached. It is to be joined to the constant value of vorticity in the eye so that the vorticity will be continuous across both regions.

3.2 Law of the Outer Region

Hoffmann and Joubert (ref. 3) have used part of their data to establish a tentative law of the outer region, the so-called defect law, in the form of $(K_0 - K)/K_1$ versus r/r_0. The data of these investigators have been examined to see if all the data which are free of the influence of starting conditions can be correlated in a defect law. The measurements near the outer edge of the vortex are least accurate, and some uncertainty arises in fairing the K curves in this region to establish the experimental values of K_0 and r_0. The curves have been faired by us putting most weight on the trend established by the data for lower r/r_0. With the values of Γ_0 and r_0, so obtained, the data have been correlated as in figure 3. A universal defect law is obtained within the indicated range of the data. From a plot of $\log [(K_0 - K)/K_1]$ versus r/r_0 the defect law is found to be

$$\frac{K_0 - K}{K_1} = 4.43 \ e^{-4.8 \ r/r_0} \tag{7}$$

We can adjust the boundary between the logarithmic region and the outer region so that ξ is continuous at $r = r_j$. From the continuity of ξ at the join, we obtain from equations (4) and (7) an implicit relationship for r_j/r_0

$$\frac{r_j}{r_0} e^{-4.8 \ (r_j/r_0)} = \frac{(0.928)}{(4.43)(4.8)}$$

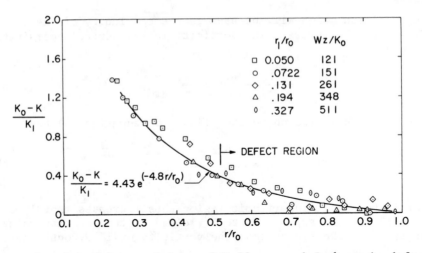

Figure 3.- Correlation of data of Hoffmann and Joubert in defect region.

which yields

$$\left(\frac{r_j}{r_0}\right) = 0.51 \tag{8}$$

This value is within the range of correlation demonstrated by figure 3.

Not only must ξ be continuous at the join, but we expect K/K_1 as obtained for the empirical correlations to be continuous there. The following additional relationship results.

$$\frac{K_0}{K_1} - 2.14 \log_{10}\left(\frac{r_0}{r_1}\right) = 4.43\, e^{-4.8\, r_j/r_0} - 2.14 \log_{10}\left(\frac{r_0}{r_j}\right) + 1$$

$$= 0.76 \tag{9}$$

Of the parameters r/r_0, r_1/r_0, and K_1/K_0 which determine the shape of the curves of K/K_1 versus r/r_0, only two are independent.

3.3 Law of the "Eye"

In the eye of the vortex, the fluid is in solid-body rotation. Thus if ω is the angular velocity of the eye,

$$K = \omega r^2 \tag{10}$$

Hoffmann and Joubert give for this relationship

$$\frac{K}{K_1} = 1.83 \left(\frac{r}{r_1}\right)^2 \tag{11}$$

However, a somewhat smaller value of the constant in equation (11) will provide for continuity of K and ξ at the join between the eye and the core. For $K = K_i$ at $r = r_i$ we have for continuity of K from equations (4) and (10)

$$\frac{\omega r_i^2}{K_1} = 2.14 \log_{10}\left(\frac{r_i}{r_1}\right) + 1 \tag{12}$$

From equation (5) we have for the eye

$$\xi = 2\omega = \frac{1}{r} \frac{\partial K}{\partial r} \tag{13}$$

Using the logarithmic law and imposing continuity of 2ω at $r = r_i$, we find

$$2\omega = \frac{0.928 \, K_1}{r_i^2} \tag{14}$$

Simultaneous solution of equations (12) and (14) gives

$$2.14 \log_{10}\left(\frac{r_i}{r_1}\right) = \frac{\omega r_i^2}{K_1} - 1 = 0.464 - 1 \tag{15}$$

so that

$$\frac{r_i}{r_1} = 0.5625 \tag{16}$$

and

$$\frac{K}{K_1} = \left(\frac{\omega r_i^2}{K_1}\right)\left(\frac{r_1}{r_i}\right)^2 \left(\frac{r}{r_1}\right)^2 = 1.47 \left(\frac{r}{r_1}\right)^2 \tag{17}$$

While this value of the constant, 1.47, is less than that suggested by Hoffmann and Joubert, it is in good accord with their data (ref. 3).

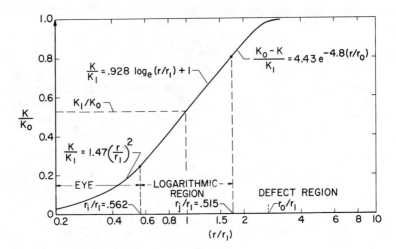

Figure 4.- Empirical circulation distribution for equilibrium turbulent vortex.

The empirical description of the circulation profile through a turbulent vortex is summarized in figure 4.

4.   ROLLING UP OF VORTEX SHEET IN STAGE I

In order to study the trailing vortices in stage III, we must calculate their characteristics in traversing stages I and II.   It is a well-known fact that a flat vortex sheet from the trailing edge of a wing tends immediately to roll up at its edges as shown in figure 5. Rolling-up calculations have been performed by a number of investigators such as Westwater (ref. 5) and Takami (ref. 6) by replacing the vortex sheet by a distribution of discrete potential vortices and computing their motions using simple cross-flow theory. Difficulty occurs in the calculations because the vortices tend to aggregate at the center of the spiral, where their mutual interactions become large. Even using very small step sizes Takami (ref. 6) was unable to define the inner turns of the rolled-up vortex, and he concluded the flow is unstable just as in the case of the planar vortex sheet.

We have carried out some rolling-up calculations for the purpose of comparison with the only measurements we could find of the wake far behind a wing that are sufficiently detailed to establish the circulation distribution. Figure 5 shows the contours of equal vorticity measured by Fage and Simmons (ref. 7) in 1925 for a rectangular wing of aspect ratio 6 at a distance of thirteen chord lengths behind the trailing edge. Contours of constant vorticity are shown forming a well defined center. Also shown in figure 5 is a theoretical rolling-up calculation for this case based on 40

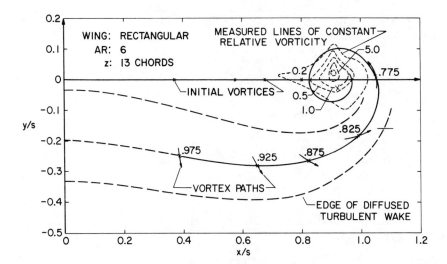

Figure 5.- Trailing vortex measurements of Fage and Simmons (1925) and 40-discrete-vortex rolling-up calculation.

equal-strength vortices at the trailing edge across the wing span. In the calculations, the vortices gravitate to the center of the spiral and get very close to each other. If two vortices come closer than a prescribed distance compatible with the accuracy of calculation based on the axial interval size, they are combined at their center of gravity and the calculation continued downstream.

Almost immediately about 45-50 percent of the vorticity rolls up into a small radius. The vortex path directions shown by arrows in figure 5 indicate velocity components inward along the spiral; thus, the remaining vorticity moves slowly into the central region. Also shown are limits of diffusion calculated approximately from the two-dimensional wake growth correlation given in Schlichting (ref. 8). These limits were taken to be equal to those for a circular cylinder having the same drag as the airfoil. Since the vortex sheet stretches as it moves inward, the diffusion band narrows approaching the center. This figure illustrates the opposing forces of convection and diffusion on the vorticity transport, the one causing inward radial motion of the vorticity and the other outward radial motion.

Experimental measurements of the velocity distributions through turbulent trailing vortices have been compared with the theory of the two-dimensional laminar unsteady vortex using a viscosity chosen to fit the data. We now show such a fit to the data of Fage and Simmons, but first the known theoretical results are presented in the form suitable for the present purpose. The well-known incompressible viscous unsteady solution for the flow due to an isolated rectilinear vortex of strength Γ_c initially concentrated along an axis is

$$\Gamma = 2\pi r v = \Gamma_c \left[1 - e^{-r^2/4\nu t} \right] \tag{18}$$

where

   v circumferential velocity

   t time

   ν kinematic viscosity

The radius r_1 corresponding to the maximum tangential velocity v_1 is given by

$$r_1 = 1.115 \sqrt{4\nu t} \tag{19}$$

and the circulation parameter K_1 at this point is

$$K_1 = 0.712 \frac{\Gamma_c}{2\pi} = 0.712 \ K_c \tag{20}$$

It is now assumed that time and distance behind the wing are interchangeable so that

$$z = Wt \tag{21}$$

an assumption not strictly valid because it neglects the existence of axial pressure gradients behind the wing. If an experimental velocity profile through the vortex is available some distance behind the wing, the distance r_1 is known. From equations (21) and (19), a value of ν, namely ν_t, which yields the correct r_1 can be determined. This determination of ν_t is independent of vortex strength. The value of ν_t so determined is an equivalent eddy viscosity which makes the laminar solution agree with the turbulent experimental results at the point of maximum tangential velocity.

The circulation distribution thirteen chords behind the wing of Fage and Simmons is shown in figure 6. In the data of Fage and Simmons the contours of constant vorticity in the crossflow planes are not circular, and the average radius of these contours has been used in calculating the abscissa of the figure. In the figure the data are compared with two other estimates of the circulation distribution. The first one, represented by the dotted line, is based on the preceding theory of the two-dimensional unsteady laminar vortex with a value of ν_t chosen to fit the data in the mean. Since the rolling-up calculations indicate that only about half of the

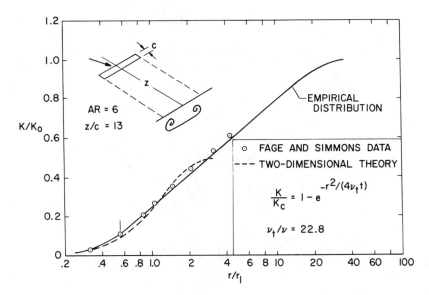

Figure 6.- Comparison of analytical circulation distributions with data of Fage and Simmons.

vorticity is initially concentrated, K/K_0 has the asymptotic value of 0.5. Also shown in the figure is the empirical distribution as given by figure 4. The fit of the empirical distribution based on the data of Hoffmann and Joubert to the data of Fage and Simmons is good. It appears that the vortex of Fage and Simmons is an equilibrium vortex free of disturbances due to the starting conditions.

It is worth noting that values of ν_t determined by matching experimental and theoretical circulation distribution will be in error if it is assumed that all the vortex circulation is initially concentrated instead of only that fraction given by rolling-up theory. Also the inward convection of vorticity shown in the rolling-up calculation is neglected in the two-dimensional theory, a fact which further contributes to error in determining the eddy viscosity from the data using that theory.

5. THEORY FOR STAGE II

In stage II an equilibrium turbulent vortex is moving downstream changing circulation distribution and axial velocity profile as it moves. We will now set up an analytical method for determining the characteristics of the vortex as it moves downstream.

5.1   Conservation of Angular Momentum

One equation of the motion can be obtained by applying the law of the conservation of angular momentum to a control volume in the vortex.

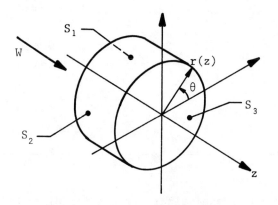

Consider the volume to be of unit thickness and of radius r_0 in which angular momentum is convected into the element on the upstream surface, S_1, and outward on the downstream surface, S_3. It is also convected radially across S_2. The angular momentum per unit time exiting from the control volume through an annulus of S_3 is

$$dH_z = (w)(vr)\rho\ 2\pi r\ dr$$

$$H_z = 2\pi\rho \int_0^{r_0} vwr^2\ dr = 2\pi\rho \int_0^{r_0} wKr\ dr \tag{22}$$

The rate at which mass flows out of the control volume across the S_2 boundary, m_2, is related by continuity to the rate it crosses the S_1 and S_3 boundaries as follows:

$$m_2 = -\frac{d}{dz} \int_0^{r_0} \rho w\ 2\pi r\ dr \tag{23}$$

Since the mass crossing the boundary S_2 has tangential velocity K_0/r_0, the transfer of angular momentum per unit time across this boundary is

$$H_r = -2\pi\rho K_0 \frac{d}{dz} \int_0^{r_0} wr \, dr \qquad (24)$$

The conservation of angular momentum for a control volume is

$$H_r + \frac{dH_z}{dz} = L \qquad (25)$$

where L is the torque on the S_2 area of unit length. The shear τ_0 at this radius is given for laminar flow by

$$(\tau_0)_\ell = \mu r \frac{\partial}{\partial r} \left(\frac{v}{r}\right) \qquad (26)$$

which with the help of the following result

$$vr = K_0; \quad r \geq r_0 \qquad (27)$$

yields

$$(\tau_0)_\ell = -2\mu \frac{K_0}{r_0^2} \qquad (28)$$

We now assume that the turbulent shear at the edge of the vortex can be represented by an eddy-viscosity law similar to that for laminar flow. The turbulent shear in question is the tangential shear, $\tau_{r\theta}$, at the edge of the vortex in the plane of the r and θ directions. It is given by

$$\tau_{r\theta} = -\rho \overline{u'v'} \qquad (29)$$

and by analogy with equation (26) at the edge of the vortex

$$\tau_{r\theta} = \nu_{t_0} \rho \left[r \frac{\partial}{\partial r} \left(\frac{v}{r}\right)\right]_{r=r_0} \qquad (30)$$

Letting $(\tau_0)_t$ be the value of $\tau_{r\theta}$ at the edge of the vortex, we thus get

$$(\tau_0)_t = -\frac{2\nu_{t_0} \rho K_0}{r_0^2} \qquad (31)$$

and the torque can be written

$$L = -4\pi\nu_{t_0} K_0 \rho \tag{32}$$

We neglect any torque due to shearing forces on areas S_1 and S_3.

Putting the results of equations (22), (24), and (32) into equation (25) yields the desired form of the conservation of angular momentum

$$\frac{d}{dz}\int_0^{r_0} (K_0 - K)wr\ dr - 2\nu_{t_0} K_0 = 0 \tag{33}$$

It is noted that only the eddy viscosity at the outer edge of the vortex is involved in this equation. This result is similar to the result of Gartshore (ref. 9) for a laminar vortex obtained by formal mathematical manipulation of the equations for axisymmetric incompressible laminar flow.

5.2   Conservation of Linear Momentum

The second equation is obtained by applying the law of the conservation of linear momentum to the vortex. Consider a single turbulent vortex formed by a generator, as shown, and enclosed by a cylindrical control surface.

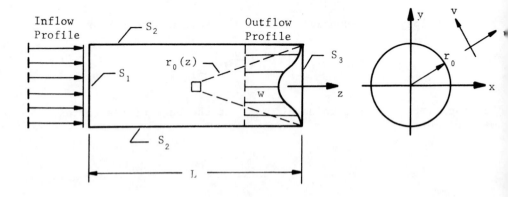

The drag of the vortex generator is equal to the negative of the net momentum flux out of the control volume.

$$D = \int_{S_1} (p_\infty + \rho_\infty W^2) dS_1 - \int_{S_3} (p + \rho w^2) dS_3 - \int_{S_2} \rho u W\, dS_2 \qquad (34)$$

Parallel flow prevails at S_1, but at the S_3 station, tangential flow associated with $K = K_0$ for $r > r_0$ occurs. At $r = r_0$, a volumetric radial flow rate exists which can be calculated using continuity.

$$\int_{S_2} u\, dS_2 = \int_{S_3} (W - w) dS_3 \qquad (35)$$

Putting this result into equation (34) yields

$$D = \int_{S_3} (p_\infty - p) dS_3 + \rho \int_{S_3} w(W - w) dS_3 \qquad (36)$$

The second integral will be evaluated for an assumed form of the axial velocity profile. To evaluate the first integral, we assume that the radial pressure gradient is not sensitive to the turbulent fluctuations and is given by

$$\frac{\partial p}{\partial r} = \frac{\rho v^2}{r} \qquad (37)$$

following reference 10, equation 2(b), for example.

We now have

$$p_\infty - p = \rho \int_r^\infty \frac{v^2}{r}\, dr = \rho \int_r^\infty \frac{K^2}{r^3}\, dr \qquad (38)$$

and

$$\int_{S_3} (p_\infty - p) dS_3 = 2\pi\rho \int_0^r \left[\int_r^\infty \frac{K^2}{\xi^3}\, d\xi \right] r\, dr \qquad (39)$$

Reversing the order of integration and noting that

$$K = K_0, \quad r \geq r_0 \tag{40}$$

we find that

$$\int_{S_3} (p_\infty - p)dS_3 = \pi\rho \int_0^{r_0} \frac{K^2}{r} \, dr + \frac{\pi\rho}{2} K_0^2 \tag{41}$$

The last term is associated with the fact that the static pressure on area S_3 is lowered by rotation of the flow outside the vortex.

The drag of the vortex generator is now

$$D = \pi\rho \int_0^{r_0} \frac{K^2}{r} \, dr + \frac{\pi\rho}{2} K_0^2 + 2\pi\rho \int_0^r w(W - w)r \, dr \tag{42}$$

5.3 Additional Equation

We note that equations (33) and (42) are in terms of integrals involving the axial velocity and circulation profiles. The circulation profile is a two-parameter family. If the axial profile is a one-parameter family, we require one more equation to solve the problem. For this purpose, we use the axial momentum equation (ref. 10, eq. (2d)) in its quasi-cylindrical approximation:

$$\rho \left(w \frac{\partial w}{\partial z} + u \frac{\partial w}{\partial r} \right) = - \frac{\partial p}{\partial z} + \frac{1}{r} \frac{\partial}{\partial r} \left(\mu r \frac{\partial w}{\partial r} \right)$$

$$- \frac{1}{r} \frac{\partial}{\partial r} (r\rho \overline{u'w'}) - \overline{\rho'u'} \frac{\partial w}{\partial r} \tag{43}$$

which we will specialize to a small cylinder near the vortex axis. As $r \to 0$, we use the following boundary condition:

$$\frac{\partial w}{\partial r} \to 0 \tag{44}$$

The remaining turbulent shear term is handled in the usual manner using the eddy-viscosity concept. The turbulent shear τ_{rz} is in the axial direction lying in the plane of the r and z directions By analogy with a two-dimensional turbulent boundary layer, we have

$$\tau_{rz} = -\overline{\rho u'w'} = \rho \nu_{t_1} \frac{\partial w}{\partial r} \tag{45}$$

where ν_{t_1} is the turbulent viscosity on the vortex axis. Neglecting the usual molecular viscosity, we obtain

$$w \frac{\partial w}{\partial z} = -\frac{1}{\rho} \frac{\partial p}{\partial z} + \frac{1}{r} \frac{\partial}{\partial r}\left(\nu_{t_1} r \frac{\partial w}{\partial r}\right) \tag{46}$$

Some light on the nature of the turbulent shear stress in a swirling jet has been obtained by Lilley and Chigier (ref. 11) from the data of Chigier and Chervinsky (ref. 12). For these jets the vortex generator has a thrust rather than a drag and the results may not be applicable to the present case. However, the results show generally that the turbulent shear stress associated with $\rho u'w'$ is an order of magnitude greater than that associated with $\rho u'v'$ in the fully developed region of the swirling jet. Also, the eddy viscosity associated with τ_{rz} does not vary greatly with radius near the axis, but falls across the jet from a maximum at the center to one-half or one-third the maximum value near the edge. In general, the turbulence is not homogeneous. Also, near the edge the values of ν_t associated with $\tau_{r\theta}$ are about half the values on the axis of ν_{t_1} associated with τ_{rz}.

In addition to the two-parameter circulation profiles, a simple one-parameter axial velocity profile with one inflection point is assumed in the theory. More complicated axial profiles can be used when adequate data to define them are available. Consider an axial velocity profile versus r of the following general shape

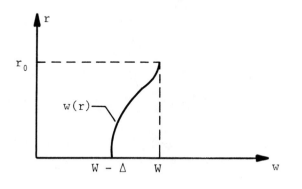

      W   free-stream velocity

      w   axial velocity in vortex

  W - Δ   axial velocity on centerline of vortex

The following boundary conditions are imposed:

$$
\left.
\begin{array}{c}
w = W \\[6pt]
\dfrac{\partial w}{\partial r} = 0
\end{array}
\right\} r = r_0 \\[18pt]
\left.
\begin{array}{c}
w = W - \Delta \\[6pt]
\dfrac{\partial w}{\partial r} = 0
\end{array}
\right\} r = 0
\quad\quad\quad\quad (47)
$$

The resulting axial velocity profile in polynomial form is

$$
w = W - \Delta \left[1 - 3 \left(\frac{r}{r_0} \right)^2 + 2 \left(\frac{r}{r_0} \right)^3 \right] \qquad (48)
$$

This profile is in fair accord with the profile for a jet and should be adequate as long as the swirl is not too high.

Certain characteristic functions are now defined which involve integrals of the circulation distribution. From the empirical relationships previously derived for the circulation distribution, we can show that the following integrals depend only on one parameter, r_1/r_0.

$$
\left.
\begin{array}{c}
f(r_1/r_0) = \dfrac{1}{r_0^2} \displaystyle\int_0^{r_0} r \left(1 - \dfrac{K}{K_0} \right) dr \\[28pt]
g(r_1/r_0) = \dfrac{1}{K_0^2} \displaystyle\int_0^{r_0} K^2 \dfrac{dr}{r} \\[28pt]
h(r_1/r_0) = 1 + 2 \dfrac{r_0^2}{K_0^2} \displaystyle\int_0^{r_0} \dfrac{K^2}{r^3} dr
\end{array}
\right\}
\qquad (49)
$$

These functions involve fairly complicated algebraic expressions. Their values and derivatives have been obtained numerically, and they are given in Table I.

TABLE I – CHARACTERISTIC FUNCTIONS AND THEIR DERIVATIVES

r_1/r_0	f	f'	g	g'	h	h'
0.002	0.2425−01	0.1748 01	0.2595 01	−0.1709 03	0.2405 05	−0.2304 08
.004	.2690−01	.1065 01	.2363 01	−0.8416 02	.7397 04	−0.3200 07
.006	.2874−01	.8086 00	.2227 01	−0.5598 02	.3751 04	−0.1052 07
.008	.3020−01	.6693 00	.2131 01	−0.4196 02	.2330 04	−0.4828 06
.010	.3144−01	.5803 00	.2056 01	−0.3357 02	.1616 04	−0.2649 06
.012	.3254−01	.5178 00	.1995 01	−0.2798 02	.1201 04	−0.1627 06
.014	.3352−01	.4710 00	.1943 01	−0.2399 02	.9366 03	−0.1079 06
.016	.3443−01	.4346 00	.1898 01	−0.2100 02	.7560 03	−0.7569 05
.018	.3527−01	.4052 00	.1858 01	−0.1867 02	.6265 03	−0.5542 05
.020	.3605−01	.3811 00	.1823 01	−0.1680 02	.5301 03	−0.4196 05
.025	.3784−01	.3356 00	.1748 01	−0.1345 02	.3732 03	−0.2334 05
.030	.3943−01	.3036 00	.1687 01	−0.1122 02	.2811 03	−0.1449 05
.035	.4089−01	.2796 00	.1635 01	−0.9620 01	.2217 03	−0.9706 04
.040	.4224−01	.2610 00	.1590 01	−0.8422 01	.1809 03	−0.6867 04
.050	.4470−01	.2336 00	.1515 01	−0.6745 01	.1293 03	−0.3864 04
.060	.4693−01	.2144 00	.1453 01	−0.5627 01	.9860 02	−0.2423 04
.070	.4900−01	.1999 00	.1401 01	−0.4827 01	.7866 02	−0.1637 04
.080	.5094−01	.1887 00	.1356 01	−0.4228 01	.6482 02	−0.1168 04
.090	.5278−01	.1797 00	.1316 01	−0.3761 01	.5475 02	−0.8679 03
.100	.5454−01	.1724 00	.1280 01	−0.3388 01	.4714 02	−0.6663 03
.110	.5624−01	.1663 00	.1248 01	−0.3083 01	.4122 02	−0.5252 03
.120	.5787−01	.1610 00	.1219 01	−0.2828 01	.3651 02	−0.4230 03
.130	.5946−01	.1566 00	.1191 01	−0.2612 01	.3268 02	−0.3469 03
.140	.6101−01	.1528 00	.1166 01	−0.2428 01	.2951 02	−0.2889 03
.150	.6252−01	.1493 00	.1143 01	−0.2268 01	.2686 02	−0.2438 03
.160	.6399−01	.1464 00	.1121 01	−0.2129 01	.2460 02	−0.2081 03
.170	.6544−01	.1437 00	.1100 01	−0.2005 01	.2267 02	−0.1794 03
.180	.6687−01	.1412 00	.1081 01	−0.1897 01	.2100 02	−0.1561 03
.190	.6827−01	.1391 00	.1062 01	−0.1799 01	.1954 02	−0.1369 03
.200	.6965−01	.1372 00	.1045 01	−0.1709 01	.1825 02	−0.1209 03

Consider equations (33) and (42) in differential form prior to introduction of the characteristic functions.

$$\frac{d}{dz}\left[\int_0^{r_0} (K_0 - K)wr\,dr\right] = 2\nu_{t_0} K_0 \tag{50}$$

$$\frac{d}{dz}\left[\int_0^{r_0} wr(W - w)dr + \frac{1}{2}\int_0^{r_0} \frac{K^2}{r}\,dr\right] = 0 \tag{51}$$

Let us also put equation (46) into the form which utilizes the characteristic function h. We can rewrite equation (46) assuming constant eddy viscosity ν_{t_1} as

$$\frac{\partial}{\partial z}\left[\frac{w^2}{2} + \frac{p}{\rho}\right] = \frac{\nu_{t_1}}{r}\frac{\partial}{\partial r}\left(r\frac{\partial w}{\partial r}\right) \tag{52}$$

which from equation (48) becomes on the axis

$$\frac{\partial}{\partial z}\left[\frac{w^2}{2} + \frac{p}{\rho}\right]_{r=0} = \frac{12\nu_{t_1}\Delta}{r_0^2} \tag{53}$$

It is easy to show that the static pressure on the axis is given by

$$\left(\frac{p}{\rho}\right)_{r=0} = \frac{P_\infty}{\rho} - \frac{K_0^2}{2r_0^2} - \int_0^{r_0} \frac{K^2}{r^3}\,dr \tag{54}$$

With the use of equations (54) and (49), equation (53) becomes

$$\left(\frac{\Delta}{W}\right) = \frac{r_0^2 W}{12\nu_{t_1}}\frac{d}{dz}\left[\frac{1}{2}\left(\frac{\Delta}{W}\right)^2 - \left(\frac{\Delta}{W}\right) - \frac{1}{2}\left(\frac{K_0}{Wr_0}\right)^2 h\left(\frac{r_1}{r_0}\right)\right] \tag{55}$$

Consider now that equations (50), (51), and (55) are to be integrated for given initial values of r_{0_i}, r_{1_i}, and $(\Delta/W)_i$ to determine the ratios r_0/r_{0_i}, r_1/r_0, and Δ/W as functions of z.

The set of three equations in terms of the characteristic functions is:

$$\frac{2\nu_{t_0} z}{W(r_{0_i})^2} - \frac{r_0^2}{(r_{0_i})^2} f(r_1/r_0) = C_1 \tag{56}$$

$$g\left(\frac{r_1}{r_0}\right) + 2\left(\frac{r_0 W}{K_0}\right)_i^2 \frac{r_0^2}{(r_{0_i})^2}\left[\frac{3}{20}\left(\frac{\Delta}{W}\right) - \frac{3}{35}\left(\frac{\Delta}{W}\right)^2\right] = C_2 \tag{57}$$

$$\left(\frac{\Delta}{W}\right) = \frac{r_0^2 W}{12\nu_{t_1}} \frac{d}{dz}\left[\frac{1}{2}\left(\frac{\Delta}{W}\right)^2 - \left(\frac{\Delta}{W}\right)\right.$$

$$\left. - \frac{1}{2}\left(\frac{K_0}{Wr_0}\right)_i^2\left(\frac{r_{0_i}}{r_0}\right)^2 h\left(\frac{r_1}{r_0}\right)\right] \tag{58}$$

If $\nu_{t_0} = \nu_{t_1}$ these equations have solutions in terms of $z\nu_t/W(r_{0_i})^2$ for given initial conditions with $(r_0 W/K_0)_i$ as parameter. The constants C_1 and C_2 are evaluated from the initial conditions and are related to the angular momentum and drag of the vortex.

6. CALCULATIVE RESULTS FOR STAGE II

6.1 Calculated Equilibrium Solutions

It is convenient to reduce equations (56) through (58) to differential form for solution on a computing machine even though the first two are already integrals of the motion. This has been done and the equations programmed to yield r_1, r_0, and Δ/W versus z for arbitrary initial values of these quantities and the initial value of Wr_0/K_0. In the first sets of calculations the eddy viscosities are taken equal at the vortex center and at its outer edge. The eddy viscosity was computed from the Owen relationship, equation (3), using a value of Λ of 0.63 based on the flight-test data of Rose and Dee (ref. 13). The flight conditions for the first calculative example are shown on figure 7(a). With these conditions, it was found that the ratio of turbulent viscosity to molecular viscosity was 3960.

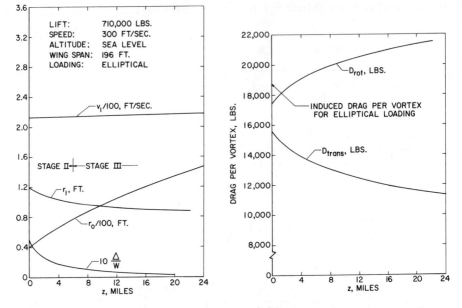

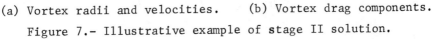

(a) Vortex radii and velocities.   (b) Vortex drag components.

Figure 7.- Illustrative example of stage II solution.

Logically the initial conditions for the stage II calculations should arise from the stage I solution. For reasons which are discussed in the next section, we find it is not possible to do this. Instead, engineering estimates were made to obtain initial conditions to illustrate the nature of the solution, and some parametric studies of the initial conditions were carried out. For the first calculative example a value of r_0 somewhat less than the wing semispan was chosen, a value of Δ/W which gives reasonable drag, and a value of r_1/r_0 compatible with the value of K_1/K_0 in equation (9). The value of K_1/K_0 was estimated from equation (20) using the approximate result from a rolling-up calculation that $\Gamma_c = 0.35\ \Gamma_0$. The values so obtained are:

$$r_0 = 40.0 \text{ feet}$$
$$r_1 = 1.2 \text{ feet}$$
$$\Delta/W = 0.05$$

Examination of figure 7(a) for the first calculative example shows that r_0 increases approximately as $\sqrt{z}$, and the value of Δ/W decreases asymptotically towards zero. The surprising result is that the value of v_1 remains nearly constant. Also, r_1 decreases only slightly for long downstream distances. Another point of interest is the partition of the drag. For the present example, the components of the drag associated with rotation and

translation are shown in figure 7(b). The initial drag in the
rotational phase has been taken not much different from the induced
drag per vortex as calculated for an elliptical loading with sufficient additional drag in the axial velocity profile to allow about
as much profile drag as induced drag. As the downstream distance
increases, the drag of the translational phase passes into the
rotational phase. It is true that the axial velocities decrease
quite rapidly as the downstream distance increases, but the vortex
radius r_0 does not increase fast enough to keep the translation
drag from decreasing. This behavior for the present turbulent
vortex differs from the behavior of a laminar vortex as predicted
by Batchelor (ref. 14), who expects a transfer of the rotational
drag to the translational phase asymptotically. The direction of
the energy transfer is sensitive to small changes in the asymptotic
behaviors of r_0 or Δ/W.

As a criterion where stage II ends, we use the condition that

$$r_0 = \frac{\pi}{4} s = s_v$$

It is thus assumed that the asymptotic centerline position of the
turbulent vortex is at the lateral centroid of the vorticity for
one wing panel, and that the outer regions of the left and right
vortices just touch at the vertical plane of symmetry of the aircraft. On the basis of the foregoing criterion, stage II ends at
about six miles behind the aircraft.

The calculations of figure 7 have been repeated for several
initial values of the parameter r_0. The calculated results are
not sensitive to r_{0_i}. The downstream distance to the position
where the two vortices just touch is shown in figure 8 as a function
of r_{0_i}. This distance, which marks the end of stage II or the
beginning of stage III depends somewhat on the value of r_{0_i}. In
fact, if r_{0_i} is assumed equal to s_v (i.e., 77 feet), then the
distance would be zero. Another significant distance in stage III
is that distance at which the outer edge of one vortex as calculated by stage II means just touches the center of the opposite
vortex. This occurs at about 23 miles, almost independent of the
initial value of r_{0_i} within reasonable limits. There is thus an
asymptotic behavior of the vortex which is not sensitive to initial
conditions.

In the foregoing calculations, the eddy viscosities at the
vortex centerline and at its outer edge were taken equal and
independent of z. There are, however, no data to support the
uniformity of the eddy viscosity across a jet vortex nor its

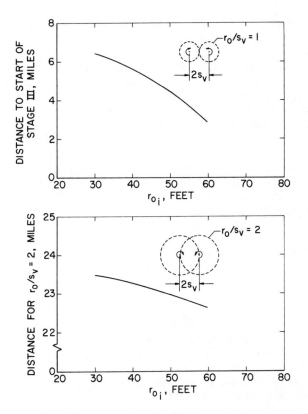

Figure 8.- Characteristic distances in stage III.

constancy with axial distance. It is of interest, therefore, to
know how sensitive the calculations are to variations in eddy
viscosity. For this purpose, we have repeated the former calculations for the case wherein the edge viscosity is zero rather than
equal to the value at the centerline. The initial conditions for
this case, shown in figure 9, are

$$
\begin{aligned}
\text{Lift:} &\quad 710{,}000 \text{ lbs.} \\
\text{Speed:} &\quad 300 \text{ ft/sec.} \\
\text{Altitude:} &\quad \text{Sea Level} \\
\text{Wing Span:} &\quad 196 \text{ ft.} \\
\text{Loading:} &\quad \text{Elliptical}
\end{aligned}
$$

$$
\nu_{t_0} = 0; \quad \nu_{t_1} = 0.645 \text{ ft}^2/\text{sec.}
$$

The vortex radii and velocities calculated for this case as
shown in figure 9(a) are directly comparable with those of figure
7(a). A direct comparison of these figures shows that the vortex
radius decreases in the second case as opposed to increasing in

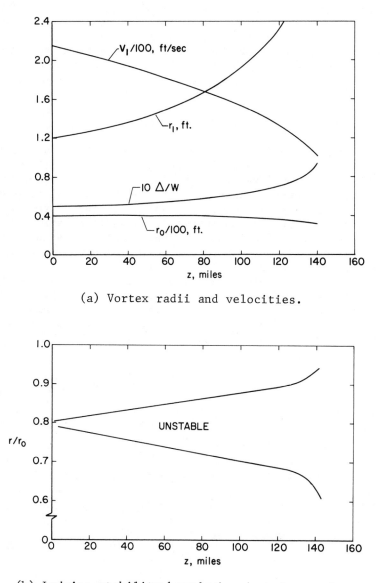

(a) Vortex radii and velocities.

(b) Ludwieg stability boundaries in outer region.

Figure 9.- Illustrative example of stage II solution with zero eddy viscosity at outer edge of vortex.

the first case. The maximum tangential velocity v_1 now decreases significantly while the axial velocity defect Δ increases. Suppressing the eddy viscosity at the outer edge of the vortex thus entirely changes the nature of the vortex behavior. It is clear that the vortex behavior does depend on the eddy-viscosity model for a jet vortex; and it will be necessary to obtain more detailed knowledge of eddy viscosity before an adequate predictive method for the jet vortex can be developed.

6.2   Stability Considerations

Suppressing the eddy viscosity at the edge of the vortex also has a significant effect on the vortex stability. For a two-dimensional inviscid rotary flow, the Rayleigh (ref. 15) criterion of hydrodynamic stability states that if the tangential velocity decreases more rapidly than inversely as the radius, instability results. Ludwieg (ref. 16) has generalized Rayleigh's result to account for stability of helical flow between two co-axial cylinders with radial gradients of the axial velocity. Assuming the annulus to be narrow, and assuming the tangential and axial velocities to be linear in the radial distance, Ludwieg has devised a hydrodynamic stability criterion which generalizes the Rayleigh criterion for the effect of the radial gradient of axial velocity. This criterion can be applied to an annulus of infinitesimal width to obtain a local stability criterion. Letting

$$C_z = \frac{r}{v} \frac{\partial w}{\partial r} \qquad\qquad C_\theta = \frac{r}{v} \frac{\partial v}{\partial r} \qquad\qquad (59)$$

we obtain stability if

$$(1 - C_\theta)(1 - C_\theta^2) - (5/3 - C_\theta)C_z^2 > 0 \qquad\qquad (60)$$

For uniform axial velocity and for a potential vortex we have

$$vr = \text{constant}; \quad \frac{\partial w}{\partial r} = 0 \qquad\qquad (61)$$

so that

$$C_\theta = -1 \qquad\qquad (62)$$

We thus have neutral stability for a two-dimensional potential vortex. For solid-body rotation, it is readily found that

$$C_\theta = 1 \qquad\qquad (63)$$

and we have neutral stability again. While a planar vortex in solid-body rotation is highly stable according to the results of Rayleigh, the result of Ludwieg is not contradictory since the least radial gradient of axial velocity can induce instability through spiral disturbances not met in the planar case. What is of importance is the effect of viscosity which might be expected to allow some radial gradient of axial velocity without instability. Kiessling (ref. 17) has considered the effect of viscosity on the results of Ludwieg, but no attempt has been made herein to apply the results of Kiessling to assess the stability of the equilibrium solutions presented. We thus draw no conclusions concerning the stability of the eye of the vortex.

The detailed experimental data of Hummel (refs. 2 and 18) of the breakdown of leading-edge vortices shed by a slender delta wing indicate that Ludwieg's theory is in good accordance with the measurements. On the basis of this finding, it seems pertinent to apply the Ludwieg theory of vortex breakdown to the logarithmic and outer regions of the trailing vortex examples just discussed.

Since the axial and tangential velocity profiles are specified in the theory for both the logarithmic region and the outer region of the vortex, we can readily make local determinations of stability in these regions. For the logarithmic region, we find

$$C_\theta = -1 + 0.928 \frac{K_1}{K}$$

$$C_z = 6 \left(\frac{\Delta}{W}\right) \left(\frac{Wr_{0_i}}{K_0}\right) \left(\frac{K_0}{K}\right) \left(\frac{r_0}{r_{0_i}}\right) \left(\frac{r}{r_0}\right)^3 \left(1 - \frac{r}{r_0}\right)$$

$$(64)$$

and for the outer region

$$C_\theta = -1 + 4.8 \left(\frac{r}{r_0}\right) \left(\frac{K_0}{K} - 1\right)$$

$$C_z = 6 \left(\frac{\Delta}{W}\right) \left(\frac{Wr_{0_i}}{K_0}\right) \left(\frac{K_0}{K}\right) \left(\frac{r_0}{r_{0_i}}\right) \left(\frac{r}{r_0}\right)^3 \left(1 - \frac{r}{r_0}\right)$$

$$(65)$$

The stability of the previous vortex solutions was investigated using these results and equation (60) at each step of the calculation for a series of radial positions. No instabilities were found for the calculative case with equal eddy viscosities on the axis and at the edge of the vortex. For the second case wherein

the eddy viscosity was zero at the outer edge of the vortex,
definite indications of instability were found in the outer
region. Figure 9(b) indicates the extent of the outer region
which is unstable.

6.3 Joining of Stage I and Stage II Solutions

A completely satisfactory method of joining the stage II
solution to the stage I solution has not been found. The conditions of constant vortex drag (axial momentum) and conservation of
angular momentum are two conditions which can be used for joining
stages I and II. However, another condition is needed in the
present theory. The possibility of making v_1 continuous is now
examined.

It has been shown in connection with the measured results of
Fage and Simmons, figure 6, that a fair fit to the measured circulation distribution in its inner part can be made using two-
dimensional unsteady viscous theory together with an experimentally
determined eddy viscosity. From equations (19), (20), and (21),
the result for the maximum tangential velocity in the vortex is
given by

$$v_1 = \frac{K_1}{r_1} = \frac{0.712 \ K_c}{(1.115) \ \sqrt{4\nu_t}} \sqrt{\frac{W}{z}} \qquad (66)$$

To the extent that the vorticity rolled up into the center of the
vortex is constant and ν_t is constant, the maximum velocity
v_1 varies inversely as z. Variations of this formula have been
used to predict the decrease in maximum tangential velocity with
axial distance. The results of the stage II theory tend to yield a
constant value of v_1 with changes in z, neglecting vortex breakdown. Accordingly, matching the results of equation (66) to the
stage II theory presents a real difficulty.

Some experimental evidence exists that suggests the existence
of a constant value of v_1 at the end of the rolling-up region.
Some flight data of McCormick, Tangler, and Scherrieb (ref. 19)
are presented in figure 10 to illustrate the possibility. The
data for both flights suggest the existence of a constant v_1
(plateau) well downstream. Data for two flights are not included;
for one flight the data do not extend far enough downstream to be
meaningful, and for the other flight a plateau is not manifest
based on one data point. One possible explanation of why the
inverse square root of z law of the two-dimensional unsteady
vortex theory does not persist downstream is that the theory does
not account for the inward convection of vorticity during the
rolling-up process nor does it account for axial gradients of the

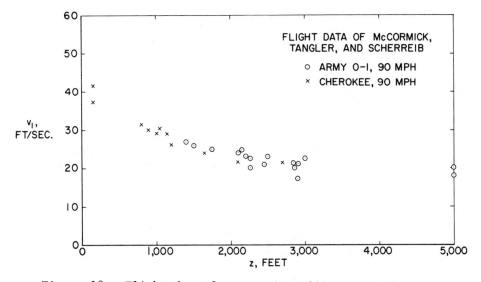

Figure 10.- Flight data for v_1 in rolling-up region.

physical quantities. A satisfactory theory for stage I must account for the fact that the vortex in this region is not fully developed, and must account for the rolling-up process as well as diffusion and axial gradients. The determination of v_1 at the end of stage I awaits such a theory or an extensive data correlation.

7. STAGE III CALCULATIONS

The beginning of stage III is taken at that downstream distance where the two axisymmetric vortices treated as if they were isolated from each other become tangent at their outer edges (r_0). Downstream of this position, the direct interaction between the flow fields of the vortices is considered. During the interaction vorticity is transported across the vertical plane of symmetry within the recirculation region shown from both the right and left vortices and is annihilated there. Additional vorticity transported across the outer $\psi = 0$ streamline is entrained by the outer flow, and is annihilated along the vertical plane of symmetry above the recirculating region. It is possible to make an estimate of the rate of decay of the strength of the individual vortices by this mechanism on the basis of some simplifying assumptions.

The first assumption concerns the shape of the outer $\psi = 0$ streamline for the stage III flow. The streamlines are those seen by an observer moving downward with the vortices. The streamlines for a pair of two-dimensional potential vortices can readily be calculated. In the notation of the following sketch

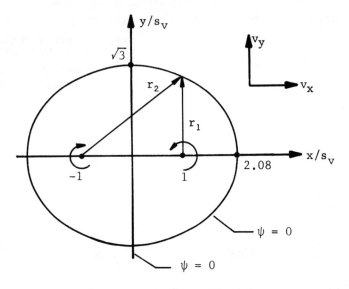

the branches of the $\psi = 0$ streamline are given by

$$\psi = -\frac{\Gamma}{2\pi} \, \log_e \left[\frac{r_a}{r_b} + \frac{1}{2} \, \frac{x}{s_v} \right] = 0 \tag{67}$$

One branch is the vertical axis of symmetry. The outer branch is
nearly an ellipse of major semi-axis $2.08 \, s_v$ and minor semi-axis
$1.732 \, s_v$ (fig. 11). To obtain the streamlines for a pair of fully
developed turbulent vortices, we will superimpose their stream
function as a first approximation. For the logarithmic region,
we find for the stream function of a clockwise vortex of strength
K_0

$$\frac{\psi}{K_0} = \frac{1.47}{2} \left(\frac{r_i}{r_1}\right)^2 \frac{K_1}{K_0} + \frac{K_1}{K_0} \, \log_e \left(\frac{r}{r_i}\right) \left[1 + \frac{0.928}{2} \, \log_e \left(\frac{r_i r}{r_1^2}\right) \right];$$

$$r_i \leq r \leq r_j \tag{68}$$

and for the outer region

$$\frac{\psi}{K_0} = \frac{1.47}{2} \left(\frac{r_i}{r_1}\right)^2 \frac{K_1}{K_0} + \frac{K_1}{K_0} \, \log_e \left(\frac{r_j}{r_i}\right) \left[1 + \frac{0.928}{2} \, \log_e \left(\frac{r_i r_j}{r_1^2}\right) \right]$$

(eq. (69) cont. on next page)

$$+ \log_e \left(\frac{r}{r_j} \right) + 4.43 \frac{K_1}{K_0} \left[E_i \left(-4.8 \frac{r_j}{r_0} \right) - E_i \left(-4.8 \frac{r}{r_0} \right) \right] ;$$

$$r_j \leq r \leq r_0 \tag{69}$$

The exponential integral Ei is given by

$$Ei(-x) = - \int_x^\infty \frac{e^{-t}}{t} \, dt \tag{70}$$

The value of the stream function for a single turbulent vortex depends only on r_1/r_0 and r/r_0 . The total stream function for a pair of turbulent vortices which are stationary in the x,y coordinate system shown above is

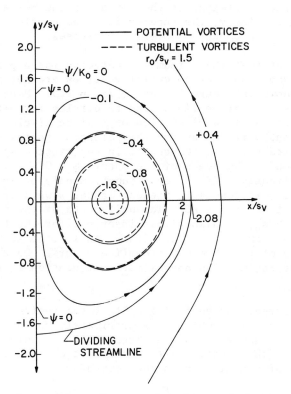

Figure 11.- Streamlines for overlapping turbulent vortices.

$$-\frac{\psi_T}{K_0} = \frac{\psi}{K_0}\left(\frac{r_1}{r_0}, \frac{\sqrt{y^2 + (s-x)^2}}{r_0}\right)$$

$$-\frac{\psi}{K_0}\left(\frac{r_1}{r_0}, \frac{\sqrt{y^2 + (s+x)^2}}{r_0}\right) + \frac{x}{2s_v} \qquad (71)$$

under the assumption of linear superposition. These equations apply downstream of the axial position for which $r_0 = s_v$. For the condition

$$\frac{r}{s_v} = \frac{2}{1 + \left(\frac{r_i}{r_1}\right)\left(\frac{r_1}{r_0}\right)} \qquad (72)$$

the outer radius of one vortex will just reach the outer edge of the edge of the eye of the opposite vortex if the single vortices are simply overlapped. The departure of the turbulent vortex from a potential vortex increases as the center of the vortex is approached. Accordingly, the superposition assumption becomes less accurate as the amount of overlap increases.

It turns out that the streamline patterns for a pair of turbulent vortices and a pair of potential vortices are not greatly different as shown in figure 11 for $r_0/s_v = 1.5$. Near the center of either vortex the streamlines tend to circles for either case, but the value of the stream function on a given circle is less for the turbulent vortex than for the potential vortex because ψ tends to a finite value at the center in the first case, but to $-\infty$ in the second. The shape of the outer $\psi = 0$ streamline is altered only slightly. If K is a function only of r, the superposition of two vortices as in present manner always produces a stagnation point at $y/s = \sqrt{3}$. The half-width of the recirculation region, $x = 2.08\ s_v$, is also almost identical for the two cases. In fact, for the range of r_0/s_v from 1 to 1.9, no appreciable change in shape of the $\psi = 0$ streamline occurs. On the basis of this result, we can make an approximate calculation of the rate of vortex decay.

The zero streamline enclosing half of the recirculation flow is moving with the flow, which is approximately two-dimensional in crossflow planes, and which has a circulation Γ which measures the vortex strength. According to the result of reference 20, the rate of change of circulation is then given by

$$\frac{d\Gamma}{dt} = \nu_t \oint_{\psi=0} \nabla^2 \vec{q} \cdot d\vec{r} \qquad (73)$$

where the vector velocity $\vec{q}$ and the vector $\vec{dr}$ are given by

$$\vec{q} = v_x + iv_y \; ; \quad \vec{dr} = dx + i \, dy \qquad (74)$$

Since we have superimposed the rotational flow fields of two turbulent vortices, $\nabla^2 \vec{q}$ will not be zero. Along the vertical axis we find

$$\nabla^2 \vec{q} \cdot \vec{dr} = \nabla^2 v_y \, dy \qquad (75)$$

On a circle of radius ρ, we find

$$\nabla^2 \vec{q} \cdot \vec{dr} = \rho \left[-\sin \theta \, \nabla^2 v_x + \cos \theta \, \nabla^2 v_y \right] d\theta \qquad (76)$$

In evaluating the integral of equation (73), the first part of the contour has been taken from $-\sqrt{3} \le y/s \le \sqrt{3}$, and the second half along a semicircle $\rho e^{i\theta}$, $-\pi \le \theta \le +\pi$, which fits the outer $\psi = 0$ streamline approximately. The relative magnitude of the integrand as given by equation (75) and (76) is shown in figure 12 for the first quadrant of the recirculating flow region. The vorticity transport across the $\psi = 0$ streamline is greatest at the origin and on the opposite side of the vortex on the x-axis. These positions are where the inner region of a turbulent vortex with its greater vorticity is closest to the $\psi = 0$ streamline.

The value of $d\Gamma/dt$ has been calculated from the start of stage III at $z = 6$ miles behind the example airplane of figure 7 to about 22 miles behind the aircraft, where the ratio r_0/s_v is

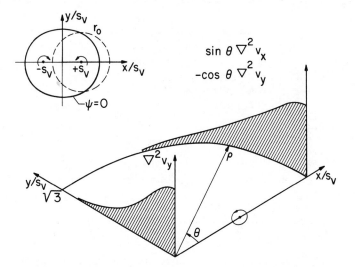

Figure 12.- Vorticity transport across the zero streamline.

1.8.   At the onset where $r_0/s_V = 1$, the two outer regions are tangent at the vertical plane of symmetry, and for $r_0/s_V = 1.96$ the logarithmic regions are tangent at the vertical plane of symmetry. Within these ranges vorticity is transferred across the $\psi = 0$ streamline from only the outer regions of the vortices. As r_0/s_V increases and the inner regions of the vortices contact the $\psi = 0$ streamline, the vorticity there increases and with it the vorticity transport.

In the form presented in figure 13 the calculated results are expressible by the equation

$$\log\left(-\frac{d\Gamma/dt}{\Gamma}\right) = \log C + 0.7 \log (z - 3.3) \tag{76}$$

which yields

$$\left.\begin{array}{c} \dfrac{\Gamma}{\Gamma_0} = e^{-k(z-3.3)^{1.7}} \\[2em] k = 0.64\times10^{-4} \\[2em] z = \text{distance behind aircraft,} \\ \text{miles} \end{array}\right\} \tag{77}$$

Within the range of z shown in figure 13, the decay of Γ is not large. From equation (77), the distance for Γ to decrease by the ratio $1/e$ is about 280 miles. This distance is not to be taken literally since the logarithmic regions start to overlap when r_0/s_V is 1.96, which occurs at about $z = 23$ miles, and the

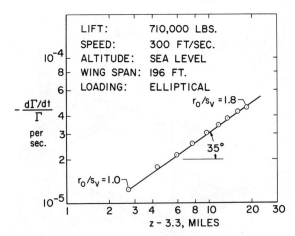

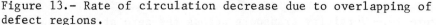

Figure 13.- Rate of circulation decrease due to overlapping of defect regions.

decay law will then change. Whether the rate of decay associated with vorticity diffusion in the logarithmic region will cause the law established in figure 13 to be too low or too high if extrapolated downstream is not known. While it is possible to carry out the present form of analysis for values of $r_0 > 1.96\ s_v$, a serious question arises concerning the validity of the linear superposition of the turbulent vortex fields to obtain the combined field. This assumption can be justified in part for overlapping outer regions where the vorticity is least, but if the logarithmic regions overlap, a better means may be required for determining the combined flow.

8. CONCLUDING REMARKS

An analysis has been made of a trailing vortex pair behind an aircraft for the case of a quiet atmosphere. The analysis provides an equilibrium solution which may be considered a limiting case from the viewpoint of vortex persistency. The three stages considered in the analysis are a rolling-up stage directly behind the aircraft, a second stage in which the vortices act independently as equilibrium turbulent vortices, and a third stage where the vortices physically interact. No development of a stage I theory has been made beyond the usual rolling-up calculative method.

A new theory has been developed for stage II using experimental correlations of the radial circulation profiles based on the data of Hoffmann and Joubert. These correlations present empirical results for the three regions of the circulation distribution; an inner region called the "eye," an intermediate region exhibiting a logarithmic law, and an outer or defect region. Also used in the theory are certain integral relations based on the constancy of vortex drag and the conservation of angular momentum. An eddy viscosity model is also required.

In stage III, a theory has been developed to estimate the rate of circulation decay during the time when the defect regions as calculated for the isolated vortices overlap. A theory for the next part of the region where the logarithmic regions overlap has not been developed, but it appears that the decay of vorticity during this period may be more significant in reducing the vortex strengths.

Example calculations for stages II and III for a heavy transport flying at sea level reveal several important results. The rate of circulation decrease in stage III due to interaction between the vortices appears not to present a promising mechanism for rapid decay of trailing vortices. In fact, barring such events as vortex breakdown, the equilibrium solutions found may represent limiting cases of persistency which could be significant

from the safety point of view. From this point of view, it is
probably important to develop the stage III theory for overlapping
logarithmic regions.

The theory of Ludwieg yields no instabilities in stages II
or III for a turbulent vortex with equal values of eddy viscosity
on the axis and its outer edge (with the possible exception of the
eye of the vortex). However, reducing the eddy viscosity at the
edge to zero significantly affected the equilibrium solution, and
it induces instability of the flow in the outer region. The state
of knowledge concerning the turbulent mechanics of jet vortices is
not yet sufficiently developed to predict their eddy-viscosity
characteristics. Further work in this area is clearly required;
in particular, detailed eddy-viscosity measurements are needed.
The effect of meteorological conditions on the outer-edge eddy
viscosity needs investigation. Further work is needed in several
other areas before the characteristics of a trailing vortex pair
behind an aircraft can be predicted. In the first place, a better
theory of the stage I region is needed since the present methods
based on rolling-up calculations or the two-dimensional unsteady
vortex solution are oversimplifications. In fact, more data are
needed in the stage I region to supplement those of Fage and
Simmons.

The stage II and III theories are based on empirical circulatio
distributions for a limited range of data. It is essential to exten
the range of the measurements over the important ranges of the independent parameters established on the basis of dimensional analysis.
Axial velocity profile measurements as well as circulation distribution measurements are needed.

There is further work to be done in the area of vortex
stability. For example, the effects of viscosity on the instability
of the eye of the vortex predicted by the Ludwieg theory needs furth
attention. Data are needed to verify the prediction of vortex
bursting in trailing vortices by the Ludwieg criterion or other
criteria. Eventually a method of predicting the characteristics
of a burst vortex will be needed if its effect on aircraft
safety is to be assessed.

REFERENCES

1. Crow, S. C.: Stability Theory for a Pair of Trailing Vortices.
    AIAA 8th Aerospace Sciences Meeting, New York, N. Y.,
    January 19-21, 1970, AIAA Paper No. 70-53.

2. Hummel, D.: Investigation on Vortex Breakdown on a Sharp-
    Edged Slender Delta Wing. Bericht 64/24 Institut für
    Strömungsmechanik Braunschweig, 1964.

3. Hoffmann, E. R. and Joubert, P. N.: Turbulent Line Vortices.
    Jour. of Fluid Mech., vol. 16, part 3, July 1963, pp. 395-411.

4. Owen, P. R.: The Decay of a Turbulent Trailing Vortex.
    F.M. 3446, British A.R.C. 25 818, Apr. 1964.

5. Westwater, F. L.: Rolling Up of a Surface of Discontinuity
    Behind an Airfoil of Finite Span. Brit. ARC R&M 1962, 1935.

6. Takami, H.: A Numerical Experiment with Discrete-Vortex
    Approximation, with Reference to the Rolling Up of a Vortex
    Sheet. AFOSR 64-1292, May 1964.

7. Fage, A. and Simmons, L. F. G.: An Investigation of the Air-
    Flow Pattern in the Wake of an Aerofoil of Finite Span.
    Reprint - Phil. Trans. Roy. Soc. London, Series A, vol. 225,
    no. 7, Jan. 1926.

8. Schlichting, H.: Boundary Layer Theory. Fourth ed., McGraw-
    Hill Book Co., Inc., 1960, pp. 602-603.

9. Gartshore, I. S.: Some Numerical Solutions for the Viscous
    Core of an Irrotational Vortex. N.R.C. (Canada) Aero. Rep.
    LR-378, June 1963.

10. Hall, M. G.: The Structure of Concentrated Vortex Cores.
    Chapter 4 in "Progress in Aeronautical Sciences," vol. 7,
    edited by D. Küchemann, Pergamon Press, 1966.

11. Lilley, D. G. and Chigier, N. A.: Nonisotropic Turbulent
    Stores Distribution in Swirling Flows From Mean Value Distri-
    butions. (To be published in International Journal of Heat
    and Mass Transfer.)

12. Chigier, N. A. and Chervinsky, A.: Experimental Investigation
    of Swirling Vortex Motion in Jets. Jour. of Applied Mech.,
    June 1967, pp. 443-451.

13. Rose, R. and Dee, F. W.: Aircraft Vortex Wakes and Their
    Effects on Aircraft. A.R.C. C.P. 795, Dec. 1963.

14. Batchelor, G. K.: Axial Flow in Trailing Line Vortices.
    Jour. of Fluid Mech., vol. 20, part 4, Dec. 1964, pp. 645-658.

15. Lord Rayleigh: On the Dynamics of Revolving Fluids. Proc.
    Roy. Soc., London (A), 93, 1916, pp. 148-154.

16. Ludwieg, H.: "Ergänzung zu der Arbeit: Stabilität der
    Strömung in einem Zylindrischen Ringraum." Z. Flugwiss.,
    9 (1961), Heft 11, pp. 359-361.

17. Kiessling, I.: Über das Taylorsche Stabilitätsproblem bei
     zusätzlicher axialer Durchströmung der Zylinder. DVL Bericht
     Nr. 290 (1963).

18. Hummel, D.: Untersuchungen über das Aufplatzen der Wirbel an
     Schlanken Deltaflügeln. Z. Flugwiss., 13 (1965), Heft 5,
     pp. 158-168.

19. McCormick, B. W., Tangler, J. L., and Sherrieb, H. E.: Structur
     of Trailing Vortices. Jour. of Aircraft, vol. 5, no. 3, May-
     June 1968, pp. 260-267.

20. Milne-Thomson, L. M.: Theoretical Hydrodynamics. Second ed.,
     The Macmillan Company, 1950, p. 510.

RESULTS OF TRAILING VORTEX STUDIES IN A TOWING TANK

John H. Olsen

Boeing Scientific Research Laboratories

ABSTRACT

Flow visualization studies were performed in a towing tank using an electrochemically activated dye. Although test Reynolds numbers were far below flight Reynolds numbers (10^4 vs 10^7), the results were strikingly similar to flight test data.

Two types of instability were observed in the tank. An instability associated with the axial flow within the core was observed to destroy the flow in the neighborhood of the core without destroying the motion far from the core. A second instability involving the mutual interaction of the two vortices was observed but somewhat masked by the first instability.

I.   INTRODUCTION

Interest in the trailing vortex wakes of aircraft has increased because of possible hazards to other aircraft, particularly at congested airports. The region of interest is the far wake (many spans) of the generating aircraft. One wishes to know how the vortex wake disintegrates for developing avoidance procedures or wake destruction methods.

This paper describes some preliminary results from flow visualization studies of vortex wakes in a towing tank. The general features of the wakes are illustrated, but precise measurements remain to be made.

II.   APPARATUS AND METHODS

A towing tank was used for the experiments because of the greater capability to control "flight" conditions and because of the cost advantage over flight tests. In the tank, the far wake could be studied easily by trading time for the large distances that would have been required in a wind tunnel. The tank was ten feet long with a sixteen inch square cross-section and plexiglas side walls.

The carriage carrying the model rode on wheels on rails parallel to the free surface. It was towed by a cable driven by a variable speed motor. The model was, in most cases, a vertical strut of rectangular plan form and NACA 0012 section which penetrated the free surface to join the towing system. Effects of surface waves were minimized by using low towing speeds so that the surface acted as a plane of symmetry. The effective wing span was, thus, twice its penetration depth.

An electrolytic dye technique described by Baker[1] was used for flow visualization. The neutrally buoyant dye was formed electrochemically on the airfoil itself in some experiments and on a grid in the path of the vortex in others. The fluid conductivity was raised by adding sodium sulphate and the electrodes were brass and copper rather than platinum as suggested by Baker. The dye was illuminated by sodium vapor light for maximum contrast. Photographs were taken at fixed intervals with a 35mm camera and all data was measured from these photographs.

The residual velocity in the tank due to convection and past runs was about 10^{-3} feet per second at the start of a new run. A thin layer of oil to prevent evaporation cooling at the free surface was necessary to achieve these low residual velocities.

III.   RESULTS

Transient Effects

The airfoil was rapidly accelerated and decelerated at the ends of the tank. The effects of these transient motions on the motion in the center of the tank was judged to be small on the basis of results shown in figures 1 and 2 showing dye shed from the moving airfoil. The starting process forms the classic starting vortex which joins the trailing vortex in a continuous way and is quickly indistinguishable from it. On stopping, a vortex is shed again, but a different result is seen. The trailing vortex has an axial velocity (see below) which must vanish at the

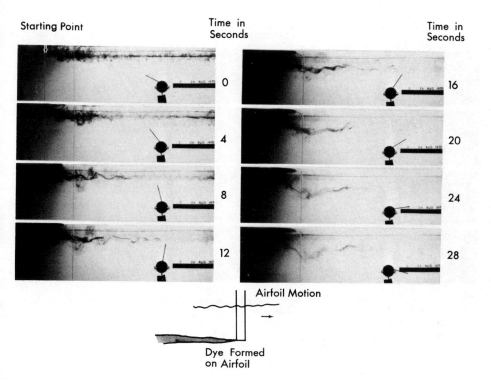

Fig. 1.   Starting Vortex and Transients.

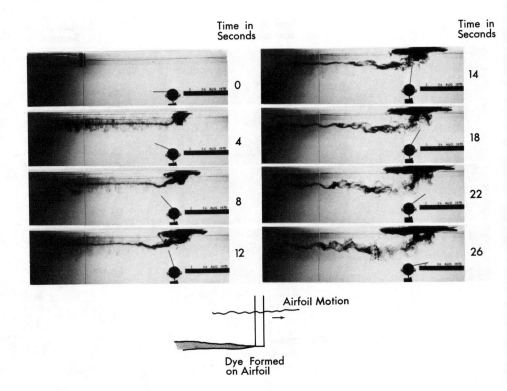

Fig. 2.   Stopping Vortex with Wave-like Disturbance
          Propagating Leftward.

free surface. Accordingly, a disturbance reminiscent of a hydraulic
jump propagates slowly along the core enlarging it and reducing the
axial velocity. The disturbance is possibly related to the vortex
bursting observed in certain steady flows.[2] A further example of
the differences between downstream and upstream transients is shown
in figure 3 where a vortex has been cut with a rigid plate. A
stopping disturbance slowly propagates opposite to the axial velocity on the left while no appreciable effect is noticed on the right.

These experiments indicate that the flow in a region far from
the transient regions is essentially unaffected by the transients
until the disturbance from stopping propagates into that region.
In all further discussion the events considered occurred before the
influences of events at the tank ends became important.

Axial Flow

Figure 4 shows the dye pattern resulting from moving the airfoil past a fixed cloud of dye formed on a grid of wires. Similar
patterns have been observed in flight tests with fixed smoke generators for flow visualization. The dye is pulled from the cloud
in the direction of airfoil motion contrary to the inviscid calculation of Batchelor.[3] A crude measurement of the peak velocities
may be obtained by measuring positions of the dye tip in successive
photographs. The method lacks precision because the fine tip of
dye fades and is difficult to locate, but certain trends can be
observed. For example, comparing figures 4 and 5 shows that the
span of the wing does not significantly affect the axial velocity.

A plot of successive dye tip positions and the wing motion is
shown in figure 6. From this and similar plots the axial velocity
at a point 60 chords behind the airfoil was computed. The results
were found to scale as shown in figure 7. The scatter at a fixed
Reynolds number is about ± 20% at worst and is probably largely due
to measurement errors. The wing aspect ratio, angle of attack, and
velocity each varied by more than a factor of three over the plot.
Conditions for each numbered data point are listed in Table 1.

Core Size

By assuming that the total vorticity shed by a wing appears in
a concentrated core and that the energy in the wake is largely due
to the circular motions around the core, Spreiter and Sacks[4] have
reached the conclusion that the core radius is

$$R = .0775 \ s \tag{1}$$

for an elliptically loaded wing of span s. As seen in figures 4 and

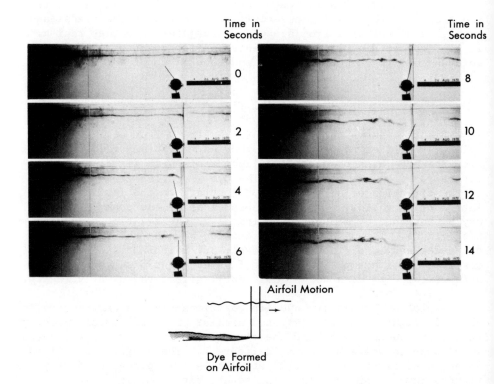

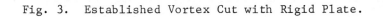

Fig. 3.   Established Vortex Cut with Rigid Plate.

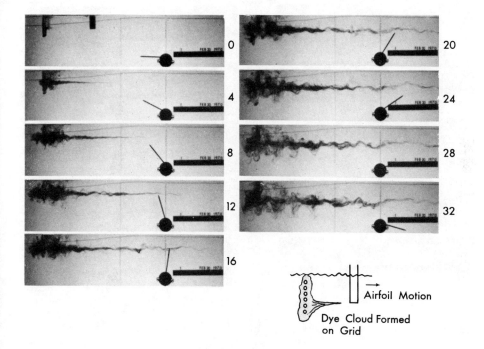

Fig. 4.   Time History of a Vortex.   Conditions: section = NACA
         0012, chord = 1", semi-span = 2", angle of attack =
         7°, speed = 1.8 ft/sec.

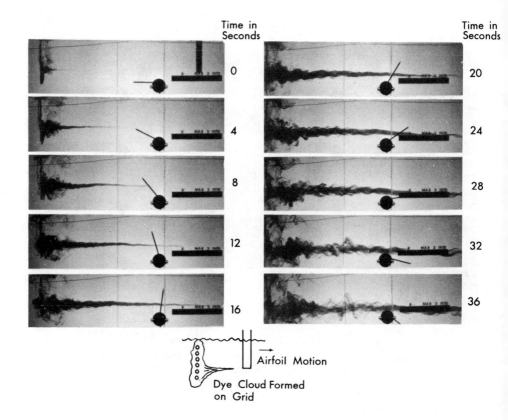

Fig. 5. Time History of a Vortex. Conditions: section = NACA
         0012, chord = 1", semi-span = 5", angle of attack = 7°,
         speed = 1.8 ft/sec.

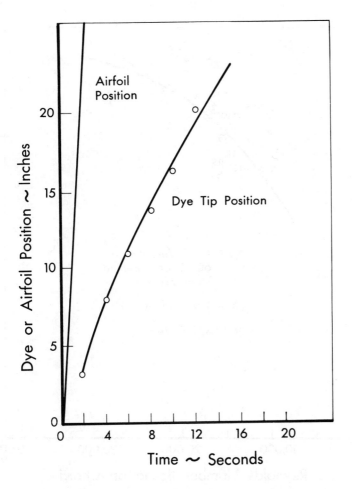

Fig. 6. Motion of Airfoil and Dye Tip.

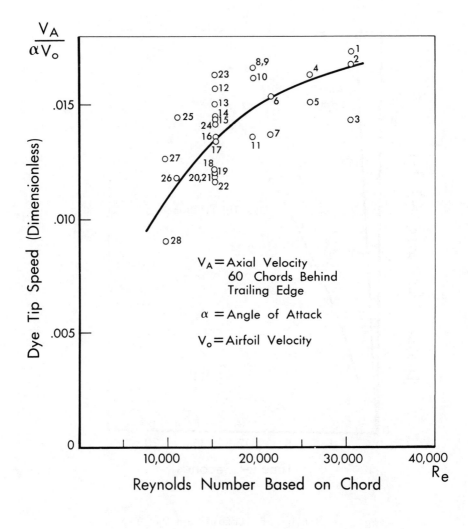

Fig. 7.   Summary of Axial Velocity Measurements.

Table 1. Experimental Parameters for Figure 7

Point Number	Speed ft/sec	Angle of Attack	Depth	Chord
1	1.79	3°	4"	2"
2	1.79	7°	4"	2"
3	1.79	5°	4"	2"
4	1.55	7°	4"	2"
5	1.55	3.4°	4"	2"
6	1.32	4°	4"	2"
7	1.32	7°	4"	2"
8	1.1	5°	4"	2"
9	1.1	7°	4"	2"
10	1.1	7°	4"	2"
11	1.1	4.8°	4"	2"
12	1.79	7°	3"	1"
13	1.79	7°	$3\frac{1}{2}$"	1"
14	1.79	7°	4"	1"
15	1.79	7°	4"	1"
16	1.79	5°	2"	1"
17	1.79	7°	5"	1"
18	1.79	7°	$2\frac{1}{2}$"	1"
19	1.79	7°	3"	1"
20	1.79	7°	2"	1"
21	1.79	7°	2"	1"
22	1.79	7°	1"	1"
23	.88	6°	4"	2"
24	.88	7°	4"	2"
25	.69	7.6°	4"	2"
26	.69	7°	4"	2"
27	1.1	10°	2"	1"
28	1.1	5°	1"	1"

5, the core size as marked by the dye moving axially increases
with span, but the diameter is far less than .31 of the half span.
Whether the core radius as defined by the profile of the circular
motions is larger or smaller than that defined by the dye is not
known.   However, core radii based on the circular motions measured
in flight tests by McCormick, Tangler, and Sherrieb[5] are smaller
than predicted by (1).   In addition, S. C. Crow pointed out in a
discussion that the calculation of Spreiter and Sacks is very sensitive to the assumption that all the shed vorticity appears in a
concentrated core.   If some of the vorticity remains diffusely
spread in the region outside the compact core, the velocity profile
may not be qualitatively very different, but the energy calculation
would require a smaller core.   Since the question is not yet settled
core size in the remainder of this paper will be taken to mean as
defined by the dye column radius.

Axial Flow Instability

A recent work by Bergman[6] has shown that a vortex flow can be
unstable when an axial jet is present in the core.   The eigenfunc-
tions of the unstable modes were not given, but estimates show the
disturbances to be confined to the neighborhood of the core.   The
radial profiles of circulation and axial velocity for the undisturbed basic flow are shown in figure 8.   The profiles were chosen
as being appropriate for dust devils, but are used here as rough
approximations to the, as yet, unknown profiles in the trailing
vortex.

The perturbations considered by Bergman[6] were of the form

$$[v_r, v_\phi, v_z] = [if(r), g(r), h(r)] \, e^{i(m\theta + k\frac{z}{r_1} + \sigma\frac{Wt}{r_1})}$$

The most unstable mode was found to be for $m = 1$.   Figure 9 gives
Bergman's[6] curves of constant growth rates ($Im[\sigma]$) as functions
of wave number, k, and the ratio of vortex to jet strength,
$\mu = \Gamma_\infty/r_1 W$, for $m = 1$.

The value of μ can be estimated for the towing tank experiments
by using the following relations

$$\Gamma = \frac{V_o C C_\ell}{2} \tag{2}$$

$$V_A = 2W = V_o \alpha f(R_e) \qquad \text{from figure 7} \tag{3}$$

$$C_\ell \approx .1\alpha \qquad \text{NACA 0012 section data} \tag{4}$$

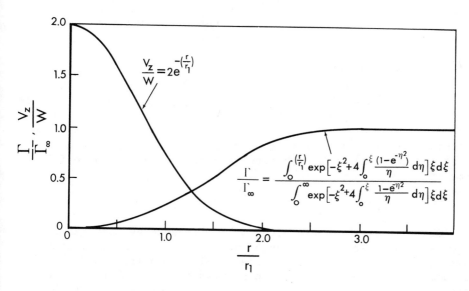

Fig. 8.   Profiles of Initial State in Bergman's
Stability Calculation.

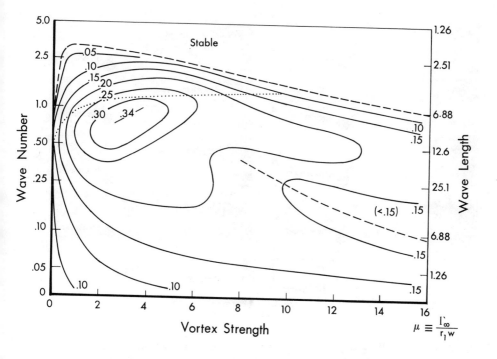

Fig. 9.   Disturbance Growth Rates for m = 1 Mode (after Bergman[6]).

$$r_1 \approx .1S \qquad\qquad \text{from measurements of dye} \qquad (5)$$
$$\text{filled core}$$

The value of μ is then

$$\mu = \frac{1}{A_r\, f(R_e)} \approx 16 \text{ for most experiments} \qquad (6)$$

At this value of μ the variation of growth rate with μ is not great and the maximum growth rate is about .15 giving an e-folding time of

$$\tau = 6.7 \frac{r_1}{W} \qquad (7)$$

Qualitative agreement with Eq. (7) is seen by comparison of figures 4 and 5 and of figures 4 and 10. The instability grows more slowly for large cores or small axial velocities and is fully visible at about 3 or 4τ. Close observations of movies of the large cloud of dye near the generating grid show that disturbances to the flow occur mainly within the core. The instability serves to enlarge the core quickly without destroying the circulation about the core at large radii.

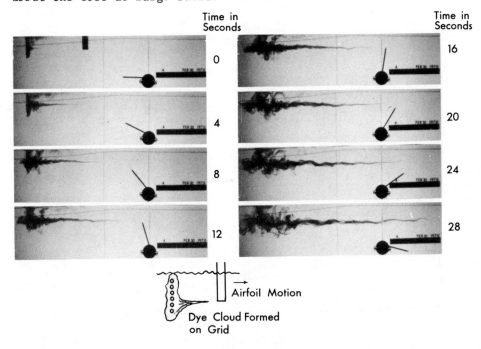

Fig. 10.   Time History of a Vortex.   Conditions: section = NACA 0012, chord = 1", semi-span = 2", angle of attack = 7°, speed = 1.1 ft/sec.

Mutual Interaction Instability

It has been shown by S. C. Crow[7] that a pair of parallel vortices of opposite circulations undergo a symmetric and nearly sinusoidal instability as illustrated in figure 11. The disturbances are on planes about 45° to the horizontal and grow by a factor of e in a time

$$T = 1.21 \frac{2\pi b^2}{\Gamma_o} \tag{8}$$

By using the relation[4] for an elliptically loaded wing that

$$b = \frac{\pi}{4} s, \tag{9}$$

the ratio between growth times for the axial and mutual interaction instabilities can be written as

$$\frac{\tau}{T} = \frac{.0143}{A_r \, f(R_e)} \tag{10}$$

Then, after noting that $f(R_e) \approx .014$ from figure 7 we can write that

$$\frac{\tau}{T} \approx \frac{1}{A_r} \tag{11}$$

in the range of the present experiments. Thus, the axial flow instability occurs first and somewhat masks the mutual interaction instability. Figure 12 shows the dye pattern left by a 1" chord 2" span wing towed horizontally and held by a vertical strut. The first irregularities to appear are very short and near the 12 core radii or 1.2 spans expected from figure 9. These disturbances remain confined to the neighborhood of the core. Then, a larger wavelength disturbance appears and leads to a final dissipation. Due to the camera angle, the closer vortex appears nearly straight, thus roughly confirming the predicted angle. The wavelength of the longer instability as marked in the fourth picture of figure 12 is about 12". The vortex separation for an elliptically loaded wing of 2" span is estimated in Reference 4 to be

$$b = \frac{\pi}{4} s = 1.57" . \tag{12}$$

The wavelength measured is found to be 7.7b compared to the maximum growth rate wavelength of 8.4b predicted by Crow[7].

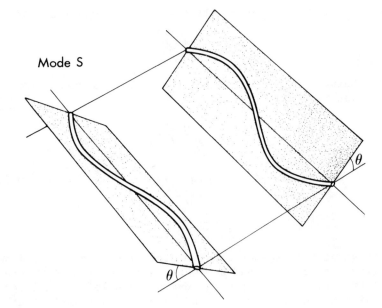

Fig. 11.   Shape of Unstable Mode in Mutual Interaction Instability
           (after Crow[7]).

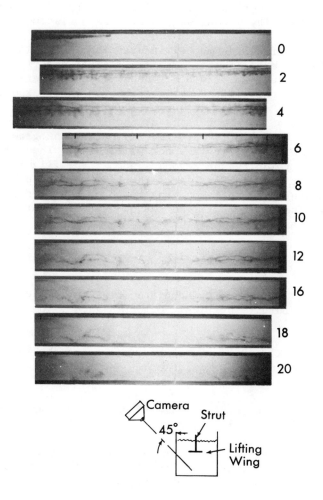

Fig. 12.   Simultaneous Occurrence of Axial Flow and Mutual
Interaction Instabilities.

IV. CONCLUSIONS

In these experiments no vortex was ever observed to disintegrate slowly by a simple diffusive process. The disintegration always was violent and spread the dye marking the core over a large region in the tank. We conclude that the disintegration is by an instability mechanism rather than a viscous decay.

Both axial flow and mutual interaction instabilities were observed in the experiments. The theoretical descriptions for both instabilities had previously been given and were found to be consistent with the limited observations of these experiments.

Questions as to the details of the instability mechanisms remain open. The velocity profiles affecting the growth rate of the axial flow instability are not well known. The relation between axial velocity and Reynolds number for flight Reynolds numbers is not known. Details of the final break-up after the disturbances have grown beyond the range of the linear theory are not known. These items will be the objects of future research on vortex wakes at BSRL.

REFERENCES

1. Baker, D. J., 1966: A Technique for the Precise Measurement of S all Fluid Velocities. *J. Fluid Mech.*, Vol. 26, Part 3, pp. 573-575.

2. Harvey, J. K., 1962: Some Observations of the Vortex Breakdown Phenomenon. *J. Fluid Mech.*, Vol. 14, Part 4, pp. 585-592.

3. Batchelor, G. K., 1964: Axial Flow in Trailing Line Vortices. *J. Fluid Mech.*, Vol. 20, Part 4, pp. 645-658.

4. Spreiter, J. R. and Sacks, A. H., 1951: The Rolling Up of the Trailing Vortex Sheet and Its Effect on the Downwash Behind Wings. *J. of the Aero. Sci.*, Vol. 18, No. 1, pp. 21-32.

5. McCormick, B. W., Tangler, J. L., and Sherrieb, H. E., 1968: Structure of Trailing Vortices. *J. Aircraft*, Vol. 5, No. 3, pp. 260-267.

6. Bergman, K. H., 1969: On the Dynamic Itability of Convective Atmospheric Vortices. Ph.D. Thesis, Department of Atmospheric Sciences, University of Washington.

7. Crow, S. C., 1970: Stability Theory for a Pair of Trailing Vortices. Boeing document D1-82-0918; AIAA Aerospace Sciences Paper 70-53.

RESULTS OF THE BOEING COMPANY

WAKE TURBULENCE TEST PROGRAM

P. M. Condit and P. W. Tracy

The Boeing Company

Commercial Airplane Division

ABSTRACT

The development of large jet transport airplanes led to concern that the turbulent wakes generated by these aircraft would create a hazard for other air traffic. The Boeing Company initiated a study of large jet airplane wake turbulence in mid 1969. The Boeing flight test program was undertaken to evaluate the behavior of trailing vortices and to obtain a direct comparison between the turbulent wakes of the 747 and a 707-320C. A fully instrumented Boeing 737-100 was used as the primary wake probing aircraft. Additional probes were made with the Boeing owned F-86 and NASA's CV990. The 737 was also flown in the turbulent wake of the 747 on approach to landing in order to assess the effects of wake turbulence near the ground. These tests produced several significant conclusions. It was found that the dynamic response of the chase airplanes (737, CV990, F-86) was essentially the same when flying in the turbulent wakes of either the 747 or 707. Thus the wakes of large jet airplanes (747 and C-5A) do not represent a significant new hazard for other air traffic. The turbulent wake was expected to descend behind the generating airplane at a constant rate until fully dissipated. It was observed, however, that the wake leveled off and never descended more than 900 feet below the generating airplane. This indicates that in level flight, airplanes should be separated vertically by approximately 1000 feet in order to avoid wake turbulence. In the approach and landing tests it was found that the wake generated in ground effect does not "roll up" into strong trailing vortices. Consequently, wake turbulence is relatively weak near the runway in the landing flare and takeoff rotation areas. Recommendations are presented for air traffic regulations which will virtually eliminate hazardous wake turbulence encounters.

1. INTRODUCTION

Vortex wake turbulence has been a subject of interest for many years with the theory dating back to the beginnings of powered flight (Reference 5). Significant experimental work on the wake was performed in the early fifties (Reference 4). The hazards of wake turbulence were recognized and published for light plane operators as early as 1952 (Reference 6). However, theoretical and experimental work did not keep pace with the rapid growth of aviation, particularly the development of large jet transports. A review of available literature in 1969 indicated that there were only a few experimental studies available for jet transport aircraft (Reference 7). Any examination of the vortex wake of large airplanes depended heavily upon extrapolation of data from relatively small airplanes. Decay of the wake was then predicted using viscous dissipation theories, which resulted in extreme vortex durations for the large airplanes.

The Boeing Company study, begun in mid 1969, led to a significant conclusion; there was strong evidence, both experimentally (References 3, 4 and 8) and theoretically (Reference 9), that the vortex wake would break up through other than purely viscous means. Brief flight tests behind the 747 indicated qualitative support for this position. Following the issuance of special separation criteria for the 747 and C5A, Boeing defined a test program which would provide data directly applicable to this problem. On February 12, 1970, the FAA instituted a three part test which included the Boeing test program.

The Boeing tests were aimed at providing a direct comparison between the 747 and a representative of the current jet fleet, the 707-320C. The encountering aircraft was the smallest Boeing jet transport, the 737-100. The NASA-Ames Research Center CV990 was also used to provide a second comparison behind the 747 and 707. Additional testing included takeoff and landing conditions with the 737 following the 747. Boeing also participated in the tower measurement portion of the FAA program by flying the 747 and 707 by the instrumented tower at Arco, Idaho.

2. TEST PROGRAM

The Boeing Wake Turbulence Test Program was aimed at a direct comparison between the 747 and a representative from the current jet fleet. A 707-320C was chosen for this role in the test program. The effects of a wake encounter were measured with Boeing 737-100 and NASA CV990 jet transport aircraft. A company owned F-86 was also used in order to obtain at least a subjective check on the effects of encountering airplane span.

Special Equipment

Based on earlier testing, it was found desirable to visualize the vortex wake. Two systems were designed for this purpose; wing tip mounted smoke grenades and oil introduced into the primary nozzle of the outboard engines. The wing tip smoke system used 12 grenades, wired to fire in salvos of 3. Flight tests indicated that all 12 grenades were required to provide sufficient smoke volume and the system was discarded as too inefficient. The engine oil system injected Corvus oil into the nozzles of the outboard engines. The smoke flow from the outboard engines was entrained into the trailing vortices. This system worked very well and was used for all tests.

Both of the Boeing chase aircraft (737 and F-86) were equipped with bore-sighted Milikan movie cameras. These cameras were calibrated so that range information could be obtained by measuring the span (or engine span) of the lead aircraft. For the landing tests, a forward looking 35 mm APAC camera was installed in the nose radome of the 737.

Conduct of the Testing

The comparative testing behind the 747 and 707 was conducted with the two lead aircraft flying formation approximately 3000 feet apart. This assured similar flight and atmospheric conditions. The chase aircraft then encountered the wakes at the same separation, making a direct comparison. Two flight conditions were used; clean (flaps and landing gear up) at 250 knots and approach flaps down at approximately 160 knots.

Takeoff and landing tests were conducted with the 737 following the 747 at spacings from 1.7 to 3.1 nautical miles. These tests were conducted to evaluate wake turbulence near the runway. An intentional encounter was set up by flying below and to the left of the glide slope on one approach. A runway crossing condition was flown with the 747 rotating prior to the runway intersection and the 737 lifting off through the intersection 68 seconds later.

Data Systems

The 737 was equipped with a standard set of flight test instrumentation. From this instrumentation, the following data was selected for display as a function of time:

>   Flight Condition
>   Indicated Airspeed
>   True Airspeed

Flight Conditions (continued)
    Altitude
    Ambient Temperature

Airflow
    Angle of Attack Vane
    Sideslip Pressure

Response
    Pitch Angle
    Bank Angle
    Yaw Angle
    Lateral Acceleration
    Normal Acceleration (cg and pilot's station)

Control Positions
    Elevator Angle
    Spoiler Angles
    Aileron Angle
    Control Wheel Position

Detailed data on the flight conditions of the 747 and 737 were also available in printed format.

Range information was the least accurate data and three independent sources were used. A visual, hand-held sight was used by the copilot in the chase aircraft to get approximate range. This sight was scaled for the outboard engines of the 747 and checked well with data from other sources. The engine smoke could be started sharply and by measuring the time from the start of smoke until it reached the chase aircraft, the range could be computed. Ground radar ranging, provided by the FAA, was used as primary range information for the 747/707 comparison with the 737 as chase plane. This information was much less accurate when the CV990 was used since the tests were conducted in Eastern Washington with reduced radar coverage.

Range information for the landing tests was obtained on the ground by timing the separation between the 747 and 737 over the runway threshold. For these tests, smoke was provided on the ground as well as from both outboard engines of the 747. Smoke grenades on poles were positioned at the threshold and at 500 and 1000 feet down the runway on both sides. Camera coverage was provided from under the approach path looking down the runway and from the side.

3.   TEST RESULTS

The test program results can be divided into four areas. The prime purpose of the testing was to provide data on the comparative strength of the 747 wake relative to a baseline, the 707. In addition, the testing provided valuable insight into the dynamics of a wake encounter and what a pilot may expect. The final two items were the position of the wake relative to the generating aircraft and the ground, and the duration of the wake. Each of these areas will be covered here, with emphasis on those that bear directly upon air traffic control separations.

747 and 707 Comparative Wake Strength

In order to obtain a direct comparison of the wake turbulence generated by the 747 and 707, the three chase aircraft (737, F-86, CV990) encountered one wake and then the other at the same separation. Figure 1 shows a typical encounter with the 737 about 3 miles behind the 747 and 707. There is little difference between these two encounters. This is also apparent in the peak values of acceleration shown in Figure 1.

Subjective comments by the pilots of all three chase aircraft agree with this analysis. The following comments are typical:

> ". . . there was not a gross difference in the characteristics of strength between the 707 and 747." - Jim Gannet, Boeing (737 Pilot)

> "I was impressed by the very little, if any, difference between the 707 wake and the 747 wake." - Paul Roitsch, Pan American Airlines (737 Copilot)

> "I couldn't really tell the difference." - John Armstrong, Boeing (F-86)

> "Moving over to the 707, again in the approach configuration, it seemed almost the same." - Fred Drinkwater, NASA (CV990)

> "They did all seem to be in the same ball park (C5A, 747, 707) and it doesn't look to me like there is any new big problem associated with the really heavy airplanes compared to the others." - Glenn Stinnett, NASA (CV990)

> ". . . I have to agree that there doesn't seem to be very much difference between the 747 and a 707." - Joe Tymczyszyn, FAA (CV990)

In order to examine both the effects of encountering aircraft span and generating aircraft weight, data from the NASA testing

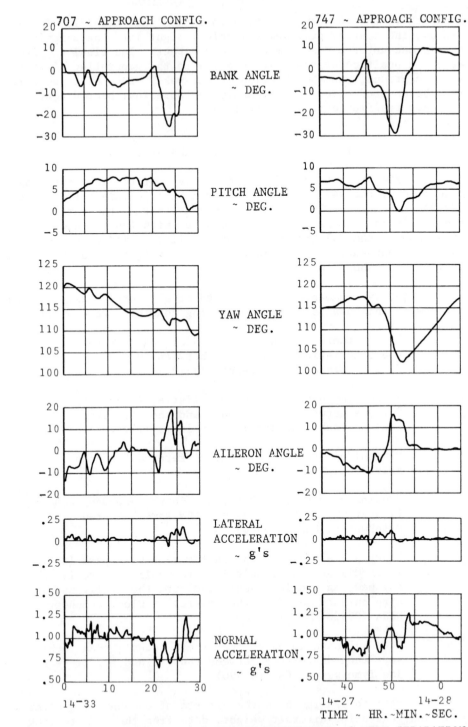

FIGURE 1. TYPICAL 737 WAKE TURBULENCE ENCOUNTERS, 3 MILE SEPARATION.

(Reference 1) was combined with the Boeing test data. Figure 2a
shows the very strong effect of span on peak induced roll rate. It
should be noted that these roll rates were measured with the pilot
attempting to remain in the vortex and maintain wings level. The
relatively weak effect of generating aircraft gross weight is shown
in Figure 2b. Although roll rates are shown here, there is a direct
correlation between roll rate and peak bank angle. As shown in
Figure 3, the peak bank angle is typically less than 2 times the
peak roll rate.

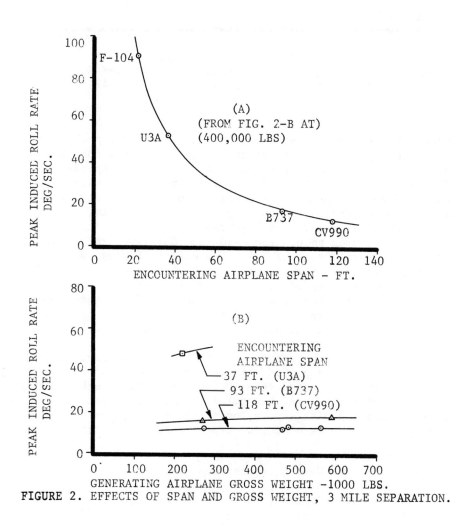

FIGURE 2. EFFECTS OF SPAN AND GROSS WEIGHT, 3 MILE SEPARATION.

It is worthwhile to compare the above results to the results
obtained by calculation. Using the empirical relationships

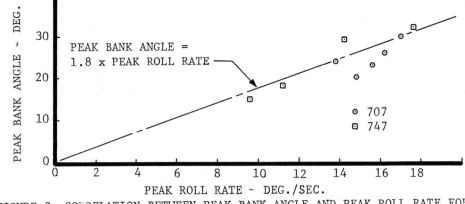

FIGURE 3. CORRELATION BETWEEN PEAK BANK ANGLE AND PEAK ROLL RATE FOR THE 737 FLYING 3 MILES BEHIND THE 707 AND 747.

developed by McCormick (Reference 10), an equation for the induced roll rate of an encountering aircraft may be derived for the special case of an airplane centered in the vortex. (Appendix B). While this is not a realistic condition, it does indicate relative effects. Figure 4a presents the calculated span effect assuming that the encountering aircraft is geometrically similar to the 737. The character of the curve is essentially the same as that observed experimentally, even though the F-104 and U3A have dramatically different planforms from the 737. The effect of gross weight was calculated using a constant wing loading, Figure 4b. A pronounced gross weight effect appears only at very low weights. In both of the above cases, the theory predicts greater effects than those observed experimentally. This is probably the result of encounter dynamics which are not included in the theoretical model.

Test results on the effect of separation were very surprising. Contrary to what was expected, little change in response was found over the separation range tested (Figure 5a) until breakup terminated the wake. Again using McCormick's empirical relationships, a predicted variation with separation is shown in Figure 5b. The theory appears to agree well with experimental observations.

Wake Encounter Dynamics

During the test program, the wake was approached from all directions in order to evaluate the effect of encounter direction on response. One item was common to all encounters – without considerable effort by the pilot the airplane would be quickly expelled from the wake. Figure 6 illustrates several encounters to indicate the type of response experienced. In no case did the pilots feel

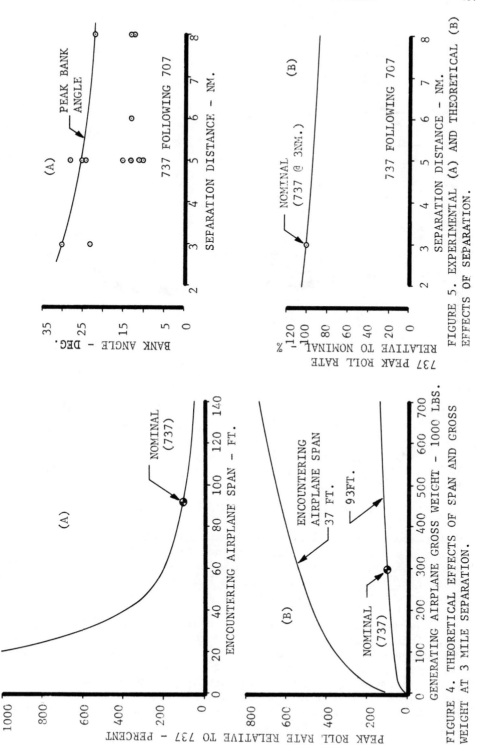

FIGURE 5. EXPERIMENTAL (A) AND THEORETICAL (B) EFFECTS OF SEPARATION.

FIGURE 4. THEORETICAL EFFECTS OF SPAN AND GROSS WEIGHT AT 3 MILE SEPARATION.

FROM ABOVE (RIGHT)   FROM ABOVE (CENTER)

FROM SIDE   **FROM SIDE (RAPID)**

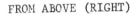

FROM BELOW (RAPID)

FROM BELOW (RIGHT)

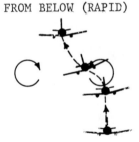

FIGURE 6. DYNAMICS OF ENCOUNTERS EVALUATED BY THE 737.

that there was a possibility of losing control. Airplane response data and pilot comments indicated that there were no structural implications over the range of separations tested (1.7 to 9 nautical miles). The accelerations shown in Figure 1 are well below those experienced in severe turbulence.

The most critical encounter is near the ground where large excursions cannot be tolerated. Figure 7 shows the 737 encountering the wake on approach to landing 1.8 nautical miles behind the 747. It was necessary for the pilot to deviate to the left and below the normal flight path to encounter the wake. As shown, the encounter occurs with very little loss in altitude and with a maximum bank angle of 28 degrees. The encounter, while dramatic, was conducted safely at 60% of the recommended range.

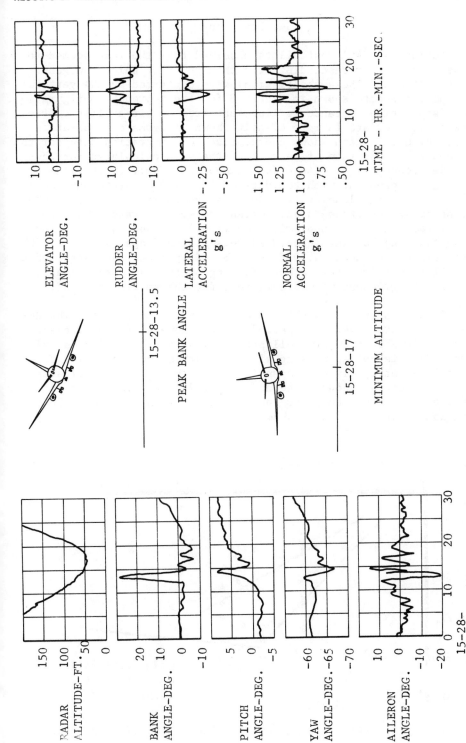

FIGURE 7. 737 WAKE TURBULENCE ENCOUNTER ON LANDING APPROACH 1.8 NM. BEHIND A 747.

Wake Position

Boeing and NASA test pilots observed that the wake seemed to "level off", never descending more than 900 feet below the generating aircraft. These observations were confirmed by the vertical wake position data shown in Figure 8 for the approach configuration. While separation distance is important from an air traffic control viewpoint, the data is more clearly shown in terms of separation time. Figure 9 again shows the approach configuration data for the 747 and 707. The rate of descent of the wake calculated for the 747 (Reference 11) is 8.44 ft/sec (about 500 fpm). This value shows good agreement with the data of Figure 9 prior to wake level off. An elliptic loading is often assumed to simplify the calculation of vortex strength and descent rate. The rate of descent calculated using this approximation is 6.31 ft/sec, which is low since the wing loading inboard increases with flaps down. The 707 wake descends at about 5.33 ft/sec, based upon the test data, which is 3.11 ft/sec less than a best fairing of the 747 data. Again, this is in agreement with predictions for the smaller aircraft.

Clean configuration data shows much the same effect. The descent rates are lower as expected for the lower lift coefficient. As shown in Figure 10, the elliptic loading approximation is fairly good. This is not surprising since the cruise configuration results in a nearly elliptic span loading. The 707 data indicates wake level off at about 60 seconds. The 747 wake levels off at near the same point (60 sec.) for the 34-1 test, with a slightly longer duration observed on test 34-2.

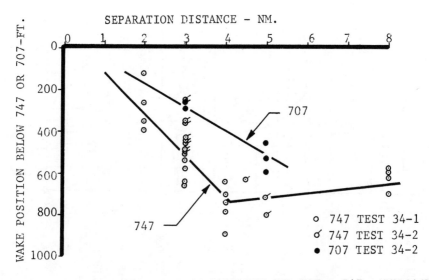

FIGURE 8. VERTICAL WAKE POSITION FOR 707 & 747, APPROACH.

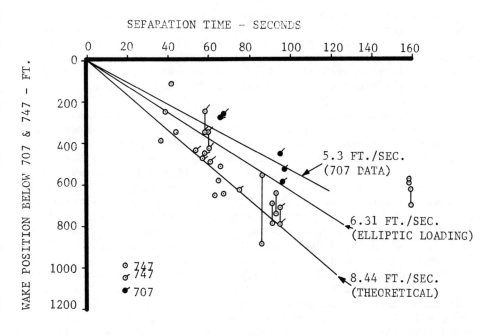

FIGURE 9. VERTICAL WAKE POSITION FOR 707 AND 747, APPROACH.

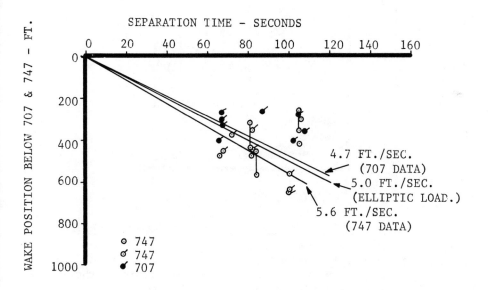

FIGURE 10. VERTICAL WAKE POSITION FOR 707 AND 747, CRUISE.

At separations greater than 5 miles the wakes were observed to oscillate vertically and laterally. Vertical oscillations were estimated at ±200 feet about a mean level altitude. The scatter in the data obscures any quantitative measure of the amplitude. These oscillations and level off of the wake appear to be directly related to wake breakup, as discussed in the next section.

The lateral position of the vortex refers to the displacement of the vortex wake laterally from the flight path of the generating aircraft due to crosswinds. Near the ground, lateral displacement is limited to dissipation of the vortices. Atmospheric turbulence directly affects the time required for vortices to dissipate. At altitudes up to approximately 5000 feet, wind speed is a reliable indicator of atmospheric turbulence level. At higher altitudes this may not be true. Data showing the relationship between vortex life and wind speed has been plotted in Figure 11. The curve from W. A. McGowan (Reference 2) is based on limited flight test data and represents a conservative upper bound for vortex life. Two observations on 747 vortex life in test 34-4 are shown in the figure and indicate the conservative nature of McGowan's curve. This curve was used to determine the maximum possible lateral displacement of the vortex wake by crosswinds, Figure 12. This figure shows that for approach conditions, the effect of wind on vortex life limits the maximum possible lateral displacement to about 0.35 nautical miles.

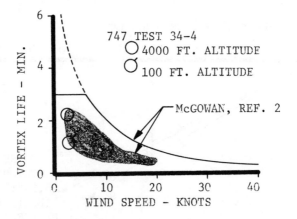

FIGURE 11.   EFFECT OF WIND SPEED ON VORTEX LIFE.

At cruise altitudes, atmospheric turbulence still affects vortex life but wind speed may not be a reliable indicator of the turbulence. Making no assumptions on vortex life, the lateral position of the wake due to crosswind is shown in Figure 13. It is seen that the high cruise speed of the generating aircraft serves to limit the lateral displacement of the wake.

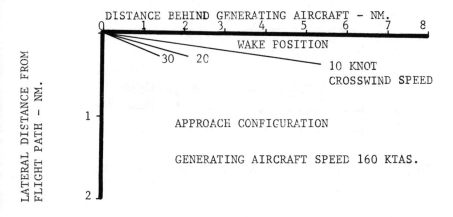

FIGURE 12. LATERAL DISPLACEMENT OF VORTEX DUE TO CROSSWIND.

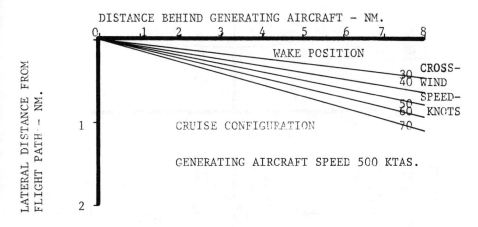

FIGURE 13. LATERAL DISPLACEMENT OF VORTEX DUE TO CROSSWIND.

Wake position near the ground was observed during test 34-4 in which the 737 flew approach and landings behind the 747. Observers were stationed at the approach end of the runway as well as in the 737. The wake of the 747 was observed descending below the approach path to an altitude of approximately 1/4 to 1/2 span (50 feet) as illustrated in Figure 14. At this point, the downward descent of the wake ceased and lateral movement of the vortices occurred. With no crosswind, the vortices moved laterally apart but with a 1 to 3.5 knot crosswind the upwind vortex was held near the ground track of the 747. (Maximum ground wind speed during the test was 3.5 knots.) This behavior is characteristic of vortices in ground effect as illustrated in Figure 15.

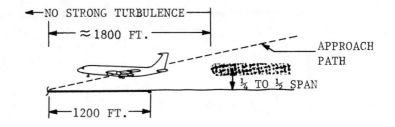

FIGURE 14.   WAKE POSITION RELATIVE TO THE APPROACH PATH.

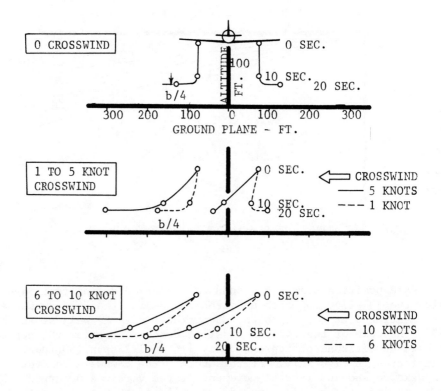

FIGURE 15.   TYPICAL EFFECT OF CROSSWIND ON LATERAL MOVEMENT OF
VORTICES NEAR THE GROUND.

One feature of the wake behavior in ground effect was not anticipated. The wake generated in ground effect (flying less than
1/4 to 1/2 span above the ground) did not form strong trailing
vortices but produced a strong outward flow. The turbulence generated in ground effect was found to be relatively weak and was
described as "light chop" by the 737 pilots. The data shows that
below 50 feet altitude the 737 never experienced more than 2° of
roll and 0.15g center of gravity normal accelerations in landings
made as close as 1.7 nautical miles behind the 747. The implication of this is that following aircraft will not encounter hazardous turbulence at liftoff or landing flare.

Wake Breakup

Most studies of vortex wake turbulence have assumed that the
wake decayed by purely viscous means (References 10, 12 and 13).
This assumption results in the conclusion that the wake from a
large aircraft will persist for an extremely long time. During
the Boeing study program, strong evidence was found which indicated
that the vortex wake would break up through other than purely viscous mechanisms. This data, both theoretical (Reference 9) and
experimental (References 3, 4 and 8), described an instability of
the vortex pair. Data gathered during the Wake Turbulence Program
further supports this position.

Figure 16 shows a time sequence of the 747 wake as defined by
the engine smoke (Reference 14). The 747 was in a takeoff configuration (flaps 10, gear up, 162 knots) for the overhead pass. As
shown, an instability develops which results in the start of breakup at 100 seconds (from a timed movie) and complete wake dissipation
at 130 seconds. Confirmation of the linear nature of the instability is shown in Figure 17 with the characteristic logarithmic
increase in amplitude (Reference 14). Data taken from this pass
correlates well with the theory presented in Reference 9. It is
interesting to note that the theory predicts a more rapid breakup
than was observed. It may be possible to excite this instability
artificially and thus hasten wake breakup. Excellent correlation
is also noted with the time observed (Figure 8) for the wake level
off. This is not surprising since the descent of the wake is due
to the interaction between the two vortices of a complete vortex
pair. Initial breakup appears to result in the destruction of
parts of the vortex pair, thus eliminating the mechanism causing
descent and resulting in wake level off.

Both the empirical and analytical (References 3 and 9) treatment of the instability conclude that the time to breakup should be
directly proportional to the square of the vortex spacing (b_v) and

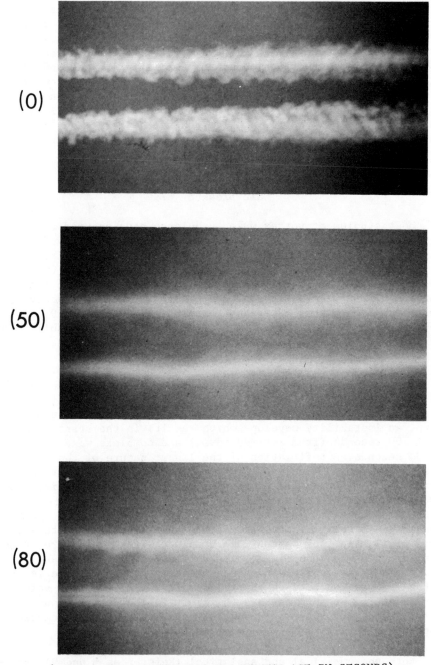

(NUMBERS IN PARENTHESIS ARE VORTEX AGE IN SECONDS)

FIGURE 16. PHOTOGRAPHIC SEQUENCE OF WAKE BREAKUP FOR A 747 IN A TAKEOFF CONFIGURATION.

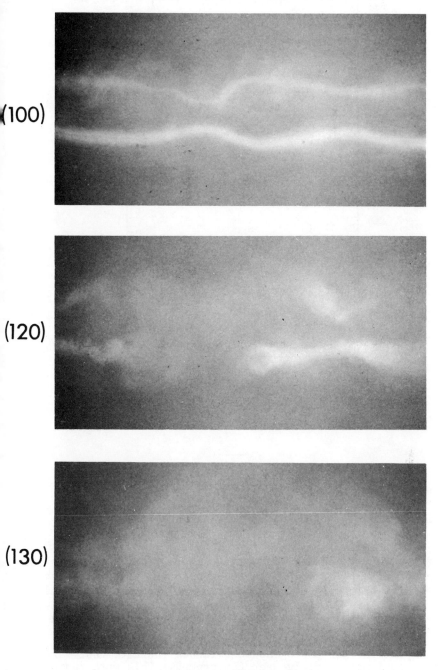

(100)

(120)

(130)

FIGURE 16. (CONTINUED)

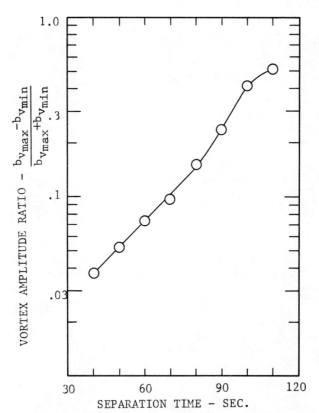

FIGURE 17. AMPLITUDE HISTORY OF THE VORTEX INSTABILITY.

inversely proportional to the circulation (Γ_∞)

$$t^* \sim \frac{b_v^2}{\Gamma_\infty} \tag{1}$$

This can be simplified to

$$t^* = K_t \frac{\rho V_\infty b^3}{W} \tag{2}$$

Data from several sources is shown in Figure 18. This data shows good agreement with $K_t = 14$. There is also an indication that K_t is a function of airplane configuration and atmospheric turbulence. Further testing is required to accurately define these effects. The relationship for t^* is particularly important when predicting wake turbulence for new configurations such as the SST.

Observations and photographic evidence indicate that wake breakup occurs more rapidly near the ground. This more rapid break-up may be the result of ground induced turbulence or the direct effect of the uneven ground plane. Future study could lead to methods for producing an even more rapid decay.

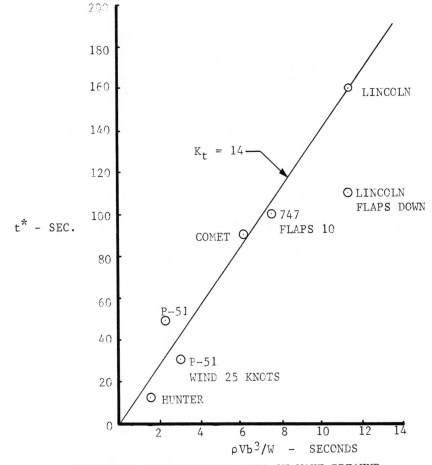

FIGURE 18. EXPERIMENTAL DATA ON WAKE BREAKUP.

4. APPLICATION OF TEST RESULTS TO AIR TRAFFIC CONTROL

The significance of the wake turbulence test program is in the application of the results to the formulation of logical air traffic control regulations. The primary consideration must be safety of flight with a secondary consideration of passenger comfort. To meet these goals, the possibility of a wake encounter must be minimized in all cases and virtually eliminated for cases where an encounter would be dangerous. The proper application of air traffic separations and procedures to appropriate categories of aircraft will achieve this result.

Aircraft Categories

Two factors which strongly influence the categorization of aircraft are flight test results and past experience. Figure 19 shows the points at which flight test data is available and indicates a measure of the effect in terms of induced roll rate. The experience factor is. shown by the shaded areas where wake turbulence is suspected as a contributing accident cause. A "critical area" is defined by the curve. The significant parameters in defining this critical area, by order of importance, are:

1. Span of the encountering airplane - b

2. Control available on the encountering airplane - P_a

3. Span loading of the generating airplane - W/b

Thus a "degree of hazard" can be approximated by:

$$(\frac{W/B}{\rho \, V_\infty})_{\text{Generating A/P}} \times (\frac{1}{b^2 \, P_a})_{\text{Encountering A/P}} \tag{3}$$

where a value of 1 represents control capability equal to the peak induced roll rate.

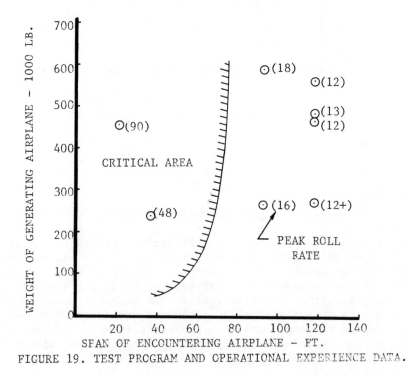

FIGURE 19. TEST PROGRAM AND OPERATIONAL EXPERIENCE DATA.

While a degree of hazard can be calculated for any aircraft
pair, this is not a reasonable procedure for current air traffic
operations. Span of the encountering airplane, while the most important parameter, may not be a convenient dimension for air traffic control. Figure 20 shows the major categories of aircraft in
terms of span and gross weight. A convenient category break is evident at a maximum takeoff weight of 75,000. This division at
75,000 pounds is proposed as a logical break satisfying air traffic
requirements, test program data, and operational experience.

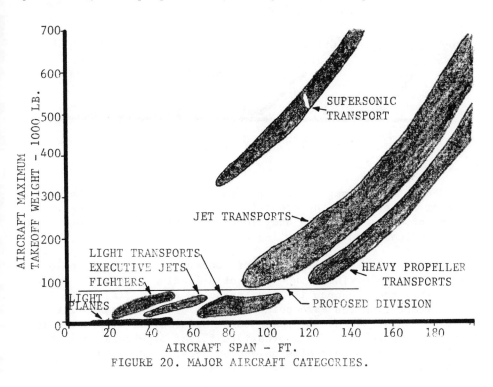

FIGURE 20. MAJOR AIRCRAFT CATEGORIES.

Flight Procedures

The possibility of a wake turbulence encounter can be dramatically reduced by the application of certain flight procedures. For
the approach and landing phase, the area of a potential wake turbulence encounter is shown in Figure 14. There is little chance of
encounter if the following aircraft flies on or above the flight
path of the leading aircraft. An encounter is possible if the following airplane is low in the runway threshold. It should be noted
that if such an encounter occurs it will be at a height of about
1/4 to 1/2 the span of the leading aircraft. As shown in the 737
landing encounter, this is sufficient altitude for recovery if the
excursions are not excessive.

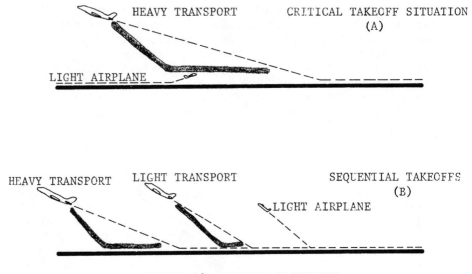

FIGURE 21. TAKEOFF PROCEDURE.

Near the ground, the lateral position and life of the wake are complex functions of the crosswind velocity. The only reasonable flight procedure would be additional caution for cases where the crosswind component is from 1-5 knots.

The only takeoff condition of concern is one which would place an aircraft under the flight path of a leading airplane. One case where this could happen is illustrated in Figure 21a. Takeoffs which start from the same point will minimize the possibility of a wake encounter because of the natural ordering of airplane sizes as shown in Figure 21b. An appropriate flight procedure would be to limit intersection takeoffs unless at least three minutes has elapsed since rotation of the lead aircraft.

Out of the terminal area, the general rule of staying on, above, or at least 1000 feet below the flight path of a leading airplane applies. Encounters away from the ground, while not pleasant, should not present a control or structural hazard at normal separations.

Separation Standards

The test program demonstrated that a small jet transport (the 737 with 93 ft. span) will not experience uncontrollable upsets in wake turbulence encounters behind large jet transports. This testing covered separations from 1.7 to 9 nautical miles. Operational experience shows that aircraft with spans greater than approximately 75 feet do not have a history of wake turbulence upsets. Based on

these facts, it may be concluded that aircraft with spans greater
than approximately 75 feet can safely operate at the standard 3
mile separation behind large jet transports. As seen in Figure 20,
the 75,000 pound division proposed in section 4, excludes only a
few of the aircraft with spans greater than 75 feet (light propeller
transports like the Convair 580). Thus, it is not necessary to increase air traffic separations beyond 3 miles for aircraft with
maximum takeoff weight greater than 75,000 pounds.

Additional separation does appear to be warranted for short
span aircraft landing behind transports. Vortex duration near the
ground appears to be on the order of 1.0 to 1.5 minutes. Even away
from the ground with takeoff flaps, 2.5 minutes is the upper bound
for the vortex life. At a final approach speed of 120 knots, a
conservative separation of 5 miles is indicated.

5. CONCLUSIONS AND RECOMMENDATIONS

The Boeing Wake Turbulence Test Program has provided significant data on the strength, position, and duration of the vortex
wake from large jet transports. The following conclusions and recommendations are made as a result of this test program and analysis
of the data.

Conclusions

The conclusions presented here are based directly upon data
gathered as part of the test program. Data from the NASA tests
(Reference 1) was used to better understand the significant results
noted. The most important of the conclusions are noted here with
other conclusions contained in the main text of the report.

1. The 747 and 707 produce a similar dynamic response in
    an airplane encountering their respective wakes.

2. Induced roll rate (and bank angle) is a strong function
    of the encountering airplane's span and a relatively
    weak function of the generating airplane's gross weight.

3. A wake encounter results in the airplane leaving the
    wake immediately. No structural or control impli-
    cations were indicated for a small jet transport
    encountering the wake of a large transport.

4. The wake was observed to move down initially and then
    level off. The wake was never encountered at the same
    flight level as the generating airplane or more than
    900 feet below the generating airplane.

Conclusions (Continued)

5. Data indicated that the wake generated in ground effect
    (1/4 to 1/2 span above the ground) does not form hazar-
    dous, tight vortices. Thus, wake turbulence is relatively
    weak near the runway in the landing flare or takeoff ro-
    tation areas.

6. Theoretical predictions gave good agreement with flight
    test data. The following sources were used for pre-
    diction:

        Wake strength – McCormick (Reference 10)
        Wake breakup – Crow (Reference 9)
        Wake position (vertical) – Spreiter and Sacks
                                        (Reference 11)
        Wake position (near ground) – Dee and Nicholas
                                        (Reference 3)

Recommendations

It is the purpose of this section to present recommendations
which will add to the safety of flight where wake turbulence is a
factor. Information from the FAA Air Traffic Service was included
in order to provide recommendations which were operationally work-
able as well as being technically correct.

Application of test data to specific runway configurations
has not been attempted in this report. As a result, recommendations for parallel or crossing runways are made only generally.
These could be extended to specific cases using data currently
available.

The following recommendations are considered to be the most
significant. It is recommended that the FAA:

1. Define two categories of airplanes with a division at
    75,000 pounds maximum takeoff weight. It should be
    noted that a division based on the span of the follow-
    ing airplane, while technically more correct, did not
    appear workable.

2. Use the standard 3 mile separation within a category
    or for heavy transports following light aircraft. Im-
    pose the following wake turbulence separation criteria
    for light airplanes following heavy transports:

        Landing – A. A minimum 5 mile separation at the
                      runway threshold.

Landing (Continued)

      B. No special separation if the light airplane is on the upwind of parallel runways spaced 3500 feet or less.

Takeoff - A. A minimum 5 mile or 2 minutes separation.

      B. Intersection takeoffs only after 3 minutes.

      C. Crossing runway takeoffs if the leading airplane has not rotated at the intersection.

3. Inform general aviation and transport pilots of the test program results. Show them graphically (films and/or simulator experience) the results of a wake encounter and procedures for avoiding the wake of another airplane.

4. Support further studies which could lead to methods, either airborne or on the ground, which would hasten wake breakup in the threshold area.

SYMBOLS

b	-	wing span, ft.
C_L	-	airplane lift coefficients
$\bar{c}$	-	mean aerodynamic chord, ft.
$c_{\ell\alpha}$	-	slope of the wing section lift curve
$\{cc_\ell\}_{Root}$	-	wing section loading at the wing root, ft.
q	-	dynamic pressure, lbs./ft.^{2}
S	-	wing area, ft.^{2}
V_∞	-	freestream velocity, ft./sec.
W	-	airplane gross weight, lbs.
α	-	angle of attack, radians
Γ	-	vortex circulation at radius r, ft.^{2}/ sec.
ρ	-	atmospheric density, slugs/ft^3

TIME ~HRS-MIN-SEC	747 ALTITUDE ① FT.	737 ALTITUDE ② FT.	① − ② FT.	747 TRUE AIRSPEED (CORRECTED) ~KNOTS	747 FLAPS-GEAR	737 TRUE AIRSPEED ~KNOTS	737 g's NORMAL @ C.G.	737 g's LATERAL C.G.	737 MAX. WHEEL ANGLE-DEG.	737 MAX. BANK ANGLE ~DEG.	SEPARATION ~N.M.	SEPARATION TIME ~MIN.-SEC.	747 GROSS WEIGHT ~1000 LB
14-22-01	7032	6450	582	178.3	20-DOWN	167	+.15 / −.1	.25	30	35	3	1-05	572.2
14-27-43	7012	6435	517	175.3	20-DOWN	163	+.28 / −.20	.05	41	19	3	1-06	558.5
14-28-47	7002	6346	656	173.6	20-DOWN	176	+.32 / −.35	.1	42	12	3	1-03	558.0
14-29-10	7012	6366	646	173.6	20-DOWN	160	+.32 / −.54	.13	82	11	3	1-07	557.8
14-29-30													
14-45-10	7006	6310	696	203.5	20-DOWN	157	+.25 / −.42	.11	82	13	4	1-31	548.8
14-45-43	7015	6227	788	204.4									
14-46-36	6997	6356	641	199.0	20-DOWN	155	+.26 / −.4	.2	77	17	4	1-33	548.0
14-47-15	7027	6286	741	201.8									
14-55-35	6998	6445	553	192.1	20-DOWN	168	+.39 / −.41	.1	60	11	4	1-26	543.0
14-56-15	7009	6120	889	190.0									
14-58-00	7007	6385	622	198.1	20-DOWN	173	+.16 / −.21	.09	62	18	8	2-39	541.7
14-58-27	6997	6296	701	196.5									
14-59-26	6988	6415	573	191.8	20-DOWN	172	+.2 / −.22	.07	61	9	8	2-38	540.9
14-59-55	7009	6415	594	190.2									
15-06-15	6990	6564	359	251.2	0-UP	254	+.30 / −.18	.30	65	30	7¼	1-45	537.7
15-06-44	6981	6674	270	250.1	0-UP	246	+.37 / −.26	.20	28	9	7¼	1-46	537.5

APPENDIX A. TABLE A-1. FLIGHT TEST DATA FROM TEST 34-1, 737 FOLLOWING 747.

TIME ~HRS-MIN-SEC	ALTITUDE 747 ~FT	ALTITUDE 737 ~FT	① - ② (CORRECTED)	747 TRUE AIRSPEED ~ KNOTS (CORRECTED)	747 FLAPS-GEAR	737 TRUE AIRSPEED ~ KNOTS	737 NORMAL g AT ~ KNOTS	737 C.G. LATERAL g	737 MAX. WHEEL ANGLE ~DEG.	737 MAX. BANK ANGLE ~DEG.	SEPARATION ~ N.MI.	SEPARATION TIME ~ MIN.-SEC.	747 GROSS WEIGHT ~1000 LB.
15-11-26	7013	6445	531	216.5	0-UP	258	+.36 / -.26	.28	42	28	6	1-24	535.9
15-12-00	6983	6525	421	212.7									
15-44-20	3145	2828	317	162.7	10-UP	162	+.25 / -.25	.17	52	16	3	1-07	524.3
15-44-35	3137	2588	545	158.6									
15-44-54	3190	2712	418	164.2	10-UP	156	+.15 / -.23	.11	77	10	3	1-09	524.0
15-45-05	3207	2837	370	170.5									
15-48-39	3098	2703	395	171.8	10-UP	196	+.35 / -.41	.17	32	17	2	0-37	522.4
15-49-00	2991	2739	252	169.8	10-UP	184	+.24 / -.36	.15	42	25	2	0-39	522.2
15-49-36	2849	2750	119	166.9	10-UP	173	+.35 / -.15	.11	20	18	2	0-42	522.0
15-53-52	2903	2553	350	165.0	20-UP	164	+.41 / -.31	.24	41	27	2	0-44	520.2
15-57-23	1620	1871	471	141.4	30-DOWN	150	+.24 / -.55	.11	55	23	3	1-12	518.6
15-57-54	1313	1552	545	139.0	30-DOWN	145	+.28 / -.26	.17	59	18	3	1-14	518.3
16-05-45	3109	2668	441	164.0	10-UP	172	+.40 / -.44	.21	71	50	3	1-03	514.3
16-08-16	3129	2765	364	152.6	20-UP	172	+.18 / -.56	.21	43	16	3	1-03	513.3

TABLE A-1 (CONTINUED). FLIGHT TEST DATA FROM TEST 34-1, 737 FOLLOWING 747.

TIME HRS-MIN-SEC	747 ALTITUDE FT ①	737 ALTITUDE FT ②	③ ①-②	747 TRUE AIRSPEED KNOTS (CORRECTED)	747 FLAPS-GEAR	737 TRUE AIRSPEED KNOTS	737 NORMAL 'g' AT C.G.	737 LATERAL 'g'	737 MAX. WHEEL ANGLE-DEG.	737 MAX. BANK ANGLE-DEG.	SEPARATION N.MI.	SEPARATION TIME MIN.-SEC.	747 GROSS WEIGHT 1000 LB.
13-53-39	8866	8390	481	268.7	0-UP	257	+.42 / -.28	.23	75	21	5.6	1-22	608.5
13-53-48	8866	8396	470	268.6	0-UP	250	+.20 / -.18	.15	84	30	5.6	1-22	608.4
13-54-22	8855	8500	355	268.5	0-UP	250	+.04 / -.22	.12	82	23	5.6	1-22	608.2
14-00-24	8863	8385	478	261.0	0-UP	268.5	-.42 / +.25	.13	38	19	5	1-06	605.9
14-00-43	8860	8306	554	261.0	0-UP	267.0	+.20 / -.50	.25	65	27	5	1-08	605.7
14-01-46	8868	8495	373	264.0	0-UP	256.0	+.12 / -.20	.20	70	19	5	1-12	605.3
14-05-45	8870	8220	650	266.0	0-UP	272.0	+.55 / -.40	.25	60	28	7.35	1-40	603.7
14-20-45	8895	8280	715	185.0	20-DOWN	194.0	+.27 / -.30	.25	60	18	5	1-33	596.8
14-20-55	8100		795										
14-23-10	8905	8280	625	187.0	20-DOWN	203.0	+.35 / -.53	.12	0	8	4.5	1-15	595.4
14-27-52	6980	6490	490	183.0	20-DOWN	176.0	+.10 / -.15	.10	80	29	3	1-01	593.1
14-28-14	6950	6600	350	183.0	20-DOWN	175.0	+.50 / -.40	.13	60	18	3	1-00	592.9
14-28-21	6520		430										
14-29-29	7000	6560	440	190.0	20-DOWN	196.0	+.13 / -.25	.25	55	19	3	0-54	592.2
14-34-43	6972	6520	452	176.0	20-DOWN	182.0	+.30 / -.37	.09	40	NA	3	0-58	589.3
14-34-47													

TABLE A-2   FLIGHT TEST DATA FROM TEST 34-2, 737 FOLLOWING 747.

747/707 GROSS WEIGHT ~1000 LB.	747/707 SEPARATION TIME ~ MIN.-SEC.	SEPARATION ~ N.MI.	737 BANK ANGLE ~ DEG.	737 MAX. WHEEL ANGLE ~ DEG.	737 LATERAL C.G. AT ~ g	737 NORMAL C.G. AT ~ g	737 TRUE AIRSPEED ~ KNOTS	747/707 FLAPS-GEAR	747/707 TAS ~ KNOTS (CORRECTED)	⓵ - ⓶	737 ALTITUDE ~ FT. ⓶	747/707 ALTITUDE ~ FT. ⓵	TIME HRS-MIN-SEC.
588.8	0-57	3	NA	80	.08	+.23 / -.15	186.0	20-DOWN	187	480	6520	7000	14-35-30
588.7	0-58	3	NA	70	.10	+.10 / -.33	184.0	20-DOWN	187	250	6740	6990	14-35-42
										350	6640		14-35-53
737 FOLLOWING 707 →													
275.4	1-07	5	24	35	.12	+.25 / -.26	265.0	0-UP	268.0	305	8550	8855	13-55-07
275.3	1-08	5	11	50	.15	+.03 / -.17	264.0	0-UP	269.0	275	8580	8855	13-55-25
275.0	1-07	5	13	62	.07	+.26 / -.22	265.0	0-UP	267.0	330	8520	8850	13-56-40
													13-56-51
274.8	1-06	5	15	65	.12	+.36 / -.35	268.0	0-UP	268.0	406	8470	8876	13-57-20
271.7	1-48	8	13	55	.17	+.42 / -.37	268.0	0-UP	271.0	364	8500	8864	14-08-34
271.4	1-45	8	22	40	.07	+.20 / -.30	275.0	0-UP	269.0	275	8580	8855	14-09-10
269.1	1-36	5	28	60	.20	+.35 / -.28	189.0	25-DOWN	189.0	590	8330	8920	14-16-34
													14-16-46
268.6	1-35	5	24	59	.15	+.27 / -.18	186.0	25-DOWN	191.0	454	8450	8904	14-18-09
268.2	1-37	5	10	55	.09	+.33 / -.15	187.0	25-DOWN	188.0	524	8370	8894	14-19-20
264.8	1-06	3	23	60	.10	+.15 / -.25	164.0	25-DOWN	187.0	280	6760	7040	14-30-54
264.2	1-07	3	30	80	.12	+.05 / -.30	163.0	25-DOWN	182.0	260	6700	6960	14-32-21

TABLE A-3. FLIGHT TEST DATA FROM TEST 34-2, 737 FOLLOWING 747 AND 707.

APPENDIX B

Derivation of an equation for the roll rate of an aircraft symmetrically encountering one of the trailing vortices.

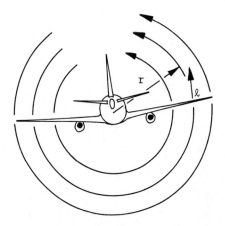

FIGURE B-1. AIRCRAFT SYMMETRICALLY ENCOUNTERING A VORTEX.

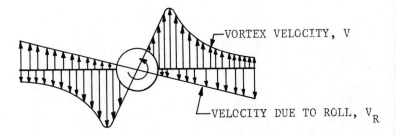

FIGURE B-2. ASSUMED VELOCITY FIELD IMPOSED ON THE AIRCRAFT.

The rolling moment R acting on the wing is

$$R = 2 \int_{o}^{b/2} r \, \ell \, dr \tag{1}$$

The section lift, ℓ, at any radius or span station is

$$\ell = c_{\ell_\alpha} \propto q\ c \tag{2}$$

then

$$R = 2 \int_0^{b/2} rc_{\ell_\alpha} \propto q\ c\ dr \tag{3}$$

The rolling moment coefficient C_ℓ is $C_\ell = \dfrac{R}{Sbq}$ so

$$C_\ell = \frac{2}{Sb} \int_0^{b/2} rc_{\ell_\alpha}{}^\alpha\ c\ dr \tag{4}$$

The velocity at any span station is

$$\vec{V} = \bar{V} + V_\infty$$
$$\bar{V} = V - r\bar{P}$$

Velocity $\bar{V}$ is composed of the tangential velocity imposed by the vortex, V, and the tangential velocity due to wing rolling, $V_R = r\bar{P}$, where $\bar{P}$ is roll rate. The angle of attack at any span station is $\alpha = \tan^{-1} \dfrac{\bar{V}}{V_\infty}$ which for small angles is approximated by $\alpha = \dfrac{\bar{V}}{V_\infty}$. The rolling moment coefficient is then

$$C_\ell = \frac{2c_{\ell_\alpha}}{SbV_\infty} \int_0^{b/2} rc\bar{V}dr = \frac{2c_{\ell_\alpha}}{SbV_\infty} \int_0^{b/2} rc(V-r\bar{P})dr \tag{5}$$

where c_{ℓ_α} is assumed constant over the span. The chord c at any span station r can be written in terms of taper ratio λ as

$$c = \frac{3\bar{c}\ (\lambda + 1)}{b(\lambda^2 + \lambda + 1)} \left\{ \frac{b}{2} + (\lambda - 1)r \right\} \tag{6}$$

or to simplify

$$c = D \left\{ \frac{b}{2} + (\lambda - 1)r \right\} \tag{7}$$

where

$$D = \frac{3\bar{c}\ (\lambda + 1)}{b\ (\lambda^2 + \lambda + 1)} \tag{8}$$

The tangential vortex velocity V at any radius r is $V = \dfrac{\Gamma}{2\pi r}$. From McCormick, Tangler and Sherrib in Reference 10:

$$\frac{\Gamma}{\Gamma_a} = \ln \left(\frac{r}{a}\right) + 1 \tag{9}$$

$$\Gamma_a = .16\ \Gamma_\infty \tag{10}$$

$$\frac{V_o}{V_\infty} = .625\ C_L \tag{11}$$

$$\frac{V_o}{V_a} = \sqrt{1 + .0065\ \frac{z}{\bar{c}}} \tag{12}$$

where Γa and $V a$ are the circulation and maximum tangential vortex velocity which are found at radius $r = a$. The circulation Γa is constant for all distances z downstream from the wing. The maximum tangential vortex velocity immediately behind the wing is V_o. The total circulation

$$\Gamma_\infty = \frac{V_\infty\ \{cc_\ell\}_{Root}}{2} \tag{13}$$

can be approximated by

$$\Gamma_\infty = \frac{K\ C_L\ \bar{c}\ V_\infty}{2} \tag{14}$$

where K is a constant such that $\{cc_\ell\}_{Root} = K\ C_L\ \bar{c}$. For the 707 and 747, $K \approx 1.0 - 1.15$. Using these relations, the radius at which maximum tangential velocity Va occurs is given by

$$a = .01275\ K\bar{c}\ \sqrt{1 + .0065\ \frac{z}{\bar{c}}} \tag{15}$$

The tangential velocity at any radius r is then

$$V = \frac{.16\ \Gamma_\infty}{2\pi r}\ \{\ln\ (\frac{r}{a}) + 1\} = \frac{A}{r}\ \ln\ r + \frac{A}{2r}\ \{2-B\} \tag{16}$$

where

$$A = .01275\ K\ C_L\ \bar{c}\ V_\infty \tag{17}$$

$$B = \ln\ \{\ (.01275\ K\bar{c})^2 + (13 \times 10^{-7})\ K^2\ \bar{c}\ z\ \} \tag{18}$$

Substituting this expression for V in terms of r and the expression for chord c in terms of r into the integral relation C_ℓ and integrating gives

$$C_\ell = \frac{2\ C_{L\alpha}\ D}{V_\infty\ Sb}\ \{\ -\frac{\bar{P}b^4}{48} - \frac{\bar{P}\ (\lambda - 1)\ b^4}{64} + \frac{Ab^2}{4}\ \{\ln\ \frac{b}{2} - 1\}$$

$$+ A\ (\lambda - 1)\ \frac{b^2}{8}\ \{\ln\ \frac{b}{2} - \frac{1}{2}\}$$

$$+ \frac{b^2 A}{8}\ \{2 - B\} + \frac{b^2}{16}\ A\ (\lambda - 1)\ (2 - B)\} \tag{19}$$

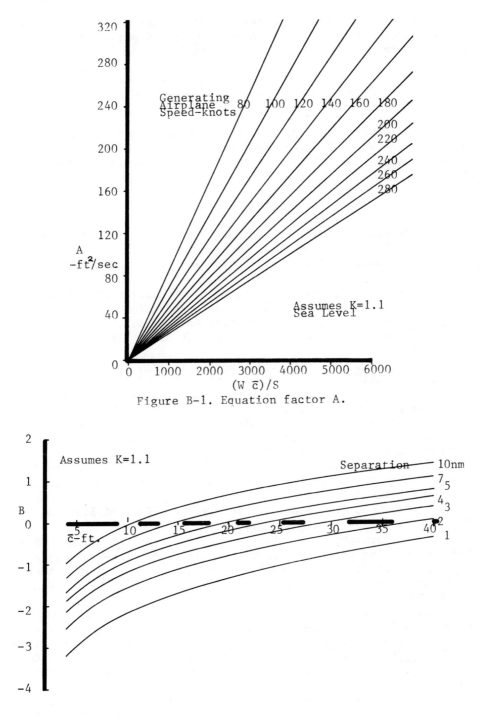

Figure B-1. Equation factor A.

Figure B-2. Equation factor B.

It is assumed that the aircraft experiences a steady roll, i.e. no roll acceleration, with no lateral control inputs. Thus, $C_\ell = 0$ and the above equation solved for the roll rate is

$$\bar{P} = \frac{12A\ (\lambda + 1)\ \{2\ \ln\ (b/2) - B + (\frac{\lambda - 1}{\lambda + 1})\}}{b^2\ \{3\lambda + 1\}} \tag{20}$$

where A and B are determined for the aircraft generating the vortices while λ and b are the taper ratio and span of the encountering aircraft. The parameters A and B are plotted in Figures B-1 and B-2.

REFERENCES

1. Andrews, W. H., "Preliminary NASA Report to the FAA on the Vortex Wake Investigation", Feb. 1970.

2. McGowan, W. A., "Trailing Vortex Hazard", SAE Paper 68220, April, 1968.

3. Dee, F. W., and Nicholas, O. P., "Flight Measurements of Wing Tip Vortex Motion Near the Ground", RAE TR68007, Jan. 1968.

4. Kraft, C. C., "Flight Measurements of Velocity Distributions and Persistence of Trailing Vortices of an Airplane", NADA TN3377, 1955.

5. Prandtl, L., Tragflugeltheoric, I Metl. Gottengen Nachiuehten, 1918.

6. Anon., "Big Plane Turbulence can Cause a Flight Hazard", Safety Suggestions No. 8, Beech Aircraft Corp., 1952.

7. Zwieback, E. L., "Trailing Vortices of Jet Transport Aircraft During Takeoff and Landing", FAA Report N64-14797, Douglas Aircraft Company, January 1969.

8. Kerr, T. H., and Dee, F. W., "A Flight Investigation into the Persistence of Trailing Vortices Behind Large Aircraft", C. P. No. 489 British ARC, 1960.

9. Crow, S. C., "Stability Theory for a Pair of Trailing Vortices", AIAA Paper 70-53, January, 1970.

10. McCormick, B. W., Tangler, J. T., Sherrieb, H. E., "Structure of Trailing Vortices", Journal of Aircraft, May-June, 1968.

11. Spreiter, J. R. and Sacks, A. H., "The Rolling of the Trailing Vortex Sheet and its Effect on the Downwash Behind Wings", J.A.S. Vol. 18, No. 1, January, 1951.

12. Bennett, W. J., "State of the Art Survey for Minimum Approach, Landing and Takeoff Intervals as Dictated by Wakes, Vortices and Weather Phenomena", The Boeing Company, D6-9892, Nov. 1963.

13. Thelander, J. A., "Separation Minimums for Aircraft Considering Disturbances Caused by Wake Turbulence", Douglas Aircraft Co., Jan. 1969.

14. Crow, S. C. and Murman, E. M., "Trailing-Vortex Experiments at Moses Lake", Boeing Scientific Research Lab. FSL TC009, Feb. 1970.

AIRCRAFT RESPONSE TO TURBULENCE INCLUDING WAKES

John C. Houbolt

Aeronautical Research Associates of

Princeton, Inc.

ABSTRACT

The nature of atmospheric turbulence and the means for establishing aircraft response is reviewed, both from discrete-gust and
spectral interpretations. Application is then made to the situation
of wake turbulence encounter to show the nature and magnitude of
the loads that result. Specific cases are treated, with encounters
perpendicular to and parallel to the wake, to bring out the main
parameters that are significant. General relations are also developed to show how the wake "gust" forces on the encountering airplane are related to the lift on the aircraft generating the wake.
It is shown that normal loads in excess of 2 g's may be produced
by a perpendicular encounter of a wake vortex, and that uncontrollable rolling moments may be caused by encounters along the axis of
a vortex.

SYMBOLS

a	slope of the lift curve; also radius of vortex core
A	aspect ratio; also structural response parameter as used in $\sigma_x = A\sigma_w$
b	wing span
c	wing chord
C_L	lift coefficient
e_A	moment arm to centroid of aileron lift

$H(\omega)$ frequency response function

I rolling moment of inertia

K_g gust alleviation factor for discrete gust

K_ϕ spectral gust alleviation factor

K_v alleviation factor for vortex

L lift

m aircraft mass

Δn incremental load factor

r radial distance

s_o nondimensional core radius, $s_o = \dfrac{2a}{c}$

S wing area

t time

V flight velocity

w tangential velocity in vortex, also vertical gust velocity

w_m maximum tangential velocity

W aircraft weight

z vertical displacement

β aileron deflection

Γ vortex strength

ε aileron efficiency factor

μ mass ratio, $\mu = \dfrac{2w}{a\rho cgS}$

ρ air density

$\sigma_{\Delta n}$ r.m.s. value of incremental load factor

σ_w r.m.s. value of vertical gust velocities

ϕ roll displacement

$\phi_x(\omega)$ power spectrum of variable x

ω circular frequency

Note:

A dot denotes $\dfrac{d}{dt}$

A subscript 1 denotes the generating aircraft

INTRODUCTION

By experience, the encounter by one airplane of the wake generated by another is known to lead to sizeable "jolting" loads or
to induced uncontrollable motions of the encountering aircraft.
The purpose of this paper is to treat this problem analytically to
establish the basic parameters that govern the problem, and to
establish what severity levels are indicated by theory for the
loads and motion.

Attention is focused mainly on two "worst" cases. One is the
encounter normal to and through the center of the trailing vortices
of the generating aircraft, Case A of figure 1; the other is a
parallel type encounter directly down the core of one of the vortices, Case B of figure 1.

The encounter is very much like the gust encounter problem of
aircraft. Because of this fact a brief review will first be given
of some of the basic response expressions that apply in the gust
case, so as to set the stage for discussing the wake turbulence
encounter problem.

GUST DESIGN APPROACHES

One of the procedures that has been used for many years for
establishing gusts loads analytically is based on a discrete-gust
encounter concept. In this concept, a discrete gust is chosen so
as to act as an equivalent replacement to the random gust experienced in nature, figure 2. The gust used for about the past 15
years has a 1- cosine shape, a maximum gust velocity U of around

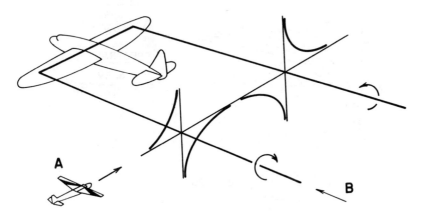

Figure 1.   Trailing Vortex Encounter

50 fps, and a gust gradient distance H of 10 chords. On the basis
of a single degree of freedom in vertical motion, theoretical calculations of the time history of response show that the incremental load factor that is produced on an airplane by this gust is
given by the equation

$$\Delta n = \frac{a\rho SV}{2W} K_g U \tag{1}$$

where K_g is a gust alleviation factor which depends on the mass
ratio μ as shown in figure 3.

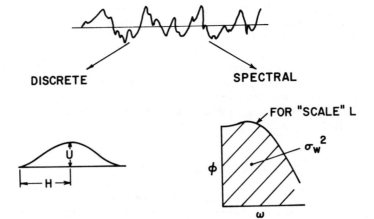

Figure 2.   Gust Design Approaches

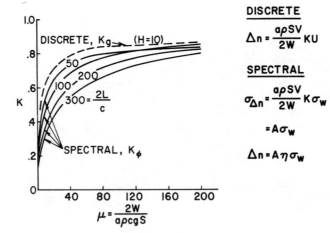

Figure 3.   Basic Elements for Establishing Gust Loads

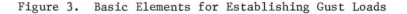

In more recent years gust encounter has been developed in terms of the more realistic power spectral methods. In this concept the random gusts are considered to be defined by a power spectrum, see right side of figure 2. The spectrum is characterized by a shape, a rms value σ_w of gust severity, and a turbulence scale length L. Calculations for response are made through means of a gust frequency response function, defined as the response due to a unit sinusoidal gust. Figure 4 shows schematically the essential steps that are involved in determining the output spectrum for response, as established from the basic input-output spectral relation

$$\phi_x(\omega) = |H(\omega)|^2 \phi_w(\omega)$$

For the case of a rigid airplane with a single degree of freedom of vertical motion, it may be shown that the spectral approach leads to the following equation for incremental rms

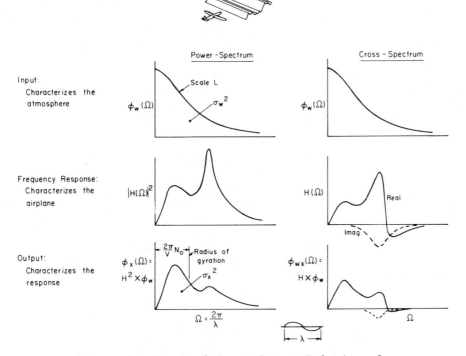

Figure 4. Spectral Input-Output Relations for
Gust Response

acceleration

$$\sigma_{\Delta n} = \frac{a\rho SV}{2W} K_\phi \, \sigma_w \tag{2}$$

This equation is noted to be analogous to the discrete-gust result. One plausible spectral design approach is to multiply this equation by some factor η and to consider $\Delta n = \eta\sigma_{\Delta n}$; we then have

$$\Delta n = \frac{a\rho SV}{2W} K_\phi \, (\eta\sigma_w)$$

If we choose $\eta\sigma_w = 54$, then we see that load levels comparable to the discrete-gust approach are indicated.

We limit our discussion of the spectral approach to just these few observations. There many other aspects of the spectral approach which are involved, such as gust severity values, proportion of time in turbulence, load exceedance curves, and flexibility effects. These aspects are not essential to the wake turbulence problem, but are discussed in detail in references 1 and 2.

VELOCITY FIELD OF THE TRAILING VORTICES

In this section we establish the essential nature of the velocity field of the trailing vortices through use of elementary theory, which is sufficiently good for present purposes. Remote from the generating airplane, the trailing wake sheet that has been left by the wing has rolled up into a pair of concentrated trailing vortices, each of which has a finite core. We assume that these trailing vortices and the circulation around the wing are represented by a horseshoe-type vortex as depicted in figure 1. In terms of the strength Γ_1 of this horseshoe vortex, the lift on the generating aircraft is given by

$$L_1 = W_1 = \rho V_1 \Gamma_1 \frac{\pi}{4} b_1 \tag{3}$$

where b_1 is the wing span. The distance $\pi/4 \, b_1$, represents the distance between the two vortex legs and follows from an assumed elliptical loading distribution on the wing.

Through the consideration of the work done by the induced drag, and the energy left in the trailing vortices, the core radius of each vortex leg can be shown to be

$$a = .086 \, b_1 \tag{4}$$

To treat the vortex encounter as a gust problem, it is desirable to express the core radius in terms of the half-chord $c/2$ of the encountering airplane, thus

$$s_o = \frac{a}{c/2} = \frac{.172\ b_1}{c}$$

which may in turn be written

$$s_o = .172\ A_1 \frac{c_1}{c} \tag{5}$$

where A_1 is the aspect ratio of the generating airplane. Non-dimensional core radii as follows are thus indicated.

$$s_o = .86 \frac{c_1}{c}, \quad A_1 = 5 \tag{6a}$$

$$s_o = 1.03 \frac{c_1}{c}, \quad A_1 = 6 \tag{6b}$$

$$s_o = 1.2 \frac{c_1}{c}, \quad A_1 = 7 \tag{6c}$$

Each vortex leg acts like a line vortex, and thus the induced velocity outside the core may be given by

$$w = \frac{\Gamma_1}{2\pi r} \tag{7}$$

where r is the radial distance from the vortex centerline. Tangential velocity is maximum at the core radius, and is given by

$$w_m = \frac{\Gamma_1}{2\pi a} \tag{8}$$

From equations (7) and (8) we may thus write the tangential velocity outside the core as

$$w = \frac{w_m a}{r} \tag{9}$$

Inside the core, solid body-type rotation is assumed; tangential velocities within the core are thus given by

$$w = \frac{w_m}{a} r \tag{10}$$

If we combine equations (3), (4) and (8), we find that w_m is given by

$$w_m = 2.355 \frac{W_1}{\rho V_1 b_1} \tag{11}$$

An alternative useful form may be shown by writing

$$W_1 = C_{L_1} \frac{1}{2} \rho V_1^2 S_1$$

Through this equation we have

$$w_m = 1.177 \frac{C_{L_1} V_1}{A_1} \qquad (12)$$

   From equations (9) and (10) we see that the encountering airplane would experience an effective vertical gust velocity field,
as depicted at the bottom of figure 5. By way of comparison the
vortex "gust" velocities that were measured with a F-106 are shown
at the top of figure 5. We see from this comparison that our
idealized representation is reasonably appropriate.

   It is of interest to compare a representative vortex "gust"
with the discrete gust used for aircraft design purposes. We consider the following case:

$$w_1 = 300,00 \text{ lbs} \qquad\qquad V_1 = 300 \text{ fps}$$
$$b_1 = 100 \text{ ft} \qquad\qquad\qquad c = 6 \text{ ft}$$
$$\rho = .00238 \qquad\qquad\qquad \text{slugs/ft}^3$$

The gust profile that applies to this example, as established by
equations (9), (10) and (11), is shown in figure 6, together with
the discrete-gust design profile. It is seen from this comparison
that rather sizeable "gust" loads can be expected from vortex encounter. It is also to be noted that, due to the very large vertica

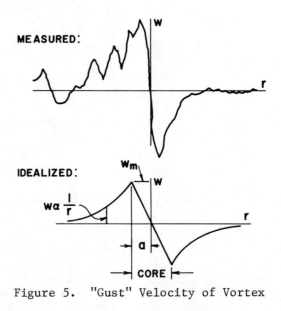

Figure 5. "Gust" Velocity of Vortex

velocities involved, vortex encounter can lead to angles of attack
in the stalling range. These nonlinear aerodynamic effects are not
considered, however, in the subsequent treatment.

ENCOUNTER NORMAL TO A TRAILING VORTEX

This section presents the treatment of an airplane encountering a trailing vortex in a direction perpendicular to the vortex
axis, Case A of figure 1. This case is truly a discrete-gust type
encounter and results similar to equation (1) can therefore be expected.

For a rigid airplane with the single degree of freedom of
vertical motion, the equation for response due to gust encounter
is (see reference 3)

$$m\ddot{z} = -\frac{a}{2} \rho SV \int_{-\infty}^{t} \ddot{z}[1 - \phi(t - \tau)] \, d\tau$$

$$+ \frac{a}{2} \rho SV \int_{-\infty}^{t} \dot{w}\psi(t - \tau) \, d\tau \tag{13}$$

where the $1 - \phi$ and ψ functions account for the lag in lift effects
due to aircraft vertical motion and due to gust penetration, respectively. By means of the simple numerical technique given in
reference 3, this equation was solved for the maximum acceleration
values that would result due to encountering vortex "gusts" of the

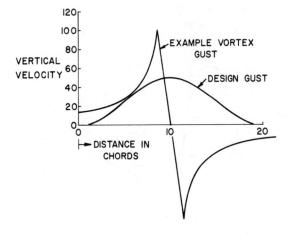

Figure 6.   Comparison of Discrete-Design Gust with
Example "Vortex" Gust

type shown at the bottom of figure 5. Results for the extreme
right-hand term of equation (13), which represents the input forcing
load due to the vortex velocities, are shown in figure 7. A smoothing of peak loads and a phase lag effect is noted due to the gust
penetration effects.

Results for maximum acceleration for the encountering aircraft
are found to be given by an equation of the same form as equation
(1), specifically

$$\Delta n = \frac{a\rho SV}{2W} K_v w_m \tag{14}$$

where the K_v values are given in figure 8, for various values of
the core radius parameter s_0. By means of equation (11), the
following instructive alternative form of equation (14) may be derived

$$\Delta n = 7.4 \ \frac{1}{A_1} \ \frac{W_1/S_1}{W/S} \ \frac{V}{V_1} \ K_v \tag{15}$$

where the subscript 1 applies to the generating aircraft, while the
symbols without a subscript apply to the encountering aircraft.
Illustrative examples indicating the severity of the loads that
might be generated by vortex encounter follow:

Case I: Suppose $A_1 = 7$

$$W_1/S_1 = 80$$

$$W/S = 10$$

$$V/V_1 = \frac{1}{2}$$

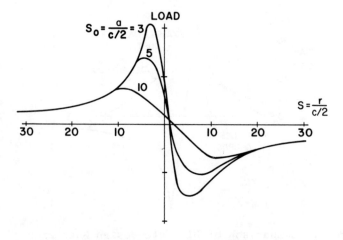

Figure 7. Gust Load of Vortex

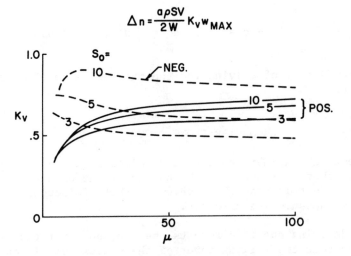

$$\Delta n = \frac{a\rho SV}{2W} K_V w_{MAX}$$

Figure 8. Load Due to Vortex Encounter

and, based on figure 7, take K = .5. With these values equation (15) indicates

$$\Delta n = \frac{7.4}{7} \frac{80}{10} \frac{1}{2} \times .5$$

$$= 2.11$$

A rather sizeable load increment due to vortex encounter is thus noted. It should be kept in mind that this increment is followed immediately by another of about the same magnitude, but of opposite sign, and then also by corresponding load pairs if the other trailing vortex is also encountered.

Case II: In this case we consider an aircraft encountering essentially its own wake. In this case W_1/S_1 = W/S, V = V_1, A_1 = A. Equation (15) indicates, taking A_1 = 7 and K = .4,

$$\Delta n = \frac{7.4}{7} \times .4$$

$$= .42$$

Thus, the "jolt" can be sizeable even for this case.

To close this section, we note that if an equation similar to (14) is used, replacing K_V by a variable K, and if the change of variables s = 2V/c t is introduced, equation (13) may be transformed to the following condensed form

$$K = \frac{1}{2\mu} \int_{-\infty}^{s} K[1 - \phi(s - \sigma)] \, d\sigma = \int_{-\infty}^{s} w'\psi(s - \sigma) \, d\sigma$$

Solution becomes that of solving for the maximum value of K from this equation.

ENCOUNTER ALONG A CORE

This section considers an aircraft following the generating aircraft such that it proceeds down the centerline of one of the vortex cores. In this case the problem is that of induced rolling motion of the encountering aircraft.

Before deriving specific results, we can show by rather elementary means that the rolling power of the vortex core is much greater than the rolling power of the aircraft. Uncontrollable rolling motions can thus be expected. The design of aircraft is usually made so as to have a minimum rolling power given by

$$\frac{\dot\phi b}{2V} = .07$$

Wing tip velocities due to a constant rolling condition as indicated by this expression are

$$w_t = \frac{\dot\phi b}{2} = .07 \text{ V}$$
$$= 14 \text{ fps for } V = 200 \text{ fps}$$
$$= 21 \quad\quad \text{for } V = 300$$

We see that these velocities are quite small relative to the rolling velocities that are present in a vortex, see figure 6. Thus it can be expected that the encountering aircraft will simply be swept around with the vortex in an uncontrollable fashion. Flight experience has shown this to be the case.

On the basis of strip theory, the equation for rolling motion may be written

$$I\ddot\phi = 2\pi\rho S_A V^2 \varepsilon\beta e_A - 2\pi\rho V\dot\phi \int_o^{b/2} cy^2 dy$$
$$- 2\pi\rho V \frac{w_m}{a} \int_o^a cy^2 dy - 2\pi\rho V w_m a \int_a^{b/2} c \, dy \qquad (16)$$

The first term on the right-hand side denotes the rolling moment due to the ailerons; the second term is the rolling moment due to rolling rate; the third term is the induced rolling moment due to the vortex core; the fourth term is the induced rolling moment due to the vortex velocities outside of the core. Aileron deflection opposing rotation is assumed. This equation may be solved for a constant rolling velocity condition ($\ddot{\phi} = 0$). Results for an assumed rectangular wing are as follows

$$\frac{\dot{\phi}b}{2V} = \left(\frac{\dot{\phi}b}{2V}\right)_A - \frac{w_m}{V} f\left(\frac{a}{b/2}\right) \tag{17}$$

where the first term on the right-hand side corresponds to the rolling power of the aircraft in free flight, while the second term is the rolling power that is induced due to the vortex. This second term depends on the ratio w_m/V and a function which depends on the ratio of the core radius to the wing semispan; this function is shown in Figure 9. From equation (12), the ratio w_m/V may be written

$$\frac{w_m}{V} = 1.176 \frac{C_{L_1}}{A_1} \frac{V_1}{V}$$

and thus the second term, defining the rolling power of the vortex, may be written

$$\left(\frac{\dot{\phi}b}{2V}\right)_V = 1.176 \frac{C_{L_1}}{A_1} \frac{V_1}{V} f\left(\frac{a}{b/2}\right) \tag{18}$$

The following example serves to show specifically the "rolling power" of the vortex. Suppose

$$C_{L_1} = 1$$
$$A_1 = 7$$
$$V_1/V = 2$$

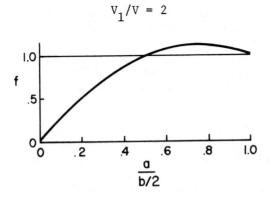

Figure 9. f function of eq. (17).

and from the sketch for f choose f = 1. Equation (18) indicates
for this case

$$\left(\frac{\dot{\phi}b}{2V}\right)_V = 1.176 \frac{1}{7} 2$$

$$= .336$$

This is seen to be nearly five times the rolling power of .07 that
is usually sought. Uncontrolled rolling motion is therefore
inevitable.

It is thus seen that if an aircraft encounters one of the more
severe trailing vortices directly along its centerline, the aircraft
will be forced into an uncontrollable rolling motion. The only hope
is that the aircraft will be thrown clear of the vortex, and that
there is sufficient altitude to effect a recovery. It can also be
seen that even if the encountering aircraft does not enter the vortex at the centerline, rather severe rolling motions are still
possible due to the large tangential velocities present in a vortex.

CONCLUDING REMARKS

This treatment of vortex encounter was limited to two specific
"worst" cases, an encounter perpendicular to a vortex, and one
along a vortex centerline. The capability of a vortex to produce
large induced loads and motions was demonstrated. We don't know
how severe the loads or motions might be for the case of an
oblique-type encounter, or for encounters which are slightly above
or below the vortex. From the nature of the results presented in
this paper, however, we can infer that loads and induced motions
can still be large even for these cases.

REFERENCES

1. Houbolt, John C.: Gust Design Procedures Based on Power
    Spectral Techniques. Technical Report AFFDL-TR-67-74, August
    1967.

2. Houbolt, John C.: Design Manual for Vertical Gusts Based on
    Power Spectral Techniques. Technical Report AFFDL-TR-70-106,
    July 1970.

3. Houbolt, John C. and Kordes, Eldon E.: Structural Response to
    Discrete and Continuous Gusts of an Airplane Having Wing-
    Bending Flexibility and a Correlation of Calculated and Flight
    Results. NACA Report 1181, 1954. (Supersedes NACA TN 3006;
    also contains essential material from TN 2763 and TN 2897.)

AIRLOADS AND MOMENTS ON AN AIRCRAFT FLYING OVER A PAIR OF INCLINED TRAILING VORTICES

W. P. Jones and B. M. Rao

Texas A&M University

ABSTRACT

When an aircraft flies across the wake of a preceding one, it is subjected to changing airloads and moments induced by the trailing vortices of the first aircraft. The purpose of this paper is to investigate the magnitude and characteristics of the time dependent aerodynamic forces so produced. Both aircraft are assumed to be in horizontal flight but the direction of flight of the second aircraft is assumed to be inclined at a small angle to the trailing vortices of the first aircraft. The airloads then will change relatively slowly with time and may be estimated with reasonable accuracy by quasi-steady aerodynamic theory without taking Wagner growth of lift effects into account. In the present study, the pilot of the second aircraft is assumed to have suffi-cient control power to maintain his aircraft in level flight. However, in practice, this condition is likely to be violated when the aircraft is close to one of the trailing vortices of the lead-ing aircraft, particularly when the latter happens to be a large transport plane. In such circumstances the following aircraft could stall and would, in any case, be subjected to big changes of lift and rolling moment as indicated by the results presented. In the development of the theory it is assumed that the trailing vor-tices are a chord length or more below the following aircraft and that the vorticity distribution over its wings will have negli-gible effect on the trailing vortices themselves. The pair of vortices will give rise to an upwash distribution over the wings of an approaching aircraft which must be balanced by an equal and opposite velocity distribution induced by the vorticity distribu-tion created over the wing. The effects of the induced velocity

523

components along the span and in the direction of flight are neglected. The problem then is one of finding the appropriate vorticity distribution at each stage of the aircraft's flight over
the trailing vortices. This can be done approximately by using
a modified lifting line theory or, more accurately, by lifting
surface theory. To illustrate the methods of analysis employed,
calculations were done for an aircraft with rectangular wings,
but only the airloads on the wings were determined. Values of the
lift, rolling moment and pitching moment coefficients, corresponding to vortex inclinations of 0, 10, 20, 30, and 60° were obtained
for a rectangular wing of aspect ratio 6 at different times during
its passage over the trailing vortices.

NOMENCLATURE

$0x, 0y, 0z$	axes of coordinates
$\xi (= x/\ell \text{ or } -\cos\Theta)$	chordwise coordinates
$\eta (= y/s \text{ or } -\cos\phi)$	spanwise coordinates
$\tau (= Ut/\ell)$	distance moved in half-chords
ℓ	semi-chord of following aircraft
A	aspect ratio
$h (=\mu\ell)$	distance of trailing vortices from plane of wing
U_0, U	velocities of leading and following aircraft
s_0, s	semi-span of leading and following aircraft
$\gamma_v (=2\pi U\ell\Gamma_v)$	trailing vortex strength of leading aircraft
λ	angle of inclination of vortex to direction of flight
$2b$	distance between trailing vortices
$Y_{01} (= y_0/\ell \text{ or } A\eta_{01})$	spanwise coordinate determining location of trailing vortex on left in Fig. 1
$W(x,y,o,t)$	downwash distribution over wing at time t
Γ	general bound vorticity distribution on wing
K	corresponding generalized circulation or doublet distribution
L	lift
$\mathcal{M}$	nose-up pitching moment
$\mathcal{L}$	rolling moment
$C_L (= L/\frac{1}{2} \rho U^2 S)$	lift coefficient
$C_\ell (= \mathcal{L}/\rho U^2 Ss)$	rolling moment coefficient
$C_m (= c\mathcal{M}/\rho U^2 S\ell)$	pitching moment coefficient
S	wing area

1.   INTRODUCTION

A theory is developed for determining the rolling and pitching moments, as well as the lift, on an aircraft as it flies over
the trailing vortices left behind by another aircraft.   The axes
of the trailing vortices are assumed to be inclined at an angle λ
to the direction of flight of the following aircraft.   As the latter
crosses the trailing vortices of the leading aircraft, it will be
subjected to airloads and moments that will change with time.   On
the basis of the assumption that the aircraft is maintained in
level flight, calculations are made of the changing airloads on
the aircraft's wings only in this preliminary study.

In the development of the analysis, it is assumed that the
trailing vortices lie in a plane below the following aircraft at
a distance greater than the radius of the core of each of the trailing vortices.   If the strength of each vortex is denoted by
$\gamma_v (=2\pi U\ell\Gamma_v)$, then the vorticity distribution induced on the wing
will be such that it produces a downwash over the wing's surface
sufficient to cancel the upwash distribution due to the trailing
vortices.   In the notation given earlier and illustrated in Fig. 1,
the downwash distribution $W(x,y,o,t)$ can be expressed in the form

$$W(x,y,o,t) = U\Gamma_v[\frac{(\xi - \tau)\sin\lambda - A\eta\cos\lambda}{[(\xi - \tau)\sin\lambda - A\eta\cos\lambda]^2 + \mu^2}$$

$$- \frac{(\xi - \tau)\sin\lambda - A\eta\cos\lambda + 2B}{[(\xi - \tau)\sin\lambda - A\eta\cos\lambda + 2B]^2 + \mu^2}] \tag{1}$$

where $x = \ell\xi$, $y = s\eta$, $t = \ell\tau/U$, $\mu = h/\ell$, $B = b/\ell$ and $A \equiv$ aspect
ratio.   The problem then is to find the vorticity distribution
over the wings of the second aircraft that would produce the above
downwash distribution.   To simplify the analysis, it has been
assumed that the aircraft has wings of rectangular plan form.

As a first approximation the effect the aircraft might have on
the trailing vortices is neglected.   Obviously, when the aircraft
flies close to one trailing vortex, there would be significant mutual interaction and the trailing vortex would probably be distorted.
However, as stated, it is assumed that when each vortex is located
at a distance equal to or greater than one chord length above or
below the aircraft, such distortion as may be produced would only
have a second order effect on the airload distribution on the aircraft.

In a previous paper,[1] the authors considered the case of a
single trailing vortex parallel to the direction of flight ($\lambda = 0$)
and the theory developed in that publication is extended in the
present paper to include the more complicated problem when $\lambda \neq 0$.
Whereas in the previous work the airloads and moments are independent
of time, in the present study they are time dependent.   Initially,

it is assumed that these time dependent loads can be estimated on
a quasi-steady basis. However, with $\lambda \sim \pi/2$, it will probably be
necessary to allow for delays in the growth of the airloads and
moments (Wagner effect).

On physical grounds, it is evident that when $\lambda = \pi/2$, only
the lift and pitching moment will be affected as the aircraft
crosses the trailing vortices. However, when $\lambda \neq \pi/2$, pitching,
rolling and yawing moments will change with time. The rate of
change for a given flight speed will increase with increasing λ
but it is probably safe to assume that for small values of λ, the
use of quasi-steady aerodynamic theory would be satisfactory while
for higher values the results can only be regarded as rough approximations.

In the present investigation, it is supposed that the aircraft
is in level flight at a small incidence with a corresponding airload distribution. Any induced loads due to the trailing vortices
would be added to this initial loading, and it is assumed that the
aircraft never becomes stalled. It is realized, however, that in
practice when the aircraft is landing or taking off behind another
aircraft, conditions might be such as to cause it to stall or roll
over when crossing one or the other of the trailing vortices left
behind by the first aircraft.

2. DEVELOPMENT OF THEORY

The task of determining the time dependent airloads on an aircraft flying across the wake of another is, in general, complicated
by the fact that the response of both the aircraft and the pilot
to the influence of the trailing vortices must be taken into
account. In the present analysis, the problem is simplified by
assuming that the aircraft is maintained in level flight and, as a
first step, the airloads on the wings only are estimated. Furthermore, the wings are taken to be of rectangular planform and of
total aspect ratio $A \geq 6$. Since the downwash distribution is prescribed by equation (1), the corresponding vorticity distribution
Γ on the wings can be estimated by using the full lifting surface
theory or by simple lifting line theory. When λ is small, the
rate of change of downwash with time is also small. Hence, the
airloads can be estimated on a quasi-steady basis without much loss
of accuracy.

To calculate the airloads on the wing, firstly, write $x = -\ell\cos\theta$
and $y = -s \cos\phi$ and let the bound vorticity distribution Γ induced
on the aircraft by the trailing vortices be represented in the
general form

$$\Gamma(\theta,\phi,\tau) = U\Gamma_v \sum_{n=1}^{N} \sum_{m=0}^{M} a_{mn}(\tau)\Gamma_m(\theta) \frac{\sin n\phi}{n}, \qquad (2)$$

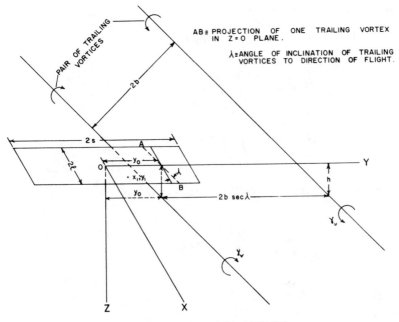

FIG. I SKETCH OF WING AND TRAILING VORTICES

where $\Gamma_0(\Theta) = 2\cot\dfrac{\Theta}{2}$, $\Gamma_1 = -2\sin\Theta + \cot\dfrac{\Theta}{2}$ and $\Gamma_m = -2\sin m\Theta$, $m > 2$ and where the $a_{mn}(\tau)$'s are arbitrary coefficients that vary with $\tau(=Ut/\ell)$. The corresponding form for the generalized circulation or doublet distribution K is then

$$K(\Theta,\phi,\tau) = \ell U\Gamma_v \sum_{n=1}^{N} \sum_{m=0}^{M} a_{mn}(\tau)K_m(\Theta) \frac{\sin n\phi}{n} \tag{3}$$

where $K_m(\Theta) = \displaystyle\int_0^{\Theta} \Gamma_m(\Theta)\sin\Theta \, d\Theta$. It then follows that

$$K_0(\Theta) = 2(\Theta+\sin\Theta), \quad K_1(\Theta) = \sin\Theta + \frac{\sin 2\Theta}{2}, \quad K_m(\Theta) = \frac{\sin(m+1)\Theta}{m+1} - \frac{\sin(m-1)\Theta}{m-1}$$

when $m \geq 2$.

When $\Theta=\pi$ at the trailing edge, $K_0=2\pi$ and $K_m=0$ for $m \geq 1$. Hence equation (3) yields

$$K(\pi,\phi,\tau) = 2\pi\ell U\Gamma_v \sum_{n=1}^{N} a_{on}(\tau) \frac{\sin n\phi}{n} \tag{4}$$

which represents the spanwise distribution of total circulation. The local lift distribution is then simply $\rho UK(\pi,\phi,\tau)$.

The corresponding lift, L, nose-up pitching moment $\mathcal{M}$ about quarter-chord and rolling moment $\mathcal{L}$ are given respectively by the following integral formulae

$$L = \rho U \int_{-s}^{s} \int_{-\ell}^{\ell} \Gamma \, dx \, dy$$

$$\mathcal{M} = -\rho U \int_{-s}^{s} \int_{-\ell}^{\ell} \Gamma(\frac{\ell}{2} + x) \, dx \, dy \tag{5}$$

$$\mathcal{L} = -\rho U \int_{-s}^{s} \int_{-\ell}^{\ell} \Gamma y \, dx \, dy$$

After substitution for Γ and integration, the above formulae yield

$$\frac{L}{\frac{1}{2} \rho U^2 S \Gamma_v} = \frac{\pi^2}{2} a_{01}(\tau)$$

$$\frac{\mathcal{M}}{\rho U^2 S \ell \Gamma_v} = \frac{\pi^2}{16} (a_{11}(\tau) - a_{21}(\tau)) \tag{6}$$

$$\frac{\mathcal{L}}{\rho U^2 S s \Gamma_v} = \frac{\pi^2}{16} a_{02}(\tau)$$

where S represents total wing area and the coefficients $a_{01}(\tau)$, etc., are to be determined.

The downwash distribution over a rectangular wing induced by the vorticity distribution Γ (or doublet distribution K) is given by

$$4\pi W(x_1, y_1, 0, t) = \int_{-s}^{s} \frac{\frac{\partial K(\ell, y)}{\partial y} \, dy}{y_1 - y}$$

$$+ \int_{-s}^{s} \int_{-\ell}^{\ell} \frac{\partial^2 K(x, y, t)}{\partial x \partial y} \frac{[(x-x_1)^2 + (y-y_1)^2]^{\frac{1}{2}}}{(x-x_1)(y-y_1)} \, dx \, dy \tag{7}$$

where $\Gamma = \frac{\partial K}{\partial x}$. For computational purposes, the above equation can be expressed conveniently as

$$4\pi W(x_1, y_1, 0, t) = \int_{-s}^{s} \frac{\frac{\partial K(\ell, y, t)}{\partial y} \, dy}{y_1 - y} + 2 \int_{-\ell}^{\ell} \frac{\frac{\partial K(x, y_1, t)}{\partial x} \, dx}{x_1 - x}$$

$$+ 4\pi U E(x_1, y_1, t) \tag{8}$$

where

$$4\pi UE(x_1,y_1t) = \int_{-s}^{s}\int_{-\ell}^{\ell}\frac{\partial^2 K(x,y,t)}{\partial x \partial y}[-\frac{[(x-x_1)^2+(y-y_1)^2]^{\frac{1}{2}}}{(x-x_1)(y-y_1)} - \frac{|y - y_1|}{(x-x_1)(y-y_1)}]dxdy$$

After substitution for K from (3) and (4) and a change of coordinates, equation (8) yields

$$\frac{W(\Theta_1,\phi_1,0,t)}{U\Gamma_v} = \frac{\pi}{2A}\ \Sigma_{n=1}^{N}\ a_{on}(\tau)\ \frac{\sin n\phi_1}{\sin \phi_1}$$

$$+ \Sigma_{n=1}^{N}\ [a_{on}(\tau) + a_{1n}(\tau)(\frac{1}{2} + \cos\Theta_1) + \Sigma_{m=2}^{M}\ a_{mn}(\tau)\cos m\Theta_1]\frac{\sin n\phi_1}{n}$$

$$+ E(\Theta_1,\phi_1,\tau)/\Gamma_v \tag{9}$$

where

$$E(\Theta_1,\phi_1,\tau) = \frac{\Gamma_v}{4\pi A}\int_{0}^{\pi}\int_{0}^{\pi}\ \Sigma_{n=1}^{N}\ \Sigma_{m=0}^{M}\ a_{mn}(\tau)\Gamma_m(\Theta)Q\ \cos n\phi\ \sin\Theta\ d\Theta d\phi \tag{10}$$

and

$$Q \equiv \frac{[(\cos\Theta - \cos\Theta_1)^2 + A^2(\cos\phi - \cos\phi_1)^2]^{\frac{1}{2}} - A|\cos\phi - \cos\phi_1|}{(\cos\Theta - \cos\Theta_1)(\cos\phi - \cos\phi_1)}$$

The value of E at Θ_1,ϕ_1 at a particular time can then be derived by numerical integration. If

$$P_m(\Theta_1,\phi,\phi_1) \equiv \int_{0}^{\pi}\ (\cos\phi - \cos\phi_1)Q\ \cos m\Theta\ d\Theta \tag{11}$$

equation (10) can be expressed in a more convenient form for computation, namely

$$E(\Theta_1,\phi_1,\tau) = \frac{\Gamma_v}{4\pi A}\ \Sigma_{n=1}^{N}\ [2a_{on}(G_{on} + G_{1n})$$

$$+ a_{1n}(G_{1n} + G_{2n}) + \Sigma_{m=2}^{M}\ a_{mn}(G_{m+1,n} - G_{m-1,n})] \tag{12}$$

where

$$G_{mn}(\Theta_1,\phi_1) = \int_0^\pi \frac{P_m(\phi,\Theta_1,\phi_1)\cos n\phi \; d\phi}{\cos\phi - \cos\phi_1} \tag{13}$$

When the G_{mn} distributions have been calculated, the function $E(\Theta_1,\phi_1,\tau)$ will be determined by (12) when the values of the arbitrary $a_{mn}(\tau)$ coefficients have been found. By the use of (1), (9) and (12), it can be deduced that they must satisfy the following equation at all points so as to insure tangential flow over the wing, namely

$$[\frac{f}{f^2+\mu^2} - \frac{f + 2B}{(f+2B)^2+\mu^2}] = \Sigma_{n=1}^N [a_{on}[\frac{\sin n\phi_1}{n} + \frac{\pi}{2A}\frac{\sin n\phi_1}{\sin\phi_1} + \frac{1}{2\pi A}(G_{on}+G_{1n})]$$

$$+ a_{1n}[(\frac{1}{2} + \cos\Theta_1)\frac{\sin n\phi_1}{n} + \frac{1}{4\pi A}[G_{1n} + G_{2n}]$$

$$+ a_{2n}[\frac{\cos 2\Theta_1 \sin n\phi_1}{n} + \frac{1}{4\pi A}(G_{3n} - G_{1n})]$$

$$+ \text{etc.}] \tag{14}$$

where $f \equiv A \cos\phi_1\cos\lambda - (\tau + \cos\Theta_1)\sin\lambda$.

If m chordwise and n spanwise modes of distribution are used, the above equation will have to be satisfied at m x n points on the wing in order to determine the a_{mn} coefficients at a given time. When this is done the corresponding bound vorticity distribution will be given by (2) and the lift and moments by (6). It should be noted that only the coefficients a_{01}, a_{02}, a_{11}, a_{21} are involved in the latter set of equations. To determine these, at least three chordwise modes of vorticity distribution will have to be used as well as a number of spanwise modes sufficiently large to insure convergence of the calculation of spanwise loading.

To evaluate G_{mn} for a range of values of m and n, the following procedure was adopted. It is evident from equations (10) and (11) that, when $\phi = \phi_1$,

$$(\cos\phi - \cos\phi_1)Q = \frac{|\cos\Theta - \cos\Theta_1|}{\cos\Theta - \cos\Theta_1}$$

$$= 1 \cdots \; 0<\Theta<\Theta_1$$

$$= -1 \cdots \; \Theta_1<\Theta<\pi \tag{15}$$

Hence, when $m \geq 1$,

$$P_m(\Theta_1,\phi_1,\phi_1) = \int_0^{\Theta_1} \cos m\Theta \, d\Theta - \int_{\Theta_1}^{\pi} \cos m\Theta \, d\Theta$$

$$= \frac{2\sin m\Theta_1}{m}$$

(16)

and $P_o(\Theta_1,\phi_1,\phi_1) = 2\Theta_1 - \pi$.

For numerical integration, $G_{mn}(\Theta_1,\phi_1)$ can be expressed in a form which eliminates the discontinuity at $\phi = \phi_1$, namely

$$G_{mn}(\Theta_1,\phi_1) = \int_0^{\pi} \frac{P_m(\phi,\Theta_1,\phi_1) - P_m(\phi_1,\Theta_1,\phi_1)}{\cos\phi - \cos\phi_1} \cos n\phi \, d\phi$$

(18)

$$+ 2\pi \frac{\sin m\Theta_1 \sin n\phi_1}{m \sin\phi_1} \, , \qquad m \geq 1.$$

and

$$G_{on} = \int_0^{\pi} \frac{P_o(\phi,\Theta_1,\phi_1) - P_o(\phi_1,\Theta_1,\phi_1)}{\cos\phi - \cos\phi_1} \cos n\phi \, d\phi$$

(19)

$$+ \frac{\pi\sin n\phi_1}{\sin\phi_1} (2\Theta_1 - \pi)$$

It is shown in Appendix I that all the P integrals can be expressed in terms of complete and incomplete elliptic integrals that have been tabulated but in practice it is easier to evaluate them by direct numerical integration.

When the aspect ratio of the wing is large, the function $E(x,y) \approx 0$ and equations (8) and (9) then reduce to their lifting line form from which the coefficients $a_{on}(\tau)$ can be determined for a range of values of τ. If the full equation, with E term included, is used, the corresponding $a_{mn}(\tau)$ coefficients may be obtained from equation (14) though the amount of computation involved when τ is changed by small steps would be very heavy. In the present paper, most of the results were obtained by using lifting line theory but a few calculations were made by lifting surface theory to estimate the degree of error involved.

3. DISCUSSION OF THE RESULTS

Most previous work concerned with the determination of airloads and moments on an aircraft flying over or into trailing
vortices is either related to the case where the aircraft flies at
right angles to the trailing vortex or along its axis. In the
first case, changes in lift and pitching moment occur while in the
second case rolling and yawing moment only are primarily effected.
For the former category, predictions of load factors and rates of
descent have been made by McGowan[2,3] for a number of airplanes and
for the latter, rolling rates have been calculated by Wetmore and
Reeder[4] using two-dimensional theory. They showed that under certain conditions design load factors and roll rates could be exceeded. To avoid such hazards a pilot must follow specified flight
procedures at landing and take-off. Several accidents, however,
have already occurred due to vortex interaction effects and accident statistics show that 50% are likely to occur at landing,
30% at take-off, and 20% at cruise.

In the present paper a preliminary study is made of the airloads and moments on an aircraft as it passes over a pair of trailing vortices, as illustrated in Fig. 1. The aircraft is assumed
to be flying horizontally in a plane at height $h \geq 2\ell$ above the
plane containing the trailing vortices which are assumed to be
inclined at an angle λ to the direction of flight. If U is the
forward speed of the aircraft, the vortices will sweep over the
span of the wing at a relative velocity of $-U\tan\lambda$. When λ is small,
the vortices move slowly across the span and quasi-steady aerodynamic theory can be used to estimate the corresponding additional
airloads.

According to Spreiter and Sacks[5] the radius R_c of the core of
a trailing vortex is 0.155 x semi-span of the generating aircraft.
The vortex strength, γ_v, is equal to $2W_o/\pi\rho U_o s_o$ where W_o, U_o and
s_o are the weight, speed and semi-span of the vortex generating
aircraft respectively. The distance between the trailing vortices
correspond to $\pi s_o/2 (=2b)$. When R_c and γ_v are known, the tangential velocity in the trailing vortex at the edge of its core is
simply given by $\dot{\gamma}_v/2\pi R_c$. Hence, this velocity is proportional to
the weight of the aircraft and inversely proportional to its air
speed and its span squared. It follows from this that super-sonic
aircraft of relatively small span such as the Concorde would have
trailing vortices with high velocities at the edge of the vortex
core. At take-off the velocity could be as high as 200 ft/sec
while for most sub-sonic transport aircraft the corresponding velocity ranges from about 54 ft/sec for the Boeing 737 to about
81 ft/sec for the Boeing 747. Because of these relatively high
velocities the problem of estimating the loads on an aircraft penetrating the trailing vortices even of the the Boeing 737 becomes a

highly nonlinear problem when one considers that the speed of the latter aircraft will be of the order of 200 ft/sec. However, in the calculations described in this paper, linearized theory was used in the hope that the results obtained would give some idea of the changing airloads when a light aircraft crossed the trailing vortices of another aircraft after the trailing vortices had been given some time to dissipate.

Calculations were done for a light aircraft of aspect ratio 6 and 30 ft span flying over a pair of trailing vortices. The distance between the trailing vortices was varied and values of 60 ft, 90 ft, and 120 ft were taken . It is assumed throughout that the trailing vortices have lost their initial strength to the extent that the maximum velocities within the vortex would be small compared with the speed of the passing aircraft. In the cases considered it is also assumed that the vortices were at a distance of at least one chord length below the aircraft. Most of the results were obtained by using lifting line theory and the additional airloads due to the presence of the vortices were calculated at small intervals of time during the aircraft's passage over them. Figure 2a shows the induced spanwise distribution of circulation when the first vortex is below the right wing tip of the aircraft at a distance of 5 ft, equal to one chord length below, for different angles of vortex inclination to the direction of flight. The induced circulation does not vary much between $\lambda = 0$ and $\lambda = 30°$. As one would expect the distribution is not symmetrical and tends to heap up towards the right wing tip. The latter effect is not so pronounced when $\lambda = 60°$ though the overall magnitude of K/γ_v is increased. When the first vortex is below mid-span, the distribution of circulation induced is roughly anti-symmetric as

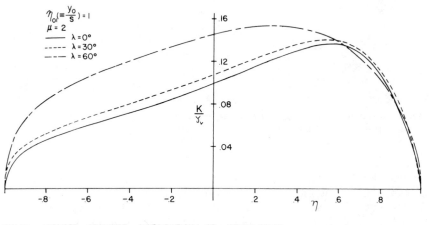

FIG. 2a   INDUCED   SPANWISE   DISTRIBUTION   OF   CIRCULATION
(First   Vortex   Below   Right   Wing   Tip)

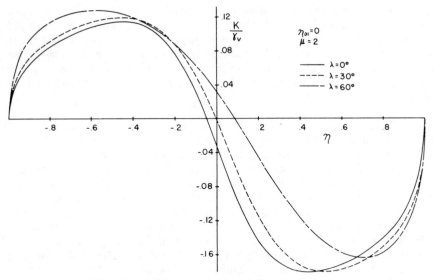

FIG. 2b   INDUCED SPANWISE DISTRIBUTION OF CIRCULATION
(First Vortex Below Mid-Span)

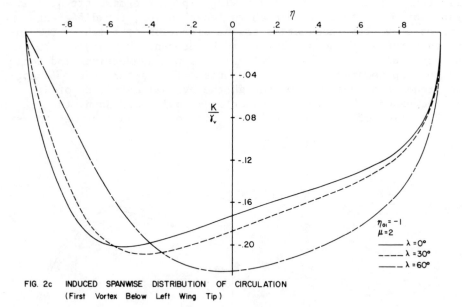

FIG. 2c   INDUCED SPANWISE DISTRIBUTION OF CIRCULATION
(First Vortex Below Left Wing Tip)

indicated by Fig. 2b.   When the first vortex is below the left
wing tip, the induced circulation is as shown in Fig. 2c and is
negative over the whole span.   The magnitude of the circulation
would of course depend on the strength, γ_v, of the trailing
vortex.

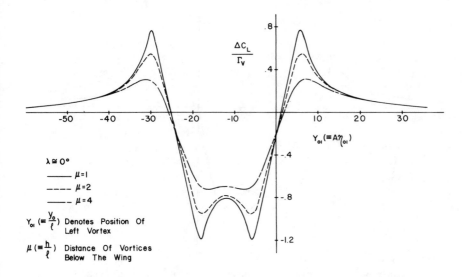

FIG. 3a   VARIATION IN INDUCED LIFT INCREMENT WITH POSITION OF TRAILING VORTICES

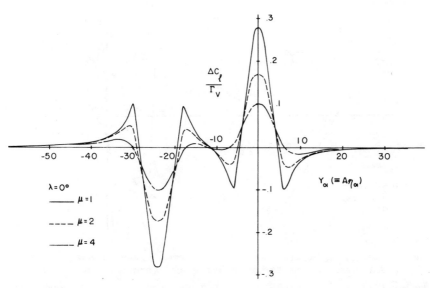

FIG. 3b   VARIATION IN INDUCED ROLLING MOMENT WITH POSITION OF TRAILING VORTICES

In Fig. 3a the ratio of $\Delta C_L / \Gamma_v$ is plotted against $Y_{01} (= A\eta_{01})$ which denotes the position of the left vortex relative to the wing (see Fig. 1). In the diagram, the wing extends along the Y_{01} axis from −6 to 6 since $A = 6$ and the chord is 5 ft. For a given Γ_v, ΔC_L increases as the left vortex approaches the right wing tip and reaches a maximum when it is roughly below the wing tip. It then drops rapidly as the first vortex of the pair of trailing vortices moves across the span of the wing. When the left vortex is below the left wing tip, the wing is between the two trailing vortices

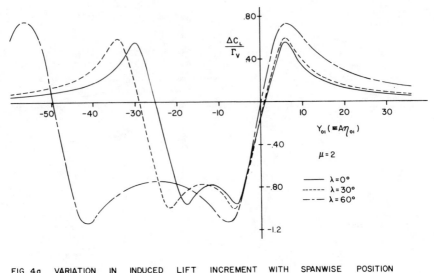

FIG. 4a   VARIATION   IN   INDUCED   LIFT   INCREMENT   WITH   SPANWISE   POSITION
          OF   TRAILING   VORTICES   FOR   DIFFERENT   ANGLES   OF   INCLINATION.

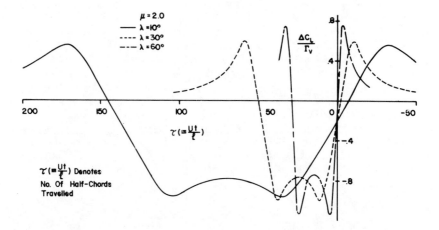

FIG. 4b   VARIATION   IN   INDUCED   LIFT   INCREMENT   WITH   DISTANCE   TRAVELLED
          FOR   DIFFERENT   ANGLES   OF   INCLINATION   OF   VORTICES.

and is subjected to a large loss of lift. This loss of lift occurs
during the time both vortices take to sweep over the wing. Calculations were also done for different positions of the trailing vortices relative to the plane of the wing and the diagram shows the
curves obtained for h = ℓ, 2ℓ and 4ℓ respectively, where ℓ is the
semi-chord of the crossing rectangular wing. The curves obtained
correspond to the case when the angle of inclination of the vortices
to the direction of flight is approximately zero. They also show
that for a given Γ_y the loss of lift would decrease if the distance
of the vortices below the wing were increased.

Similarly the variation of induced rolling moment with the position of the trailing vortices is shown in Fig. 3b. As the left vortex approaches the aircraft there is firstly a tendency to roll in the negative sense but when the vortex passes under the wing, there is a sharp increase in the rolling moment which reaches a maximum approximately when the left vortex is below mid-span and then decreases as it moves towards the left wing tip. As the right hand vortex moves under the wing from right to left, the situation is reversed and the rolling moment has a minimum when the vortex is below mid-span. On the scale used in the diagram the distances between the vortices is 24. The diagram also shows that the rolling moment decreases appreciably as the distance of the vortices below the wing is increased.

A study was also made of the variation of the induced lift increment with spanwise position of the trailing vortices for different angles of inclination. Between the curves for $\lambda = 0°$ and $30°$ shown in Fig. 4a, there is not much difference, but for $\lambda = 60°$ the corresponding curve is quite different. It should be remembered, however, that in the latter case some allowance should be made for Wagner effect in the growth of loading, as the changes in loading take place much more rapidly for this case than when $\lambda \sim 0$. It should also be noted that the negative and positive peaks for the $\lambda = 60°$ case are larger than those of the other two cases. All curves correspond to a condition where it is assumed that the aircraft is at a height of 5 ft, one chord length above the plane of the trailing vortices. In Fig. 4b, $\Delta C_L/\Gamma_v$ is plotted against a parameter $\tau(=Ut/\ell)$. This clearly shows that the changes in airload take place much more quickly for the higher angles of vortex inclination. If the aircraft is flying at 250 ft/sec, the distance 0 to 50 on the τ axis would be covered in half a second for this particular case. Similarly, Figs. 5a and 5b show the variation in induced rolling moment with spanwise position of trailing vortices and with distance traveled for different angles of inclination of the trailing vortices to the direction of flight. The induced rolling moment shows a very sharp increase as the aircraft crosses the first vortex and exhibits a peak value roughly when the left vortex is below mid-span. The magnitude of the induced rolling moment does not appear to be sensitive to the angle λ but for $\lambda = 60°$ the peaks in the $\Delta C_\ell/\Gamma_v$ are very much sharper. Figures 6a and 6b show the variation in induced pitching moment with the position of the left vortex and with the distance traveled when $\lambda = 10°$ and $\mu(\equiv h/\ell)=2$. Whereas all of the other curves previously shown were obtained on the basis of simple lifting line theory, lifting surface theory was used to calculate the induced pitching moment for the cases illustrated. These do not reveal any marked effect but, when they are compared with the ΔC_L curves shown in Figs. 4a and 4b they show that there is a slight tendency for the center of pressure of the induced airload to move forward of quarter-chord as the vortices cross under the wing.

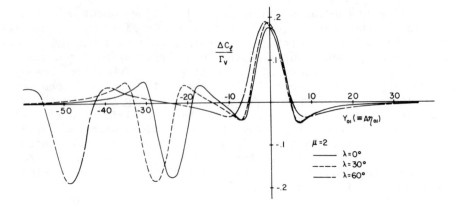

FIG. 5a   VARIATION IN INDUCED ROLLING MOMENT WITH SPANWISE POSITION OF TRAILING
VORTICES   FOR   DIFFERENT   ANGLES   OF   INCLINATION

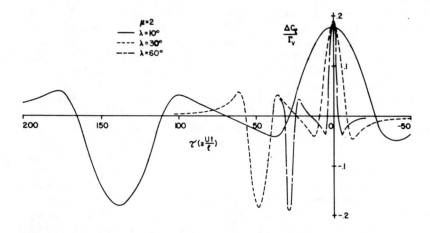

FIG. 5b   VARIATION   IN   INDUCED   ROLLING   MOMENT   WITH   DISTANCE   TRAVELLED
FOR   DIFFERENT   ANGLES   OF   INCLINATION   OF   VORTICES.

Since most of the results given were deduced on the basis of
lifting line theory, some check results were derived on the basis
of lifting surface theory. The lifting line solutions were obtained
by assuming twenty unknown $a_{mn}(\tau)$ coefficients and satisfying the
tangential flow condition, represented by Eq. (9) with E = 0, at
39 spanwise points by the least squares method. For comparison,
solutions by lifting surface theory were obtained using 27 collocation points corresponding to 3 chordwise and 9 spanwise stations.
The chordwise stations were taken to be at 1/4, 1/2, and 3/4 chord
and the spanwise stations were located at $\phi_1 = \pi/8,\ \pi/4,\ 3\pi/8,\ 7\pi/16$

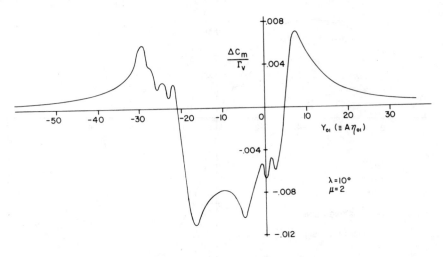

FIG. 6a   VARIATION   IN   INDUCED   PITCHING   MOMENT   WITH   POSITION   OF
         LEFT   VORTEX.

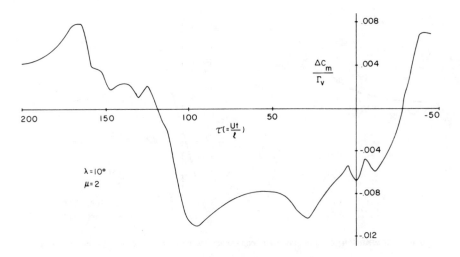

FIG. 6b   VARIATION   IN   INDUCED   PITCHING   MOMENT   WITH   DISTANCE   TRAVELLED

$\pi/2$, $9\pi/6$, $5\pi/8$, $3\pi/4$, and $7\pi/8$.   Figure 7 shows a plot of K/γ_v
against η for a particular case when the first vortex is one chord
length below mid-span and inclined at an angle $\lambda = 10°$.   The maximum difference between lifting line and lifting theory solutions in
this case is about 15%.   The difference in the total lift increment
and in the induced rolling moment, however, is much less as indicated in Fig. 8 and 9.

All the above calculations refer to the case when the perpendicular distance between the trailing vortices is assumed to be
60 ft.   Similar calculations were also done for vortices separated
by distances of 90 ft and 120 ft respectively.   The results obtained

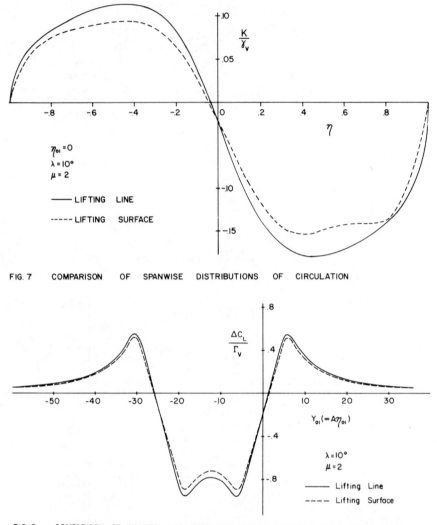

FIG. 7   COMPARISON   OF   SPANWISE   DISTRIBUTIONS   OF   CIRCULATION

FIG. 8   COMPARISON OF INDUCED LIFT INCREMENT WITH   POSITION OF LEFT   VORTEX

show no marked difference from those given in the present paper
except that, for a given trailing vortex strength, the loss of lift
induced when the aircraft is midway between the vortices tends
to be smaller, as one would expect, when the distance between
the vortices is increased. The most marked effects occur when the
aircraft crosses one or the other of the trailing vortices as has
been already indicated by the results shown.

No attempt was made to calculate the actual load factors and
rolling rates as the trailing vortex strength $\gamma_v (=2\pi U \ell \Gamma_v)$ was not
specified in this study. For a light transport aircraft weighing
about 100,000 lbs and having a span of 90 ft, the vortex strength,

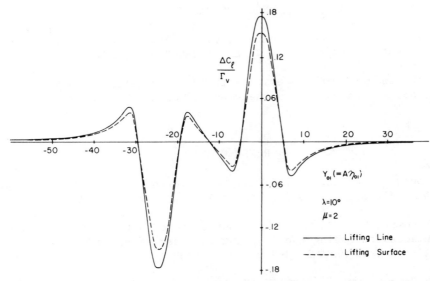

a few minutes after take-off, might be about 500π ft^2/sec. If the following aircraft of chord 2ℓ($= 5$ ft) were flying across the trailing vortices at a speed of 250 ft/sec, the corresponding value of Γ_v would be 0.4. The added induced lift ΔC_L and rolling moment ΔC_ℓ could then be readily deduced from the results given. For instance, Fig. 8 indicates that $\Delta C_L \simeq -0.32$ when the following aircraft is between the trailing vortices. If this aircraft were flying initially at a C_L of 0.2, the total lift coefficient would become negative, indicating a complete loss of lift which could have disastrous consequences if this were to happen fairly near the ground.

Further work is planned to determine the airloads and moments on particular aircraft when flying over trailing vortices of known strength. In the present preliminary study, the aircraft was assumed to be maintained in level flight with the pilot in full control. In practice, however, the pilot would not be able to maintain control when passing closely over the trailing vortices of a much heavier aircraft and further work is required to study the response of an aircraft with and without pilot control. Figure 4b, however, indicates that any attempt by the pilot to maintain control of his aircraft would tend to impose heavier loads on the aircraft structure because of the rapid changes in airloads that take place while crossing the vortex. For instance, $\Delta C_L/\Gamma_v$ drops from 0.8 to -1.0 in less than 0.1 seconds when $\lambda=60°$ and the change would be even more rapid when crossing at right angles. Similarly, Fig. 5b shows the big changes in rolling moment that take place which could produce rapid roll of a light aircraft that the pilot might not be able to control.

ACKNOWLEDGEMENTS

This work was supported by the U.S. Army Research Office - Durham under Project Themis Contract DAHC04-69-C-0015.

The authors wish to express their appreciation to Mrs. Kay Boyd for her assistance in the preparation of this paper for publication.

REFERENCES

1. Jones, W. P. and Rao, B. M., "Wing-Vortex Interaction," Texas A&M University, Project Themis Aero Report, Jan. 1970.

2. McGowan, W. A., "Calculated Normal Load Factors on Light Airplanes Traversing the Trailing Vortices of Heavy Transport Airplanes," NASA TN D-829, March 1969.

3. McGowan, W. A., "Trailing Vortex Hazard," SAE Transactions, Report No. 68-220, Vol. 77, 1968, pp. 740-753.

4. Wetmore, J. W. and Reeder, J. P., "Aircraft Vortex Wakes in Relation to Terminal Operations," NASA TN D-1777, April 1963.

5. Spreiter, J. R. and Sacks, A. H., "The Rolling Up of the Trailing Vortex Sheet and Its Effect on the Downwash Behind Wings," Journal of Aeronautical Sciences, Vol. 18, Jan. 1951.

6. Jahnke, E. and Emde, F. Funktionentafeln. 1933.

APPENDIX I

Evaluation of $P_m(\Theta_1, \phi, \phi_1)$ Integrals

The integral P_m is here defined by

$$P_m(\Theta_1, \phi, \phi_1) = \frac{1}{g} \int_0^\pi \frac{(D-1)\cos m\Theta \; d\Theta}{\cos\Theta - \cos\Theta_1} \tag{1}$$

where $D = [1+g^2(\cos\Theta - \cos\Theta_1)^2]^{1/2}$ and $gA|\cos\phi - \cos\phi_1| = 1$.

It can readily be proved that

$$P_{m+1} + P_{m-1} = 2P_m \cos\Theta_1 + \frac{2}{g} \int_0^\pi (D-1)\cos m\Theta \; d\Theta \qquad (2)$$

so that

$$P_1 = P_o \cos\Theta_1 + \frac{1}{g} \int_0^\pi D \; d\Theta - \frac{\pi}{g} \qquad (3)$$

and

$$P_{m+1} = 2P_m \cos\Theta_1 - P_{m-1} + \frac{2}{g} \int_0^\pi D \cos m\Theta \; d\Theta; \; m>0$$

The above recurrence relations can then be used to evaluate the P_m integrals once P_o has been determined and provided $X_m \equiv 1/g \int_0^\pi D \cos m\Theta \; d\Theta$ is known for all the values of m needed. The latter integrals can be expressed in terms of complete and incomplete elliptic integrals which have been tabulated but because of the complicated nature of the corresponding formulae obtained, it is easier to evaluate them numerically. The integrands of the X_m integrals have no singularities and, hence, the integrals may be evaluated numerically without difficulty.

The integral P_o can be expressed in terms of tabulated elliptic integrals by the use of the transformation

$$\cos = \frac{(1+c)\cos\psi + 1-c}{1+c + (1-c)\cos\psi} \qquad (4)$$

where

$$c = \left[\frac{1+g^2(1-\cos\Theta_1)^2}{1+g^2(1+\cos\Theta_1)^2}\right]^{1/2}$$

After considerable reduction, it was found that

$$P_o(\Theta_1,\phi,\phi_1) = - \frac{2}{g \sin\Theta_1} (F_1 E_{\psi_1} - E_1 F_{\psi_1}) - 2B \qquad (5)$$

where the elliptic integrals

$$F_1 = \int_0^{\pi/2} \frac{d\psi}{\Delta} \quad , \quad E_1 = \int_0^{\pi/2} \Delta d\psi$$

$$F_{\psi_1} = \int_0^{\psi_1} \frac{d\psi}{\Delta} \quad , \quad E_{\psi_1} = \int_0^{\psi_1} \Delta d\psi$$

and $\Delta \equiv (1 - k^2 \sin^2 \psi)^{1/2}$

The parameter k is defined by

$$k^2 = \frac{1}{2} \left(1 - \frac{1 - g^2 \sin^2 \Theta_1}{m_0^2} \right) \tag{6}$$

where

$$m_0^2 = [(1 + g^2(1 - \cos\Theta_1)^2)(1 + g^2(1 + \cos\Theta_1)^2)]^{1/2}, \quad \ldots \quad m_0 > 0$$

$$= \frac{g}{k(1-k^2)^{1/2}}$$

The last term of equation (5) is given by

$$B = \frac{\pi}{2} + F_1 F(k', \chi) - F_1 E(k', \chi) - E_1 F(k', \chi) \tag{7}$$

where

$$F(k', \chi) = \int_0^{\chi} \frac{d\chi}{\Delta(k', \chi)}$$

$$E(k', \chi) = \int_0^{\chi} \Delta(k', \chi) \, d\chi$$

$$\Delta(k', \chi) = (1 - k'^2 \sin^2 \chi)^{1/2}, \quad k'^2 = 1 - k^2$$

and $\cos\chi = \dfrac{1 - c}{1 + c}$

Values of all the integrals involved in (5) can be found tabulated in Ref. 6.

The integrals X_o, X_1 ... X_m are considered next and, as a matter of interest, are expressed in terms of elliptic integrals

$$X_o = \int_0^\pi [A^2(\cos\phi-\cos\phi_1)^2 + (\cos\theta-\cos\theta_1)^2]^{1/2} \, d\theta \qquad (8)$$

$$= \frac{2m_o E_1}{g} + 2B\cos\theta_1$$

When $\phi \simeq \phi_1$, X_o reduces to

$$X_o \simeq \pi\cos\theta_1 + 2\sin\theta_1 - 2\theta_1\cos\theta_1 \qquad (9)$$

It can also be shown that

$$X_1 = \int_0^\pi [A^2(\cos\phi-\cos\phi_1)^2 + (\cos\theta-\cos\theta_1)^2]^{1/2} \cos\theta \, d\theta$$

$$\qquad (10)$$

$$= \frac{m_o}{g} [F_1(\cos\theta_1 -\cos\chi) - E_1\cos\theta_1 - \frac{(1+g^2)B}{m_o g}]$$

When $\phi \simeq \phi_1$, similarly, it can be shown by integration that

$$X_1 \simeq \theta_1 - \frac{\pi}{2} - \sin\theta_1 \cos\theta_1 \qquad (11)$$

which also may be deduced from the general formula.

The corresponding expression for X_2 is

$$X_2 = - \frac{2B\cos\theta_1}{g^2} + \frac{2m_o E_1}{3g} (\frac{2}{g^2} + \sin^2\theta_1)$$

$$\qquad (12)$$

$$+ \frac{2m_o F_1}{3g} [\cos^2\theta_1 - \cos\theta_1\cos\chi + \frac{2}{m_o^2} (2\cos\theta_1\cos\chi - 1)]$$

It can be readily deduced that, when $\phi \simeq \phi_1$,

$$X_2 \simeq \frac{2}{3} \sin^2\theta_1 \qquad (13)$$

AIRCRAFT WAKE TURBULENCE CONTROLLABILITY EXPERIMENT

R. P. Johannes

Air Force Flight Dynamics Laboratory

Wright-Patterson Air Force Base, Ohio

The development of extremely large, high gross weight air-craft over the past several years has greatly increased the significance of aircraft wake turbulence. The increased magnitude of this turbulence and its effect on other aircraft is of major concern to the aircraft community. This concern is evidenced by the very fact that this symposium is being held for the purpose of considering the wake turbulence phenomenon and developing methods to cope with its effects. From a practical, operational point of view, perhaps the two most critical areas of concern are the induced structural loads and vehicle upset which can result from wake vortex encounter.

The Air Force Flight Dynamics Laboratory has recently completed an advanced development program, titled, "Load Alleviation and Mode Stabilization (LAMS)", which addressed the problem of providing active control of the structural responses of a large flexible aircraft. The primary goal of this effort was to show that the fatigue damage rate due to turbulence encounter could be significantly reduced through the active control to minimize the loads and vary the structural responses. Reference 1 summarizes the results of this phase of the program. This technology is directly applicable to the control of structural loads induced by wake vortex encounter as well.

Near the end of the LAMS program, a flight test experiment was conducted which addressed the other problem area, that of vehicle upset due to wake turbulence encounter. This experiment is the subject of this paper. The object of the experiment was to evaluate the effectiveness of a properly mechanized automatic control system in minimizing aircraft upset and improving the controllability which can result from wake turbulence encounter. The practical application that was addressed was to evaluate the potential for improving the paradrop mission capability of the Military Airlift Command (MAC). The paradrop mission is typically one wherein the requirements for precision control are quite stringent and wherein the turbulence environment is most severe due to the large number of aircraft flying at slow speeds, high angles of attack, and at essentially the same altitude. Any improvements that could be obtained in the ability to control the vehicle during this mission phase would be highly desirable for application to future aircraft.

The test aircraft which was used for this experiment is the LAMS test vehicle, NB-52E 56-632, which is shown in Figure 1. This airplane was initially the early model B-52 loads aircraft, and consequently, is highly instrumented for in-flight structural loads measurements. It will also be noted that the vehicle is equipped with a gust boom which is capable of measuring the characteristics of the environment that the aircraft is flying through. During the LAMS program, the aircraft was rather extensively modified to include hydraulic powered actuators on all control surfaces. Figure 2 shows a schematic of the vehicle and the modifications that were made. These modifications

LAMS TEST AIRCRAFT
FIGURE 1

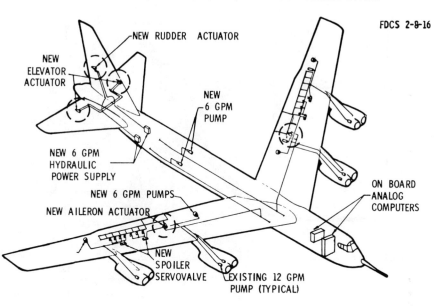

FIGURE 2

included the addition of high performance hydraulic actuators on the rudder and elevator surfaces as well as high performance actuators on the aileron surfaces and spoilers in the wings. Since the basic airplane did not initially have hydraulic powered surfaces, the hydraulic capability on board required augmentation through the use of electric motor driven pumps.

The next two figures show the response characteristics of the actuation equipment which was developed for the aileron actuator and the spoiler actuator, respectively, to provide quantitative insight into the meaning of "high performance". Figure 3 shows the response of the aileron actuator. You will note that the frequency bandwidth at the 3 db point is on the order of 7 cycles per second. During the development of this device, it was found that the requirement for mechanical, as well as electrical input, provided a limiting factor on this frequency response capability. The dotted envelopes show the initial design goals and the solid lines indicate the performance that was achieved. It will be noted that the performance does not fall within the envelope that had been projected, however, performance was found to be satisfactory. Figure 4 shows the

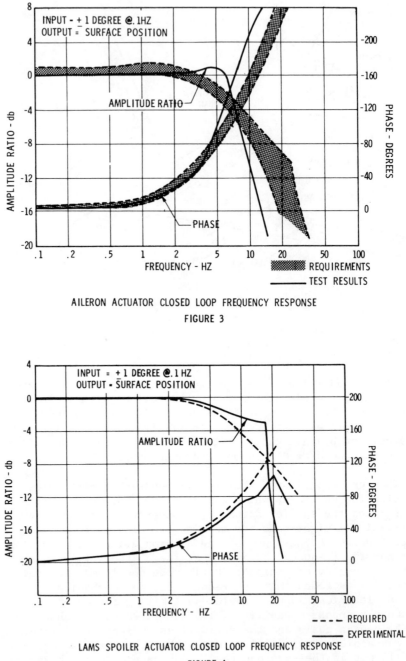

INPUT - +1 DEGREE @.1HZ
OUTPUT = SURFACE POSITION

AMPLITUDE RATIO

PHASE

FREQUENCY - HZ

REQUIREMENTS
TEST RESULTS

AILERON ACTUATOR CLOSED LOOP FREQUENCY RESPONSE
FIGURE 3

INPUT = +1 DEGREE @.1 HZ
OUTPUT - SURFACE POSITION

AMPLITUDE RATIO

PHASE

FREQUENCY - HZ

REQUIRED
EXPERIMENTAL

LAMS SPOILER ACTUATOR CLOSED LOOP FREQUENCY RESPONSE
FIGURE 4

same type of data for the spoiler actuator. In this case, the
device is a pure fly-by-wire commanded device, and as you can
see the frequency response is significantly higher. The thrust
and stroke capabilities are similar to that of the aileron
actuator and the performance is truly significant when it is
noted that this actuator is a straightforward, relatively simple
modification of the existing, off-the-shelf spoiler actuator
used on the B-52 fleet aircraft. In this case, the performance
goals were exceeded, as shown by the solid lines.

Other modifications included the addition of an extensive
on-board analog computational capability and the conversion of
the left pilot station from mechanical to a non-redundant fly-
by-wire system. These modifications greatly increased the
versatility of the aircraft for test purposes. The analog
capability is provided by two Electronic Associates, Inc. TR-48
desk top computers which had been hardened for the airborne
environment. Figure 5 shows one of these computers. Appropriate
use of interconnect patch panels in conjunction with the analog
computers allows extreme versatility in the mechanization of
various control schemes with minimal effort. The structural
mode control system, or LAMS computer, is shown in Figure 6.

FORWARD COMPUTER

FIGURE 5

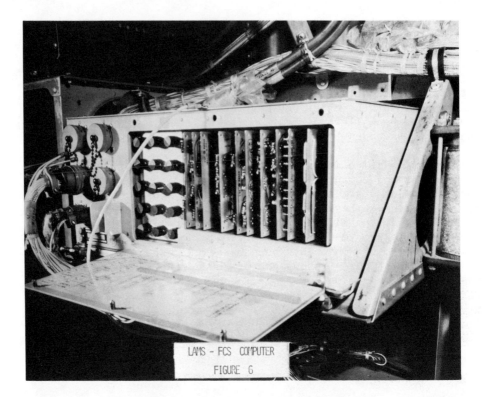

LAMS - FCS COMPUTER

FIGURE 6

You will note that this is a hard wired computer which the pilot
has the capability of engaging or disengaging, in conjunction
with various other control functions which may be generated by
the on-board computers.

Previous experience with the LAMS vehicle and various levels
of natural turbulence had indicated that the LAMS controller
itself provided significant improvements in the handling of the
test aircraft over that of the basic B-52 vehicle. However,
since the task to be considered in this experiment was formation
flying in the presence of wake turbulence, it was felt that
some type of an outer loop autopilot mode would be most beneficial.
A Pitch and roll axis outer loop control function was implemented
to aid the pilot in controlling the vehicle for this control
task. Figure 7 shows a block diagram of this outer loop system.

The roll axis system can be described as a roll command
system in which the commanded roll angle is proportional to
pilot's control wheel deflection. A roll rate limiter is used
to act as a buffer between a pilot's wheel position and the roll
command signals to smooth out the aircraft response. The pitch
axis system is an attitude hold system which has a pitch rate

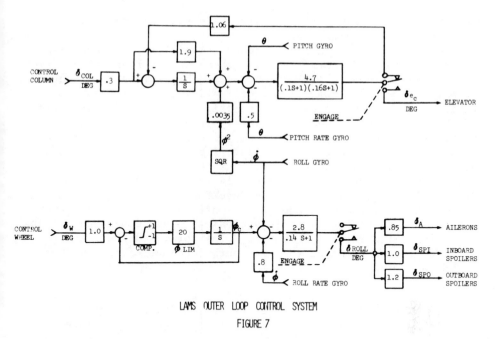

LAMS OUTER LOOP CONTROL SYSTEM

FIGURE 7

command input capability. In order to provide flight path control during turns a cross coupling term proportional to bank angle squared is included.

This control system which is referred to as the LAMS Outer Loop (LOL), system was mechanized on the two Electronic Associates, Inc. TR-48 computers which are installed on the test vehicle. This approach provided a convenient and inexpensive method of implementing the experimental control system and allowed sufficient flexibility to evaluate gain and time constant variations during the in-flight parametric investigations. It should be noted that all available control surfaces including both in-board and out-board spoiler panels are being commanded by this outer loop system. The yaw axis is controlled by the basic LAMS controller which provides appropriate damping. The gains shown in the figure are those which were determined to be nominal for a 270,000 to 300,000 pound aircraft.

The in-flight experiment itself consisted of two flights of approximately three hours duration each, wherein the B-52 test aircraft was flown in formation with a C-141 and explored the wake turbulence generated. The first half of the first flight was conducted at 15,000 feet at airspeeds ranging from 180 to 280 knots indicated airspeed and consisted of a series of control system checkouts which included a gain search, manual flutter transients, doublet inputs, and controllability checks. The loop gains were varied from nominal to 150% of nominal. This initial checkout was accomplished without the support of the C-141 aircraft. This initial checkout indicated that the original nominal gain setting which has been determined analytically provided the best handling qualities for the flight conditions to be evaluated. After the initial system checkout on the first flight, the C-141 aircraft rendezvoused with the test vehicle and the wake envelope was determined qualitatively with the C-141 stabilized at 15,000 feet altitude in smooth air. The procedure for this qualitative evaluation was to note the encounter and departure from the wake turbulence using the LAMS B-52 while ascending, descending, and passing laterally behind the C-141 vehicle. This provided the estimates of the depth and width of the wake turbulence at given distances behind the C-141.

Since the test aircraft was not equipped with appropriate radar ranging equipment, a crude, but effective sighting device was used. Figure 8 shows a diagram of the sighting template which was used for the ranging with the C-141 aircraft. As indicated, this was calculated to provide two range measurements for an

C-141 SIGHTING TEMPLATE

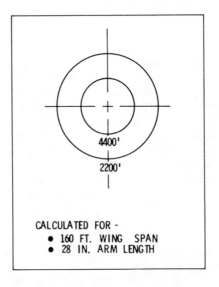

FIGURE 8

aircraft with a wing span of 160 feet. The device was used merely
by the pilot holding it at arms length ahead of him, sighting
on the aircraft and estimating the distance. Although as mentioned
previously, this is a rather crude device it was inexpensive, easy
to use and provided range measurements with sufficient accuracy
for our purposes. The initial testing of the LAMS outer loop system
in the C-141 wake was at an altitude of 15,000 feet at 250 knots
indicated airspeed. This was done to evaluate the upset hazard
before proceeding to the lower altitude wake test. The final
testing was accomplished at altitudes of 1000 feet and 5000 feet
at 180 knots indicated airspeed. The test results which were
obtained from the experiment are shown on the next few charts.

Figure 9 shows a qualitative evaluation of the C-141 vortex
field at a range of 2200 feet. This figure indicates that the
width of the field is approximately 180 feet and that the depth
is approximately 150 feet. These overall dimensions for the vortex
field appears to be in agreement with those which were predicted
by theory. However, subsequent reduction of the data obtained
from the gust boom during the test has indicated that significantly
higher velocities and smaller core diameters than had been predicted were present. This particular area of discussion is beyond
the scope of this paper, however, more detailed information is

C-141 WAKE DIMENSIONS
TRAILING VORTEX SHEET

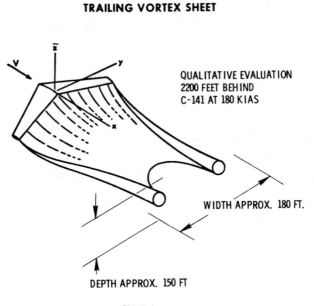

QUALITATIVE EVALUATION
2200 FEET BEHIND
C-141 AT 180 KIAS

WIDTH APPROX. 180 FT.

DEPTH APPROX. 150 FT

FIGURE 9

available in Reference 3. In order to obtain an indication of the
performance of the LAMS outer loop system, data was taken with
both the system engaged and with the aircraft unaugmented. Figure
10 shows the excursions in roll attitude with both system on and
off. In this case, the pilot's task was to fly a straight and
level in formation with the C-141 aircraft at a given range. This
figure shows the undesired roll attitude excursions. It can be
seen that these excursions vary from approximately ±10° with the
system off to approximately ±2 1/2° with the system on. This is
considered to be a significant reduction in undesired roll attitude
excursions. Figure 11 shows control wheel activity as the pilot
is attempting to maintain the wings level attitude with both system
on and system off. Here we can see that the control wheel deflections go from approximately ±60° to ±10°, which is again a very
significant reduction in pilot activity. This increased performance
is attributed to increased roll response to pilot commands and/or
wake turbulence upsets.

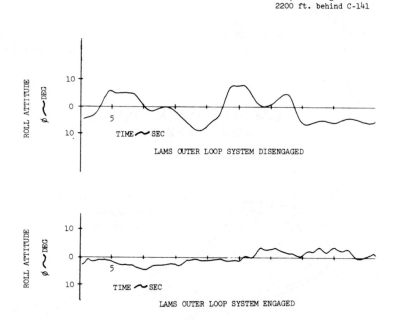

LAMS TEST VEHICLE ROLL ATTITUDE
IN C-141 WAKE TURBULENCE

FIGURE 10

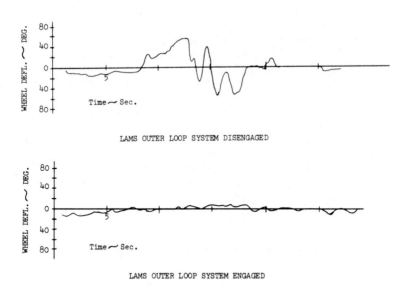

180 KIAS
280,000 Lb. Gross Wt.
2200 Ft. Behind C-141

LAMS OUTER LOOP SYSTEM DISENGAGED

LAMS OUTER LOOP SYSTEM ENGAGED

LAMS B-52
PILOT ACTIVITY IN C-141
WAKE TURBULENCE

FIGURE 11

A limiting factor in this type of approach is the rolling moment capability of the aircraft. In order to evaluate the applicability of these test results to the paradrop mission, an analytical comparison of the B-52 at the test condition and the C-141 at the paradrop flight condition was performed. Figure 12 shows the results of this analysis and indicates that at the C-141 paradrop condition the maximum steady state roll rate available is almost identical to that of the B-52 in the test condition. Further, the roll rate after one second is also very compatible. This would imply that such a system mechanized on the C-141 aircraft would be capable of providing similar results over the unaugmented vehicle.

In order to gain an insight into the acceptability of these advanced control functions to operational pilots, the second flight in the two flight experiment was made for the purpose of demonstrating the system to Air Force pilots from the Military Airlift Command. Perhaps the best way to present the results of this demonstration is to give appropriate excerpts from the post flight debriefing.

Lt Col X – "I have flown quite a bit of combat. That is, the old tactical flying, in formation, and on occasion I have been in the wake turbulence and it's rather awesome to be in there. I wasn't prepared for the ease with which we could fly through this wake turbulence with this system. To me it's real impressive. I'll tell you, I'm frankly very impressed. It would be a tremendous thing if we would have it on our airplanes. I could find no fault at all with it, frankly."

Major Y – "Those are my sentiments exactly. I feel the same way as Lt Col X. I felt one time we were in about a thousand feet (trailing position relative to the C-141) and we were sitting right there in it (wake turbulence) with no control problems at all .... So it's very impressive. I think it's wonderful."

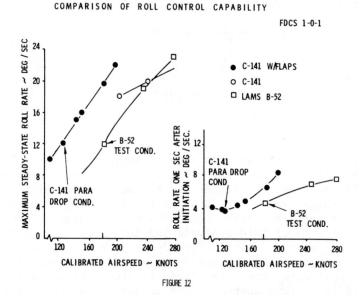

FIGURE 12

The conclusions that were drawn as a result of conducting this experiment were that control systems for large aircraft which take into account the structural characteristics of the vehicle, such as the LAMS system, can be mechanized with appropriate outer loops which will significantly reduce the upset and minimize pilot activity due to wake turbulence encounter. The improved controllability that is provided gives significant improvement in the precision with which formation flying can be accomplished in the presence of wake turbulence and this improved capability should be also beneficial to other precision control maneuvers such as landing. Admittedly, this experiment was somewhat less than rigorous in many aspects, however, I feel that there can be no doubt that improved control can do much to minimize many of the problems of wake turbulence encounter. The results of this experiment definitely support the desirability of further development in this area as new and larger vehicles are developed and become operational in the future.

REFERENCES

1. AFFDL-TR-68-158, "Aircraft Load Alleviation and Mode Stabilization (LAMS)", Burris, P. M. and Bender, M. A., April 1968.

2. Boeing Document D3-8100, "Demonstration of the LAMS-FCS in the Wake of a C-141", August 1969.

3. Boeing Document D3-8357, "C-141 Wake Turbulence Evaluation", May 1970.

FOG FORMATION AND DISPERSAL BY TRAILING VORTICES

P. Baronti and S. Elzweig

Advanced Technology Laboratories, Inc.

Jericho, New York

ABSTRACT

A model is presented to explain why fogs can be maintained for extended periods of time. The model considers the fog as a dynamical system whereby droplets depletion through fall is balanced by a continuous process of evaporation and condensation. The controlling parameters of the process are indicated. Possible means of fog dispersal by the velocity field induced by trailing vortices are suggested.

1. INTRODUCTION

Recent experiments indicate that helicopters may be successfully used for fog dispersal.[1,2,3] The explanation is quite intuitive. The vortex downwash forces the air above the fog to mix with the moisture laden air below; then if the air above the fog is sufficiently dry or warm, the decreased humidity of the resulting mixture permits evaporation of the fog droplets. The extent and the depth of the clearing, depends upon the entrainment characteristic of the helicopter wake and the stability structure of the fog layer.

The analysis of the flow field induced by vortex downwash and the ensuing mixing and evaporation of masses of humid air at different temperatures is conceptually simple. For the case of a hovering helicopter, either with or without ground effects, a rather adequate determination of the region that can be cleared, may be obtained by representing the helicopter downwash by a jet and by utilizing standard mixing analyses.[4]

Two conclusions may, however, be drawn from the experiments. Firstly, fog dispersal is achieved by mixing large masses of air, and thus at the expense of large energy requirements. Secondly, a more meaningful prediction of the effect generated by the vortices, in terms of extent and duration of the clearing, requires a good knowledge of the stability structure of the fog layer.

Investigators were long puzzled as to why fogs stay up and do not disappear through fallout of the droplets to the ground because of gravity. Several explanations are frequently offered such as sustainment of droplets by updrafts, replenishment of falling droplets by droplets diffusion or by continuous condensation on new nucleii. But none of these explanations is quite satisfactory. If, therefore, a better understanding of the phenomenon can be gained, the effort may be rewarding in that it may provide basic information on the structure of the fog layer and indicate more rational means for fog dissipation.

The purpose of this study is to propose a model for the self sustainment of fog, to show that the mechanism by which fog stays up is the result of a delicate balance in fog properties, and to indicate how aircraft wakes may be utilized to upset this balance and achieve fog dispersal.

Most of the material contained herein is taken from Reference 5. The reader is referred to this report for more details.

2.   THE FOG MODEL

The fog is assumed to have already formed, to be stationary and infinite and uniform in the horizontal direction. Therefore the problem is one dimensional with gradients allowed only in the vertical direction. The model is representative of fogs which, once formed, have large lateral dimensions and which change slowly with time. The model is postulated as follows:

(a)   Because of gravity and diffusion the droplets reach the ground. At the ground the water evaporates and the water vapor diffuses upward because of a concentration gradient. The heat of vaporization is provided by the heat transfer at the ground. The coefficient of diffusion will be, in general, turbulent.

(b)   At any fog height, the rising vapor produces supersaturation. As a consequence part of the vapor condenses either on the already existing droplets or on available nucleii to produce new droplets. The heat of condensation is carried away by local temperature gradients and radiation.

(c) All the vapor arriving at the top of the fog condenses. The heat released by condensation is transferred from the fog top by radiation and conduction.

A schematic diagram of the assumed model is shown in Figure 1.

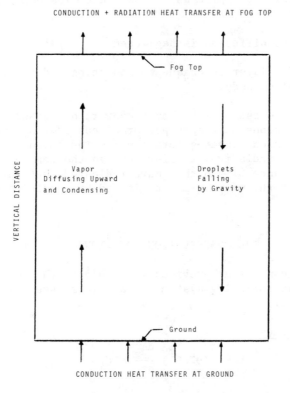

Fig. 1. Schematic Diagram of Fog Dynamics

Although many refinements are available within the framework of the model, several assumptions are now introduced in order to simplify the mathematical description and expedite numerical computations. The assumptions are:

(d) At any fog height the process of condensation from the rising vapors occurs rapidly and the amount of condensation is determined by the local temperature. The, the saturation conditions will define the amount of vapor concentration to be used for the determination of vapor diffusion.

(e) At any height, with the exception of the fog top, condensation occurs on droplets already present at that particular location.

(f) At the fog top the vapor condenses on available nucleii to form new droplets, all of which are taken to be the same size. The number of activated nucleii and thus of water droplets, is assumed to be known.

(g) Accretion (droplet collision and coalescence) is not considered.

(h) Droplet diffusion is neglected.

(i) The fog layer is opaque to radiation and the fog top radiates as a black body.

Some of these assumptions may appear rather drastic. Yet, parallel computations have been performed considering the effects of droplet diffusion, super-saturation (as indicated in References 6 and 7), and of radiative transfer through the fog layer (as indicated in Reference 8) and they have indicated that the qualitative features of the model are not altered by retaining these assumptions.

3. THE CONSERVATION EQUATIONS

Under the conditions previously postulated the fog model is described by the following system of equations (see Reference 5 for more details):

Air Continuity

$$\rho_a v = \rho (D^L + D^T) \frac{d\alpha_a}{dz} + C_1 \tag{1}$$

Total Continuity (air, water vapor and water droplets)

$$\rho v + \rho_p v_p = C_2 \tag{2}$$

where C_1 and C_2 are two integration constants.

Momentum For the gaseous mixture, air plus vapor, the momentum equation may be simplified to yield the hydrostatic equation

$$\frac{dp}{dz} = -\rho g \tag{3}$$

With regard to water droplets, their fall velocity may be taken, for all practical purposes, as

$$v_p = v_t + v \tag{4}$$

Energy   The energy equation of the mixture composed of air, water vapor and water droplets may be written as

$$\rho \, v \, L - k \frac{dT}{dz} - k^T \, \Gamma_s = C_3 \tag{5}$$

where C_3 is an integration constant, where $k \equiv k^L + k^T$, and where Γ_s, the saturation lapse rate,[9] is defined here as

$$\Gamma_s = \frac{\Gamma}{1 + \frac{L}{C_p} \frac{d(\rho_v/\rho_a)}{dT}}$$

The term $k^T \, \Gamma_s$ accounts for the change of temperature with height of parcels of mixture, as they move vertically.

The constants of integrations C_1, C_2 and C_3, can be immediately determined. In fact

$$C_1 - C_2 = 0$$

since, neither air nor water is allowed to enter the system, and

$$C_3 = \dot{q}_g$$

where $\dot{q}_g$ is the heat transfer rate to or from the ground, as can be seen by an energy balance at the ground.

If steady state is to exist the energy supplied by the ground must be balanced by the energy released at the fog top. The energy flux into the fog layer is $\dot{q}_g$. The energy flux from the fog top, assuming the top behaves as a black body radiator, is $\sigma T^4 - k^+ (dT/dz)^+$, where $(dT/dz)^+$ and k^+ are the temperature gradient and the heat conductivity above the fog top. Thus

$$\frac{dT^+}{dz} = \frac{\sigma T^4 - \dot{q}_g}{k^+} \tag{6}$$

provides the temperature gradient above the fog in terms of ground heat transfer. It will be seen below that such gradients are, in general, positive and very steep, and correspond to conditions of stable stratification above the fog.

The conservation equations must be implemented by a description of the turbulent transport coefficients. For the gaseous mixture, the usual assumption is made of unity Lewis number, so that $(\rho D) = k/C_p$. Furthermore, the point of view is taken that k^T is determined by the condition of heat transfer at the ground.

Indeed, it is the heat transfer from the ground that controls the rate of vaporization, the temperature distribution through the fog and the rate of energy transfer from the fog top. These, in turn, determine the degree of turbulence in the fog for windless conditions. In performing the integration of the equations (in the following section) it was found convenient to follow the inverse procedure of assuming an eddy conductivity (of the order of $10^4 \div 10^5$ times the laminar value,[10] and on the basis of this value to determine the heat transfer at the ground. Computations have been performed by taking constant values of k^T and values of k^T that are dependent on the vertical distance z. A quoted[11] expression for k^T is

$$k^T = \rho\, C_p \left[- 32 \frac{g}{T}\, (0.4z)^4 \left(\frac{dT}{dz} + \Gamma_s \right) \right]^{\frac{1}{2}} \tag{7}$$

The present analysis, for steady state conditions, provides results leading to values of k^T that are consistent with Equation 7.

To complete the description of the system, the conservation equations are implemented by a condition of phase equilibrium of the form

$$\rho_v = \frac{m_v}{R} \left(\frac{a}{T} + b + cT \right) \tag{8}$$

where a, b and c are the constants of a vapor pressure-temperature fit, by the equation of state

$$\rho = \left(\frac{\rho_a}{m_a} + \frac{\rho_v}{m_v} \right) RT \tag{9}$$

and by the relationship

$$\rho_p = \frac{4}{3} \pi\, r_p^3\, \rho_L\, n_p \tag{10}$$

Thus, the system of Equations 1, 2, 3, 4, 5, 8, 9 and 10 matches the number of unknowns, v, v_p, T, p, p_v, ρ_a, ρ_v and r_p.

These equations may be integrated numerically. The procedure used here specifies the fog height, T, p_a, r_p and n_p at the fog top. These conditions suffice to initialize the problem, and to determine the heat transfer from the top. A marching scheme is then established by which fog properties are determined at each height.

4.   NUMERICAL RESULTS AND DISCUSSION

Some typical numerical results are presented in this section
to illustrate the main features of the model. The results are
then related to available experimental data and interpreted in the
light of this data.

For the first set of computations the fog has been taken to
be 100 ft. high and the following properties assumed at the fog
top:

$$T = 40°F = 4.44°C$$
$$\rho_a = 1 \text{ atm}$$
$$n_p = 100/cm^3$$
$$r_p = 10 \text{ } \mu$$

Three different values of turbulent conductivity, $k^T = 10^5 k^L$,
$5 \times 10^4 k^L$ and $10^4 k^L$, have been allowed.

Presented in Figure 2, is the temperature distribution through
the fog layer, for each value of k^T. The corresponding values of
ground heat transfer and temperature gradients above the fog top
where $k^+ = k^L$, see Equation 6, are tabulated below:

k^T(cal/cm-sec°K)	$\dot{q}_g$(cal/cm²sec)	(dT/dz)$^+$(°F/ft.)
5.0	4.45x10⁻⁴	0.76x10⁴
2.5	5.60x10⁻⁴	0.75x10⁴
0.5	6.42x10⁻⁴	0.74x10⁴

The main characteristics of these results are: heat transfer from
the ground into the fog, large temperature inversions at the top,
temperature slopes that become increasingly negative with increasing values of k^T.

No experiment is known to the authors where simultaneous
measurements of profiles of temperature, liquid water content,
and particle density, as well as of heat transfer rates, have been
performed. Some experimental data does, however, exist. From
this data the following considerations may be made.

(a)   The model requires that heat transfer be provided from
the ground to the fog. No such measurement, when fog is present,
is known to the authors. Measurements of temperature profiles in
the ground are, however, reported in the literature. What can be
inferred here, is the possibility that such conditions may also
prevail when fog is present to give heat transfer rates of the
order of 5×10^{-4} cal/cm²sec, quoted above.

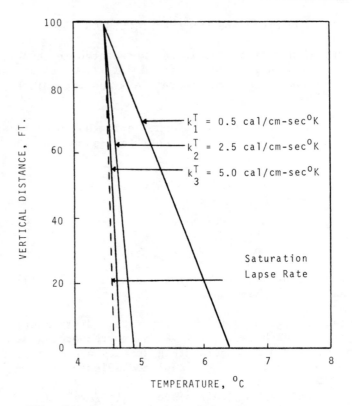

Fig. 2.   Temperature Profiles Through Fog Layer

(b)   Conditions of unstable stratification are reported to
exist within the fog layer, as shown in Figure 3, taken from Reference 13.   This condition may be recovered by the present analysis
as shown typically in Figure 2, where the temperature distribution
is compared with the saturation lapse rate for the case of $k^T = 5$
cal/cm sec °K.   Moreover, the value of k^T used to obtain this
temperature distribution, is consistent with the values that may be
obtained from Equation 7.

(c)   A sharp temperature inversion at the fog top results from
the present analysis.   The result is consistent with experimental
evidence[13] which indicates stable stratifications above fog layers
and rather uniform fog tops.

(d)   Measurements of droplets number density, average drop
diameter and liquid water content through a fog,[14] are shown in
Figure 4.   The present analysis predicts the distribution of liquid
water content shown in Figure 5, and droplets radii that change
only slightly with height.   However, the experimental distributions

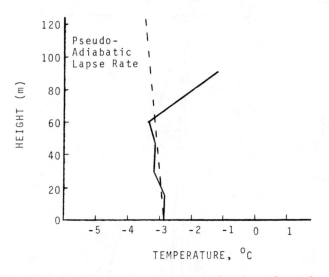

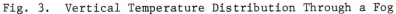

Fig. 3.   Vertical Temperature Distribution Through a Fog

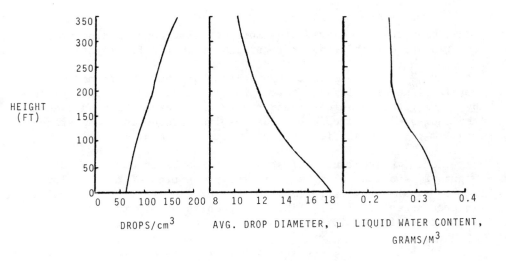

Fig. 4.   Properties of Fog in a Valley

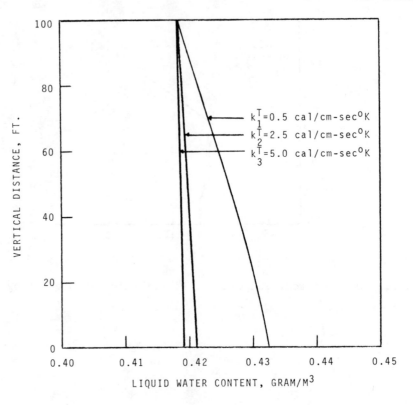

Fig. 5. Liquid Water Content Through Fog Layer

could have been recovered if a more detailed condensation process
and radiative heat transfer were allowed in the fog layer.

Additional numerical computations have been performed by
varying some of the parameters entering the analysis, for instance
by varying the turbulent conductivity with height, and by changing
the radius of the droplets. The results, presented in Reference 5,
conform to the basic features of Figures 2 and 5.

Some important considerations may now be made. As shown by
Figure 2, substantial changes in the properties of the fog layer
may be obtained by variations in eddy conductivity or, equivalently,
in ground heat transfer. This, in turn, seems to indicate the possibility of fog modification by ground heat transfer control.

An interesting situation arises if it is assumed that a vertical air flow is blown into the fog layer, either from the ground
upward or from the top downward. This case can be dealt by the
present analysis if, to retain the one dimensionality of the

problem, the ground is assumed to act as either a surface source or
a sink.   In the analysis, the only modification is that the air mass
flux, identified by the integration constant C_1, becomes non-zero.
The temperature profiles for the two air velocities of about
$\pm$ 0.5 cm/sec, and for T = 40°F, p = 1 atm, n_p = 100/cm, r_p = 10 μ
and k^T = 5 cal/cm sec °K are presented in Figure 6.   The changes
introduced into the fog layer by these relatively small air fluxes
are apparent from the figure.   These changes may not be consistent
with actual boundary conditions at the ground and at the top and,
thus, may result in fog modification.

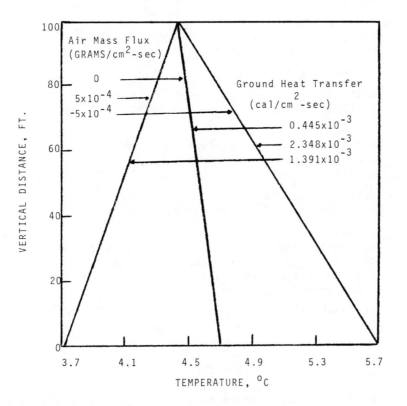

Fig. 6.   Effect of Vertical Air Velocity on Fog Properties

The possibility of modifying fog properties and thus enhancing
fog dispersal by trailing vortices becomes now evident.   The flow
field induced by the vortices will in fact alter the turbulence and
the stability structure of the fog layer.   Furthermore, vortex
downwash will diminish the upward diffusion of the water vapor and
increase the droplets fall rate.   All of these purposes may be accomplished by an aircraft flying relatively high above a fog layer,
thus affecting large extents of fog.

NOMENCLATURE

C_p	coefficient of specific heat
D^L, D^T	diffusion coefficient, laminar or turbulent
g	acceleration of gravity
k^L, k^T	thermal conductivity, laminar or turbulent
L	heat of vaporization
m_a, m_v	molecular weight of air or vapor
n_p	particle number density
p	total pressure
P_a	air partial pressure
P_v	vapor pressure of water
$\dot{q}_g$	ground heat transfer
R	universal gas constant
r_p	particle radius
T	temperature
v	velocity of gaseous mixture, mean value
v_t	terminal velocity
v_p	particle velocity
z	vertical coordinate
α_a, α_v	mass fraction of air or vapor, $\alpha_i = \rho_i / \rho$
Γ	adiabatic lapse rate
Γ_s	saturation lapse rate
ρ	density of gaseous mixture, mean value
ρ_p	liquid water content in fog
ρ_L	water specific gravity
σ	Stefan-Boltzmann constant

REFERENCES

1. Hicks, J. R., "Experiments on the Dissipation of Warm Fog by Helicopter-Induced Air Exchange over Thule AFB, Greenland," Special Rept. 87, U. S. Army Material Command, Cold Regions Research and Engineering Laboratory, Hanover, N. N., August 1965.

2. Plank, V. G., "Clearing Ground Fog with Helicopters," Weatherwise, Vol. 22, No. 3, June 1969.

3. Plank, V. G. and Spatola, A. A., "Cloud Modification by Helicopter Wakes," Journal of Applied Meteorology, Vol. 8, No. 4, August 1969.

4. Elzweig, S. and Baronti, P., "Preliminary Analysis of Fog Dispersal by Helicopters," ATL TM 157 September 1970.

5. Ferri, A., Baronti, P., and Elzweig, S., "A Dynamical Model of Fog," ATL TR 159 and AFOSR 70-2292TR, September 1970.

6. Fletcher, N. H., "The Physics of Rainclouds," Cambridge University Press, 1966.

7. Amelin, A. G., "Theory of Fog Condensation," Translated from Russian by the Israel Program for Scientific Translations, Jerusalem 1967.

8. Zdunkowski, W. G., Mecham, D., and Trask, D. C., "An Investigation of the Prediction and Stability of Radiation Fog," Technical Report ECOM-0213-F, September 1969.

9. Brunt, D., "Physical and Dynamical Meteorology," Cambridge University Press, 1952.

10. Godske, C. L., Bergeron, T., Bjerknes, J., and Bundgaard, R. C., "Dynamic Meteorology and Weather Forecasting," American Meteorological Society and Carnegie Institute of Washington, 1957.

11. Fleagle, R. G. and Businger, J. A., "An Introduction to Atmospheric Physics," Academic Press, 1963.

12. Lettau, H. and Davidson, B., "Exploring the Atmosphere's First Mile," Pergamon, New York 1957.

13. Fleagle, R. G., Parrot, W. H., and Barad, M. L., "Theory and Effects of Vertical Temperature Distribution in Turbid Air," Journal of Meteorology, 9, pp. 53-60, 1952.

14. Quarterly Progress Report, "Project Fog Drops. Investigation of Warm Fog Properties and Fog Modification Concepts," prepared by Cornell Aeronautical Laboratory, Inc., under NASA Contract No. NASA-156, Cal. Rep. No. RM-1788-P-22, February 1969.

AN ESTIMATE OF THE POWER REQUIRED TO ELIMINATE TRAILING VORTICES BY SUCTION

J. Menkes and F. H. Abernathy

University of Colorado and Harvard University

ABSTRACT

The fullest utilization under instrument flight rule (IFR) conditions of closely spaced parallel runways might very well depend ultimately on our ability to eliminate or at least to control the hazards of wake turbulence. The actual hazard is due to the possibility that the disturbance created over one runway will drift in an uncontrolled manner to a neighboring one.

A descending aircraft is affected by the interaction of its vortex system with the ground; if the ground effect could be modified, the vortex reaction would be affected. Such a modification is induced by suction. Suction in effect "removes" the ground and thus lets the vortices drift downward rather than horizontally. The vortex motion is thus influenced without relying on viscous dissipating mechanisms. The conceptual scheme involves placing one or more ditches which house the suction blowers between the runways.

The computer simulation of this scheme indicated the following main points.

1) The most adverse condition is presented by a weak vortex and winds of the order of 6 mi/hr or more.

2) The most favorable situation arises in the absence of wind and when the vortex is very strong, e.g., due to the take-off of a Boeing 747. Under those circumstances, the vortex drifts almost solely under its own induced motion into the ditch.

3) In most other cases one can expect a suction velocity of 10 ft/s or less to clear the runway of any vortex located at 100 feet above ground in 70 seconds or less.

4) The horsepower required is governed by the exit port rather than the suction inlet.

5) A conservative estimate is 2000 hp for a ditch with a surface area of 1×10^5 ft^2, two exit ports of about one-third this size, and a suction velocity of 10 ft/s.

6) The suction velocity over the runways is approximately 1 ft/s.

7) Increases in the horsepower do not substantially decrease the time required to suck down the vortex.

8) A secondary benefit derives from the possibility of using suction to clear the field of snow and ground fog.

A written version of this paper was not available. For further details on this work see "An Estimate of the Power Required to Eliminate Trailing Vortices by Suction", by Joshua Menkes and F. A. Abernathy, Institute for Defense Analysis Research Paper P-514.

PANEL DISCUSSION

Arnold Goldburg, Moderator
Boeing Scientific Research Laboratories

Steven C. Crow
The Boeing Company

Paul R. Owen
Imperial College

Walter S. Luffsey
Federal Aviation Administration

Coleman duP. Donaldson
Aeronautical Research Associates of Princeton, Inc.

Norman N. Chigier
NASA-Ames Research Center

Arnold Goldburg, Moderator: Each panel member was asked to address the question whether the research presented at this Symposium was relevant to the problems which the airframe manufacturer and the airways users and operators will face in the 1970-1980 period and beyond; or whether the research results and the research directions pointed towards questions which were not of interest or pointed towards questions which have already been resolved by direct flight experience. Each panel member spoke to this question and then the discussion was opened to questions and comments from the floor.

Steven C. Crow: I shall divide my comments into two parts, the first having to do with the scientific questions raised at the Symposium and the second with practical issues.

It must have been apparent to everyone that vortex experiments are lagging theory. The theoretical stability problem appears to be well in hand, with respect both to a simple vortex pair[1] and to the interacting jet vortices treated by Widnall and

Parks. Yet we have not resolved the most elementary questions
about axial flow in the cores of real airplane vortices, whether,
for example, the axial flow is toward the airplane, away from it,
or both. We have seen pictorial evidence that could support any
of those alternatives. Again, a member of our panel has published a theory of vortex turbulence,[2] and we have seen Donaldson
present another at this meeting. But the motion pictures suggest
that the cores of trailing vortices are laminar, and the question
remains open whether a simple vortex can sustain turbulence at
all. The experiments of Hoffmann and Joubert,[3] often cited here,
may have dealt with turbulence shed from their vortex generator
rather than with turbulence originating within the vortex core.

Experiment must answer two questions before theoretical work
on trailing vortices can proceed much further. First, what is the
structure of vortex cores immediately after they form behind a
wing, say five or ten spans downstream? Second, does a simple
vortex amplify or suppress turbulence within its core? The first
question can be answered by careful velocity-vector measurements
carried out in the near wake of a model wing. Fortunately, we
have heard about some new vector anemometers, which should reduce
the labor. The answer to the second question will rest on hot-
wire measurements of turbulence in vortex cores. Either a survey
of turbulence intensity can be carried out along the axis of a
vortex in a long tunnel, of the sort under construction by ARAP,
or a study of the turbulence energy budget can be carried out at
a fixed station. Either way, a decision could then be reached
whether turbulence grows or decays within a vortex core.
Professor Owen has reminded us about the difficulty of impaling a
vortex on a hot-wire, but there seems to be no alternative if we
intend to understand core diffusion any better than we do now.

I want to turn to the practical issue of trailing vortices as
an aeronautical hazard. The question of how to assess the danger
of trailing vortices came up several times but, I think, was not
wholly resolved. Judging from Condit and Tracy's film,* we would
agree that a 737 can be in a dangerous predicament within the in-
tact wake of a 747. On the other hand, a 707 is probably not
endangered by the wake of another 707. How do we cast those intuitive feelings into a quantitative definition of danger? Because
danger lies in the interaction between two airplanes, the definition
must involve properties of both the generating airplane g and the
following airplane f:

$$\text{Danger} = D(g,t) \quad .$$

---

* See Condit and Tracy, these Proceedings.

One definition that I find appealing is the following. Suppose airplane f momentarily falls into the eye of a vortex from airplane g and is subjected to a roll rate P. Suppose further that the maximum aileron-induced roll rate of airplane f is P_{max}, generally about 30°/sec during approach to landing. Then the danger D is defined as being proportional to P/P_{max}. One can easily show that P is proportional to Γ_g/B_f^2, where Γ_g is the circulation around the vortices of airplane g, and B_f is the wingspan of airplane f. Γ_g in turn is proportional to $(BV\,C_L/\pi A_R)_g$, where the span B, speed V, lift coefficient C_L, and aspect ratio A_R all pertain to the generating airplane.[4] Then the definition of danger becomes

$$D(g,f) = \frac{\left(BV\,\dfrac{C_L}{\pi A_R}\right)_g}{(B^2 P_{max})_f} \quad ,$$

where P_{max} is to be expressed in radians per second. Any coefficient could be placed in front of the fraction on the right hand side, of course, but it happens that $D(g,f)$, as defined, is almost exactly unity for the 747-737 event described by Condit and Tracy:

$$D(747,737) = 0.99 \quad .$$

The value D = 1 can therefore be taken as marginally dangerous. In the case of one 707 approaching after another,

$$D(707,707) = 0.20 \quad .$$

In general, a vortex encounter will tend to be dangerous (D $\approx$ 1) when the generating airplane has a large span, high lift coefficient, and low aspect ratio, and when the following airplane has a short span and low roll authority. Because it enters the definition of danger quadratically, the span B_f of the following airplane is the most crucial parameter.

In the event an encounter is too dangerous, the airplanes involved must be spaced apart far enough for natural processes to dissipate the vortices, or some artificial means must be found to destroy them. Several methods for destroying vortices were suggested at the Symposium, but the methods involve either massive airport installations or undefined performance penalties for the generating airplane. I should like to conclude my comments by suggesting an alternative method, namely artificial excitation of the vortex instability treated in Ref. 1 and expanded upon here by Widnall and Parks.

Condit and Tracy showed you some pictures and data that Earll Murman and I acquired during the Moses Lake flight test conducted by the FAA and Boeing.[5] The pictures showed the 747's vortices undergoing the now familiar symmetric instability, and the plot of displacement amplitudes showed that the phenomenon was indeed an exponential instability as predicted by theory, until the instability accelerated due to nonlinear interaction, shortly before the wake became incoherent. The implication of the experiment is that linear instability controls the duration of a vortex wake. The air above Moses Lake was fairly stable, so there was little turbulence to initiate the vortex instability, and the wake lasted 120 sec. The duration could have been cut short if the most rapidly growing mode of instability could have been excited by oscillating the 747's control surfaces.

The wavelength of the fastest growing mode is about 6.8B, equivalent to a period of oscillation τ of about 6 sec for a 747 during approach to landing. One way to excite the instability would be to execute a porpoising motion, but a 6-sec-period vertical oscillation would not sit well with the passengers. The accompanying figure depicts a better means of exciting the instability. The outboard ailerons oscillate symmetrically, while the inboard ailerons (flaperons) oscillate in opposition to keep the total lift of the wing constant. The horizontal stabilizer trims any pitching motion. The effect of the oscillations is to slosh the lift distribution cc_ℓ (local chord c times local lift coefficient c_ℓ) in and out as shown in the plot above the sketch of the airplane. The integral of cc_ℓ and thus the net lift remain constant, so the passengers feel little if any motion.

The spacing between trailing vortices is given in Ref. 4 as

$$b = \frac{C_L S}{(cc_\ell)_r} \quad ,$$

where C_L is the overall lift coefficient, S is the wing area, and $(cc_\ell)_r$ is the root loading. If the control surfaces cause $(cc_\ell)_r$ to vary in the manner

$$(cc_\ell)_r = (cc_\ell)_{ro} + \Delta(cc_\ell)_r \sin \frac{2\pi t}{\tau} \quad ,$$

then the vortex separation will vary as

$$b = b_o \left\{ 1 - \frac{\Delta(cc_\ell)_r}{(cc_\ell)_{ro}} \sin \frac{2\pi t}{\tau} \right\} \quad .$$

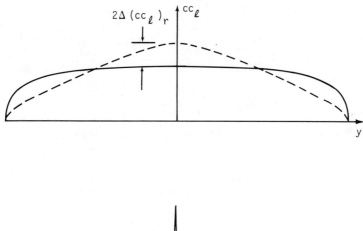

$2\Delta\,(cc_\ell)_r$ $\quad cc_\ell$

y

SLOSHING THE LIFT DISTRIBUTION BY
AILERON AND FLAPERON MOTION.

If the resulting instability grows as e^{at}, and the wake is assumed
to disintegrate when linear theory indicates that the vortices
touch, then the duration of the wake is given by

$$T = \frac{1}{a} \log_e \frac{(cc_\ell)_{ro}}{2\Delta(cc_\ell)_r} \quad .$$

According to Ref. 1,

$$a = 0.43 \left(\frac{V}{B} \frac{C_L}{\pi A_R}\right)$$

for the fastest growing mode of instability. In the case of a 707
during approach to landing, for example, $a = 0.061$ sec^{-1}, whereas
$a = 0.100$ sec^{-1} for the wake of a 747. If a five percent oscillation of $(cc_\ell)_r$ could be imposed by manipulating the 747's control
surfaces, then the trailing vortices would destroy themselves by
mutual induction in a time

$$T = \frac{1}{0.10} \log_e \frac{1}{2 \times 0.05} = 23 \text{ sec} \quad ,$$

compared with 120 sec required for natural disintegration in the
Moses Lake tests.

At the moment most airplane control systems are wired together so
that the ailerons move only antisymmetrically. That will prove
no problem in the future, when control is accomplished through
"fly-by-wire" stability augmentation systems. All that will be
required for wake disintegration is an appropriate circuit in the
stability augmentation system, to be activated on approach and
climbout. Exciting the large-scale instability is a satisfying
way of abating the vortex problem, since the energy of the vortices
themselves propells their destruction.

References

1. Crow, S. C., "Stability Theory for a Pair of Trailing Vortices,"
    AIAA Journal, Vol. 8, No. 12, December 1970, pp. ..... .

2. Owen, P. R., "The Decay of a Turbulent Trailing Vortex," The
    Aeronautical Quarterly, Vol. 21, February 1970, pp. 69-78.

3. Hoffmann, E. R. and Joubert, P. N., "Turbulent Line Vortices,"
    Journal of Fluids Mechanics, Vol. 16, No. 3, 1963, pp. 395-411.

4. Spreiter, J. R. and Sacks, A. H., "The Rolling Up of the Trailing
    Vortex Sheet and Its Effect on the Downwash Behind Wings,"
    Journal of the Aeronautical Sciences, Vol. 18, No. 1, January
    1951, pp. 21-32.

5. Crow, S. C. and Murman, E. M., "Trailing-Vortex Experiments at
    Moses Lake," Boeing Scientific Research Laboratories Technical
    Communication 009, February 1970.

<u>Paul R. Owen</u>: The physical features of the flow in a wake to which
this Symposium has so very effectively drawn attention are (i) the
presence of an appreciable axial flow along a vortex; (ii) vortex
bursting; (iii) the persistence of the core which appears to grow
rather gradually with increasing distance downstream from the wing;
and (iv) the ultimate instability of the trailing vortex pair and
the formation of loops.

    Having isolated these phenomena, it is evident, however, that
our understanding of certain of them is still vague and the accompanying physical explanations which have been proposed are contro-
versial in places: for example, the mechanism of vortex bursting
(although personally I am convinced by the Brooke-Benjamin theory)
and the existence of turbulence in a vortex core (which Coleman
Donaldson asserts is a consequence of the axial flow, yet which I
suggested could be sustained by the swirl alone).

    The danger of the situation from a practical point of view, as
emphasized by Maltby,* is that wake turbulence presents a flight
hazard and, since the standards of safety in civil aviation are so
stringent, any statement about the flow in the wake must be quantitatively reliable. We seem to be far from satisfying that requirement.

    When we ask what aerodynamic research should be done to clarify
the phenomena, and perhaps reveal others hitherto undetected, we
surely must be prepared to accept a strong dependence on experiment:
and much of that performed in flight. In saying this, I do not wish
to imply that there is no scope for analytical investigations: on
the contrary, the consequences, for instance, of perturbing an isolated vortex core, whether it be rectilinear or in the form of a
loop, remain to be examined; but the type of perturbation to be
imposed might well have to await inference from experimental observations.

    The opportunities for research offered by the conventional wind
tunnel appear to me to be restricted to the near wake, because the
length of working section normally available makes any but the
youngest trailing vortices inaccessible. Nonetheless, wind tunnels

---

*   See Bisgood, Maltby & Dee, these Proceedings.

could provide more information about the origin of the axial flow
component and its direction in relation to wing planform, attitude
and Reynolds number. They could also elucidate the conditions under
which vortex bursting occurs as well as describe the structure of
the vortex system and its dependence on wing planform. In the
latter respect, it may be pertinent to draw attention to the double
structure of the vortex shed from the tip of a slender wing, which
appears to result from a convolution of the leading and trailing
edge vortices. It would be surprising if the downstream development of such a double vortex were found to resemble in detail that
of the vortex shed from a straight or non-slender sweptback wing.
The wind tunnel could be used to study, at least, the initial behaviour of the two types of vortex systems.

As a further problem to which the wind tunnel, or some simpler
derivative from it, could be applied is that of the mechanism by
which turbulence in a vortex may be sustained. A crucial aspect,
to which I have already referred, is whether the presence of an
axial flow is a necessary condition. If that question can be resolved, as the experiments proposed by Coleman Donaldson promise to
do, the structure of the turbulence remains to be investigated, as
well as its evolution from what might be an initial laminar flow.
In this latter respect, I can mention some tentative explorations of
a trailing vortex shed from the tip of a rectangular wing made by
Dr. Harvey and me some years ago at Imperial College. They suggested the existence of turbulent slugs travelling along the core
at a moderate distance downstream from the trailing edge. But I
must emphasize the word "tentative," because we never pursued the
investigation to the stage where we could be sure that we were
detecting genuine turbulent slugs rather than a spurious intermit-
tency generated by the vortex straying from our measurement probe.
We happened to be using hot-wires, but similar difficulties would
have been encountered with any other kind of material probe:
which were to approach the vortex with sufficient care and stealth
to prevent it from eluding the probe and shyly straying to a
neighbouring part of the stream.

An attractive complement to the wind tunnel is the water
channel or towing tank which, in conjunction with dye or particle
injection and photography, offers the possibility of studying the
entire development of the wake, including the behaviour of the
individual vortex cores. The technique is not without its difficulties; in the case of the towing tank, its dimensions must be
such that not only wall and free surface constraint, but the influence of the starting and stopping motions of the model, especially
as regards the development of the axial flow in the vortex, must be
acceptably small; in addition, the towing mechanism must provide
the model with a uniform velocity during the major phase of its
travel. These are mechanical problems which careful design can

overcome. More fundamental are the problems associated with making
flow observations of a non-visual kind in water with respect to a
reference frame fixed in the undisturbed fluid, and the Reynolds
number limitation imposed by freedom from cavitation on the model.
There is little prospect of removing the latter limitation, except
by the use of an enormous tank; nonetheless, in the absence of
alternative means of studying trailing vortex behaviour in the
laboratory at large distance from the wing, I think that it is
worth accepting and effort concentrated on devising methods of
studying the flow quantitatively--and by "flow," I mean particularly
that in the core and perturbations to it. Perhaps one will have to
fall back upon visual methods, involving the use of small, neutrally-
buoyant particles.

       In general, however, the problem of the far-wake--its break-up,
formation of vortex loops, the residual motion induced by them and
the mechanics of their decay--is one that is likely to yield quantitative information most readily to flight experiment. In addition, it is only in such experiment that the effect of atmospheric
density stratification is automatically represented. But it is
easy to speak of flight experiment; its quantitative achievement
demands great ingenuity in the formulation of test programmes and
techniques. Of one thing I am convinced: and that is that the time
has come to do something more revealing than the observation from
the ground of the pattern of smoke injected into a trailing vortex
from cannisters festooned on pylons; for the pylons inevitably dis-
turb the vortex and one cannot help wondering whether the behaviour
one observes, dramatic though it may be on occasions, is the consequence of such a disturbance or is a genuine feature of the vortex
flows. Surely, the most effective and unobjectionable way of providing flow visualisation is by injecting the smoke or other material
into the vortex cores from the aeroplane itself. But whereas visual
observation has provided, and will no doubt continue to provide,
valuable information, the ultimate problem of assessing the strength
of the vortex-induced motion far from the ground remains one which is
likely to be answered by experiments of the type pioneered by Rose
and Dee at Farnborough, in which a light aircraft was flown through
the wake of a heavier one and its response to the disturbance provided
by the wake carefully measured. In that sense, I am advocating a
deterministic approach to the problem; statistics, I cannot help
feeling, will accumulate by themselves.

       But having said that, one is presupposing that the aerodynamics
of the interaction between a trailing vortex and an aeroplane encountering it are known. When the vortex contains a purely swirling
motion, the problem can be properly formulated and solutions to
varying orders of approximation may in principle be found. But what
the evidence presented at this symposium has made abundantly clear
is that, more often than not, the swirl is accompanied by a strong

axial motion, except possibly in the far-wake, and I have no im-
pression of the consequences of the axial motion on a wing exposed
to it having been worked out, even on the simplest assumptions
about the radial distribution of that motion within the vortex core.

Walter S. Luffsey: In summarizing the symposium, I must say that I
am left with two main thoughts: (1) There is much good work already
done or being done in wake turbulence research--unfortunately the
work is disjointed and redundant in several areas, and (2) We have
heard a great deal about studying wake characteristics and some
about the effect of wakes on other aircraft; but I still didn't
perceive a clear definition of hazard, i.e., when is the wake of
particular aircraft a hazard to a particular aircraft?

    I think these two thoughts really give directions for future
research, the theme for this panel. Regarding the disjointed work
effort, Boeing and AFOSR have taken a very meaningful step by
sponsoring this Wake Turbulence Symposium. Certainly some redir-
ection of effort will result from the exchanges here, and the published proceedings will provide an updating of the state-of-the-
art, the first since 1964. I suggest that continued, frequent
(perhaps annual) symposia be instituted to provide an arena for
exposing redundant activities and optimizing study results.

    At least this would provide a defacto method of centralizing
project exchange and offer the potential to establish a cohesive,
unifed program in wake turbulence study. Further, I suggest that
parties interested in sponsoring such follow-on symposia indicate
their willingness to either Dr. Goldburg here at Boeing, or perhaps
more properly to Mr. Milton Rogers of AFOSR. Optimally, I think
that a Government organization charged with basic research respon-
sibility such as NASA or OSR should take on the leadership in organizing a National, perhaps international, program effort in Wake
Turbulence.

    Now to my second thought, the need to establish a clear
definition of hazard. I have generally described the hazards related to wake turbulence in three levels of seriousness. First,
the "coffee spilling" hazard: that is, an encounter resulting in
passenger discomfort. This hazard is serious to passengers and of
concern to the airlines, but from the Air Traffic Control system
viewpoint is of little or no concern. On the other hand, the two
additional hazard levels; namely "loss of operational control" by
an encountering aircraft in close proximity to the ground or other
aircraft and "structural failure" of an encountering aircraft, are
of grave concern to all users and operators. If these descriptions
are adequate, then it seems that a clear definition would rest in
comparison of two sets of quantities: (1) the wake behind any
generating aircraft quantitatively characterized as a function of

time and all significant variables to some tolerable limit and (2) the effect of these wakes on <u>any</u> encountering aircraft as a function of all encounter modes and all entry conditions (not just along track, roll relationships). Given this comparison, an optimum application in the operating system can be determined.

In hypothesizing what could be done, several applications come to mind that singularly or in combination might at least minimize exposure to wake turbulence hazards. These applications fall in some five general areas:

1. ATC-Separation - To many, this seems to be the solution-- the only solution. Not so to me. Perhaps this is the easiest and quickest, but to my way of thinking, the least effective means of minimizing hazardous exposure. Variable separation standards are very difficult to apply and efficient use of available airspace would require a set of several ATC separation standards to incorporate wake turbulence considerations.

2. Flight Procedures - We may apply wake turbulence knowledge to flight operations, requiring that aircraft be flown on paths or profiles so that wakes will be generated in, or quickly move into, areas of no concern to other aircraft. Also, potential encounters may be avoided by requiring aircraft to fly paths or profiles which avoid hazardous wakes of preceding aircraft. Here, the Agency has taken several actions to aid in educating pilots in avoidance techniques. We have disseminated two Advisory Circulars to all airmen of record, produced a training film having general availability through loan or purchase, and we currently provide latest information in the frequently updated Airman's Information Manual. All of these publications are now being revised by our Flight Standards Service to incorporate the findings of the most recent testing.

3. Airport Design - Here, there are two questions deserving consideration: (a) For a given airport, what are acceptable runway/taxiway configurations and airport operating practices which minimize exposure to hazardous wakes, and (b) for design of future airports, how can we optimize runway/taxiway configurations to prevent hazardous exposure? Based on current knowledge, I believe much can be done in this area.

4. Sensing/Processing/Display - It has been proposed that sensing the presence and severity of vortices for processing and displaying of essential information to controllers and pilots will alleviate much of the existing operational problem with wake turbulence. I submit that this is the least practical of all applications. I don't believe sensing of

vortices in finest of detail will provide a meaningful operational solution to the problem. If we assume availability of
very inexpensive, perfect sensing devices giving detailed information on the character of aircraft wakes in airspace to
be transited by other aircraft, and the capability to process
this information in any way desired, to whom will we display
the information? Where? For what purpose? What does it mean?
etc.

The approach in this area having more potential seems to
me to be sensing and processing of essential data such as wind
vector gradients, temperature gradient or any parameter proven
conducive to maintenance of severe and persistent vortices in
flight paths of succeeding aircraft. Having sensed the proper
information, processing should derive necessary values used for
determining hazard relationships for any aircraft pair, predic-
tively. Given predicted "probable" exposure to a wake hazard,
optimum solutions should be selected and presented to the responsible agent, either the controller or pilot, well in advance.
Solutions fifteen to twenty minutes before a problem occurs
would undoubtedly allow more efficient system treatment than a
tactical solution based on sensing vortices of a preceding
aircraft when a following aircraft is 60 seconds from touch-
down. In the latter case, the resolving maneuver, a "go-round"
or missed approach, is also potentially hazardous and normally
creates severe system interaction. On the other hand, an
earlier, predictive assessment allows for solutions such as
path stretching and assignment of alternate runways. These
solutions have minimum adverse system interactions.

5. Assisted Dissipation - I have used this term frequently, but
the feedback evident to me indicates possible misunderstanding.
Perhaps the more descriptive terminology is "Assisted Deconcentration" of vortex energy--methods to redistribute the energy
concentrated in vortices to a form having characteristics of
random turbulence, thus little or no effect on encountering
aircraft. I believe deconcentration can be accomplished to a
degree either by proper aircraft and engine treatments or by
certain treatments on the airport surface for critical areas of
concern such as parallel runway operations.

Aircraft treatments might include, for example, automatically fluctuating flaperons at an optimum frequency to vary
the peak of the spanwise lift distribution and create an oscillatory mode in the shed vortex such as Dr. Crow discussed
before me. Such treatment might cause beneficial vortex interactions without significant adverse effects in terms of drag
or other operating considerations.

Engine treatments might include compression bleed directed to the wing-tips and into the center of vortex cores. By pulsating this compressed air at the proper frequency the resulting alternate increases and decreases of pressure at the core centers could lead to earlier bursting, hence earlier dissipation.

Other treatments such as stepped flaps, blown flaps, blown tips, and oscillating foils come to mind, but at least the two examples cited deserve consideration by research aerodynamicists.

Suggested airport surface treatments have run the gamut from suction trenches between parallel runways and adjacent to single runways, directed compressed air across critical portion of runways, to placement of obstructions between runways in the path of vortices. We have observed apparent dissipation of vortices contacting towers and guy-wires in our testing, and in the NASA test, shown here, the vortex contacting a 4'×8' sheet of phywood appeared to become discontinuous. Also, Dr. John Olsen's water tank tests showed rapid dissipation when the vortex was allowed to interact with several plates (or obstructions in the path). These observations indicate the possibility to deconcentrate vortex energy; we need to determine optimum techniques having practical applications. At many of today's airports, to purchase necessary additional real estate for new parallel runways meeting a one mile spacing criterion may prove far more expensive than using existing real estate with vortex dissipators between runways.

I would not be adverse to consideration of such far-fetched schemes as closed-loop water or alcohol sprays adjacent to runways to effectively create obstructions which are frangible.

Regarding the FAA Research and Development Program, it is my understanding that we will concentrate on the latter two areas, i.e., Sensing/Processing/Display and Assisted Dissipation while recognizing the need for continuation of more basic efforts by other organizations.

Having discussed the areas of possible applications, I now feel it germane to ask two questions and make a statement. I believe the questions will ultimately have to be answered to fully understand the vortex phenomenon, and as yet, no satisfactory answers are evident.

1. What is the influence of other lifting surfaces (their vortices), such as the horizontal stabilizer and the vertical fin, on wing tip vortices?

2. What are the significant atmospheric variables which
affect generation and persistence of vortices? Are there
isolatable dependencies?

My statement is in the form of a suggestion complementing Dr.
John Houbolt's recommendation that we use a standardized base-line
for referencing vortex intensity or characteristics. I agree, but
suggest that we use the Boeing 707 vortex as the unity reference.
We have done this in most of our work, and supporting data is
readily available from our previous testing.

I would now like to extend, publicly, my heartfelt thanks to
our host and co-sponsors, and to those who made presentations.
Your efforts assured a most successful symposium. Finally, I wish
to compliment the attendees who so aptly contributed from the
floor. Your diligence is exemplary.

Coleman DuP. Donaldson: I agree with Paul Owen. One should be extremely careful in interpreting the results obtained from smoke experiments downwind of a tower. When smoke is introduced into a
vortex in this way, the vortex has always been cut by the tower.
A result of this cutting is an inrush of fluid towards the core of
the vortex. The question then arises as to whether what is observed in a vortex after it has been cut by a tower has anything to
do with the flow in a vortex which has not been cut by a tower.

The second comment concerns an experiment I made by which the
question might be resolved. One could, in either a wind tunnel or
a small towing track, create a trailing vortex system and then move
a model of a tower across this vortex system. The relative motion
of the wing and the tower should be the same as in flight experiments. If smoke is introduced at the wing and has rolled up into
a trailing vortex, one could observe whether cutting this vortex
with a tower had a major effect on the character of the vortex
system.

My third suggestion: The effect of hot exhaust rolled up in
the turbulent vortex may have a very stabilizing effect on the core
of the vortex similar to that found in an arcjet. Would it not be
desirable, in order to investigate the effect, to perform vortex
wake experiments in a tunnel or towing range in which the lower side
of the airfoil creating the vortex was highly heated? I believe
this experiment would give some insight into the nature of this
effect on a trailing vortex system.

Another suggestion is the use of a glider for some of the flyby
experiments so as to eliminate the effect of thrust upon the vortex
system.

Finally I suggest that it is highly desirable, by means of instrumentation designed to minimize the effects of vortex cutting, to measure the turbulence in the core of a vortex as it passes the instrumentation. In regard to this experiment, I suggest the use of sound measurements to observe the spectra of noise resulting from the turbulence in the vortex, both before and after the vortex passes the tower. This would, it seems, be another means of determining the extent of the effect of vortex cutting on the character of the vortex.

Let me summarize my comments with a warning that I did not feel the result of smoke experiments should be accepted blindly as an indication of the character of aircraft trailing vortices until such time as it has been proven that the phenomena being observed were not strongly dependent on the fact that a leakage flow into the vortex core has been induced by the cutting of the vortex.

Norman Chigier: Many of the questions raised during the discussion at this symposium will only be resolved on the basis of accurate measurement of the flow field within the trailing vortices. Measurements we have made in the 7- by 10-foot wind tunnel at Ames show the vortex core, in the region close to the wing (less than 40 chord lengths), to be highly turbulent. An explanation is required as to the apparent anomaly between a turbulent vortex core which should have a high rate of diffusion and the fact that vortices persist for several miles. The rotating force field may well have a stabilizing effect which prevents radial diffusion and may even cause damping of turbulence. If the modified Richardson number, based on the ratio of centrifugal forces to turbulent shear forces is sufficiently large, damping of turbulence can be expected.

Because of the high intensity of turbulence, account will have to be taken of these turbulent forces when attempting to predict loads on aircraft penetrating into vortices.

CONCLUSION

From the papers and discussions presented at the Symposium,
the editors here attempt to summarize the principal conclusions
reached at the Symposium.

(i) Phenomena previously regarded as mere curiosities, e.g.
axial flow along the vortex axis, have emerged as important
tools for understanding or controlling vortex wakes. The
two types of instability occurring within the wake, axial
flow and mutual interaction, may one day be used for promoting early disintegration of the hazardous twin vortex
structure.

(ii) The influence of the atmospheric variables of stability
and turbulence level on the wake behavior has been examined,
but is not yet fully understood. High turbulence levels are
thought to accelerate wake disintegration by both increasing
dissipation and exciting the natural wake instabilities.
High atmospheric stability causes variations in the descent
path of the wake, in some cases the effect is similar to
the spreading in ground effect.

(iii) Determining the interaction between the organized vortex
wake of one aircraft and the flight of another is the most
important practical problem of the symposium. The interaction
depends, of course, on the parameters of the generating aircraft as related to the wake structure and on the parameters
of the following aircraft. Analytical models have shown that
the two most important parameters are the circulation of the
vortex and the span of the following aircraft.

AUTHOR INDEX

SUBJECT INDEX

Aeronautical research, 1, 4
AFCRL, 291
AFOSR, 2, 4, 6, 39, 55, 153, 226
Aircraft encounters, 127, 238, 254, 480, 482, 511, 517, 520, 523
  acceleration, 518
  loads, 532
  moments, 532
  roll rates, 532
Aircraft properties, 268
Aircraft response, 509, 513
Aircraft type
  Army O-1, 445
  B-47, 2, 245, 291, 292, 293, 294
  B-52, 291, 293, 548
  B-707, 94, 265, 473, 480, 481, 484, 487
  B-737, 267, 473, 478, 481, 483, 500, 503, 532, 578
  B-747, 94, 230, 265, 366, 473, 480, 484, 486, 487, 497, 500-503, 532, 578, 580, 582
  B-2707, 2
  C-5A, 2, 230, 256, 265, 287, 473
  C-47, 118, 124
  C-123, 295
  C-141, 553, 555
  Cessna 180, 294
  Cessna 210 & 310, 287
  Cherokee, 445
  Comet, 174, 245, 493
  Concorde, 2
  Convair 990, 127, 238, 260, 265, 287, 473
  DC-3, 294
  DC-8, 265
  DC-9, 256, 265, 287
  Devon (AVRO), 172
  F-86, 473
  F-104, 287, 479
  F-106, 516
  Hawker Siddeley Vulcan, 81

Aircraft type (continued)
  Hunter 6, 176, 493
  Lear jet, 24, 238, 265, 287
  "Light", 533
  Lincoln (Hawker Siddeley), 172
  Morane-Saulnier "Paris", 187, 205
  Piper Apache, 295
  Piper Cherokee, 94
  SAAB A37 Viggen, 137
  T-33, 125
  Twin Bonanza, 295
Air Force, 2, 267
Airport design, 587
Air Traffic Control, 493
Amplification rates, 382, 467
Anemometer, 57
Aspect ratio, 531
ATC-separation, 587
Atmospheric stability, 42
Axial flow, 99, 284, 338, 358, 459, 578, 583
  instability, 432
Axial momentum, 432
Axial velocity, 90, 133, 327, 332, 459, 463
Axisymmetric flow, 21

Balloons, wake marking, 295
Bank angle, 481
Biot-Savat formula, 12
Bjerkness equation, 45
Boeing, 267
  Vertol, 141
Boundary layer probe, 76
Boussinesq approximation, 54
Breakdown, vortex, 144, 146, 198, 202, 230, 290, 291, 366, 489
Buoyancy, 44, 51, 52, 290, 297
Bursting, 244, 260, 452, 583

Chaff, 97, 101, 104, 107
Circulation, 45, 51, 399, 401, 424, 433, 448, 533, 540
Complex potential of vortex, 12

599